Fundamentals of
Communication Systems

Fundamentals of Communication Systems

Contributors

Abbas Majdabadi, Saeid Marjani et al.

AURIS
Reference

www.aurisreference.com

Fundamentals of Communication Systems

Contributors: Abbas Majdabadi, Saeid Marjani et al.

Published by Auris Reference Limited
www.aurisreference.com

United Kingdom

Fundamentals of Communication Systems

ISBN: 978-1-78154-923-0

British Library Cataloguing in Publication Data
A CIP record for this book is available from the British Library

Printed in the United Kingdom

Exclusively distributed by CBS Publishers & Distributors Pvt. Ltd.

Sales & Distribution Rights only for India, Pakistan, Bangladesh, Sri Lanka, Nepal and Bhutan.This book is not to be sold outside these territories.

Contents

vi

List of Abbreviations

PSA	Algorithm of Parallel Substitutions
AM	Amplitude modulation
ASK	Amplitude shift keying
ADC	Analog-to-digital converter
BS	Base station
BPSK	Biphase shift keying
BER	Bit Error Rate
BCC	Broadcast channel with confidential messages
CSI	channel state information
CD	Chromatic dispersion
CWT	Continuous wavelet transform
CP	Cyclic prefix
DGD	Differential-group-delay
DAC	Digital-to-analog converter
DFT	Discrete Fourier Transform
DWT	Discrete WT
DFB	Distributed feedback
DOD	Distribution of Direction of Departure
DB-GSCMs	Double Bounce Geometry-based Stochastic Channel Models
EDFA	Erbium doped fiber amplifier
FEC	Forward error correction
4G	Fourth Generation
FDD	Frequency division duplex
FM	Frequency modulation
GSCMs	Geometry-based Stochastic Channel Models
GI	Guard interval
IM/DD	Intensity modulation with direct detection
IF	Intermediate frequency
LEDs	Light Emitting Diodes
LOS	Line of Sight
LO	Local oscillator
LTE	Long Term Evolution
LPFs	low pass filters
MPF	Microwave photonic filter
MCM	Multicarrier modulation
MMF	Multimode fiber
MIMO	Multiple Input Multiple Output
MISO	Multiple-input single-output
NTSC	National Television System Committee
OPLL	Optical injection locking and optical phase-locked loops
ONU	Optical network unit

OSNR	Optical signal-to-noise ratio
OSA	Optical Spectrum Analyzer
OWC	Optical Wireless Communication
OTR	Optical-to-RF
OEIC	Optoelectronic integrated circuits
PSK	Phase shift keying
PMD	Polarization mode dispersion
QAM	Quadrature amplitude modulation
RF	Radio frequency
RCI	Regularized Channel Inversion
RTO	RF-to-optical
SNR	Signal-to-noise ratio
SMF	Single mode fiber
SITO	Single-input two-output
3G	Third Generation
TITO	Two-input two-output
VCSEL	Vertical-cavity surface-emitting laser
VLC	Visible Light Communication
WP	Wavelet packet
WTs	Wavelet transforms

List of Contributors

Abbas Majdabadi
Laser and Optics Research School, Nuclear Science and Technology Research Institute (NSTRL), Tehran, Iran

Saeid Marjani
Department of Electrical Engineering, Ferdowsi University of Mashhad, Mashhad, Iran

Masoud Sabaghi
Laser and Optics Research School, Nuclear Science and Technology Research Institute (NSTRL), Tehran, Iran

Ben Zid Maha
École Nationale Supérieure d'Ingénieurs du Mans (ENSIM), Le Mans, France

Raoof Kosai
École Nationale Supérieure d'Ingénieurs du Mans (ENSIM), Le Mans, France
Laboratoire d'Acoustique de l'Université du Maine, (LAUM), UMR CNRS 6613, Le Mans, France

Giovanni Geraci
School of Electrical Eng. & Telecommunications, The University of New South Wales, Australia
Wireless and Networking Technologies Laboratory, CSIRO ICT Centre, Sydney, Australia

Jinhong Yuan
School of Electrical Eng. & Telecommunications, The University of New South Wales, Australia

Taner Cevik
Department of Computer Engineering, Fatih University Istanbul, Turkey

Serdar Yilmaz
Department of Electrical and Electronics Engineering, Fatih University Istanbul, Turkey

Yan Zhang
College of Electronic and Information Engineering, Nanjing University of Aeronautics and Astronautics, Nanjing, Jiangsu 210016, China
Nanjing College of Information Technology, Nanjing, Jiangsu 210046, China

Feng-fan Yang
College of Electronic and Information Engineering, Nanjing University of Aeronautics and Astronautics, Nanjing, Jiangsu 210016, China

Weijun Song
Nanjing College of Information Technology, Nanjing, Jiangsu 210046, China

Rongxing Duan
School of Information Engineering, Nanchang University, Nanchang 330031, China

Jinghui Fan
School of Information Engineering, Nanchang University, Nanchang 330031, China

Edward G. Kostsov
Institute of Automation and Electrometry, Russian Academy of Sciences, Russia

Sergey V. Piskunov
Institute of Computational Mathematics and Mathematical Geophysics, Russian Academy of Sciences, Russia

Mike B. Ostapkevich
Institute of Computational Mathematics and Mathematical Geophysics, Russian Academy of Sciences, Russia

Alejandro García Juárez
University of Sonora, Department of Physics Research, México

Ignacio Enrique Zaldívar Huerta
National Institute of Astrophysics, Optics and Electronics. Department of Electronics, México

Antonio Baylón Fuentes
Inst. FEMTO-ST, Université de Franche-Comté, Besançon, France

María del Rocío Gómez **Colín**
University of Sonora, Department of Physics, México

Luis Arturo García Delgado
University of Sonora, Department of Physics Research, México

Ana Lilia Leal Cruz
University of Sonora, Department of Physics Research, México

Alicia Vera Marquina
University of Sonora, Department of Physics Research, México

Y. Ben-Ezra
Department of Electrical Engineering, Holon Institute of Technology, Holon, Israel

B.I. Lembrikov
Department of Electrical Engineering, Holon Institute of Technology, Holon, Israel

Preface

In telecommunication, a communications system is a collection of individual communications networks, transmission systems, relay stations, tributary stations, and data terminal equipment (DTE) usually capable of interconnection and interoperation to form an integrated whole. The components of a communications system serve a common purpose, are technically compatible, use common procedures, respond to controls, and operate in union. The book, Fundamentals of Communication Systems, introduces the basic techniques used in modern communication systems and provides fundamental tools and methodologies used in the analysis and design of these systems. In first chapter, we investigate threshold characteristics of a single mode 1.55 μm InGaAsP vertical cavity surface emitting laser (VCSEL) with two different optical confinement structures. Second chapter provides an insight into multi-user MIMO systems. We firstly present some of the main aspects of the MIMO communication. We introduce the basic concepts of MIMO communication as well as MIMO channel modeling. In third chapter, we conclude physical layer security for multiuser MIMO communications, especially focusing on the case when multiple malicious users are present in the network, and they can eavesdrop on each other. In fourth chapter, we explore the fundamentals and challenges of indoor VLC systems. Basics of optical transmission such as transmitter, receiver, and links are investigated. Moreover, characteristics of channel models in indoor VLC systems are identified and theoretical details about channel modelling are presented in detail. Fifth chapter gives an approach on exploration of spatial diversity in multi-antenna wireless communication systems. Sixth chapter presents four different integer sequences to construct quasi-cyclic low-density parity-check (QC-LDPC) codes with mathematical theory. The chapter introduces the procedure of the coding principle and coding. Seventh chapter presents a comprehensive study on the evaluation of data communication system (DCS) using dynamic fault tree approach based on fuzzy set. Using a 3D (multilayer) IC structure is proposed in eighth chapter in order to solve most pressing problems appeared in constructing computing systems. Ninth chapter presents about wired/wireless photonic communication systems using optical heterodyning. The main goal of these systems is to reduce infrastructure cost and to overcome the capacity bottleneck in wireless access networks, allowing, at the same time, flexible merging with conventional optical access networks. In last chapter, we discuss the fundamentals of CO-OFDM systems; the WPT based OFDM systems are considered; PMD influence on the CO-OFDM dual-polarization transmission is reviewed; the structure and the properties of CO-OFDM system based on DT-CWT and DT-CWPT are described, respectively.

Chapter 1

THRESHOLD CHARACTERISTICS ENHANCEMENT OF A SINGLE MODE 1.55 μM INGAASP PHOTONIC CRYSTAL VCSEL FOR OPTICAL COMMUNICATION SYSTEMS

Abbas Majdabadi[1], Saeid Marjani[2*], Masoud Sabaghi[1]

[1]Laser and Optics Research School, Nuclear Science and Technology Research Institute (NSTRL), Tehran, Iran

[2]Department of Electrical Engineering, Ferdowsi University of Mashhad, Mashhad, Iran

ABSTRACT

In the present work, we investigate threshold characteristics of a single mode 1.55 μm InGaAsP vertical cavity surface emitting laser (VCSEL) with two different optical confinement structures. The device employs InGaAsP active region, which is sandwiched between GaAs/AlGaAs and GaAs/AlAs distributed Bragg reflectors (DBRs). The optical confinement introduced by the oxide aperture or a single defect photonic crystal design with holes etched throughout the whole structure, is compared with previous work. Photonic crystal VCSEL shows 30.86% and 57.02% lower threshold current than that of the similar oxide confined VCSEL and previous results, respectively. This paper provides key results of the threshold characteristics, including the threshold current and the threshold power. Results suggest that, the 1.55 μm InGaAsP photonic crystal VCSEL seems to be the most optimal one for light sources in high performance optical communication systems.

INTRODUCTION

In recent years, the vertical cavity surface emitting lasers (VCSELs) have attracted extremely [1] . VCSEL is one of the key light source used in high performance optical communication systems where single mode operation, high output power, high speed modulation and low manufacture cost are necessary [2] . High optical gain in the active area, high temperature, low threshold current and high thermal conductivity in the reflecting mirrors are the main difficulties in developing VCSELs which are used in the field of optical spectroscopy [3] . When all lateral modes except one are suppressed, a fully single mode, VCSEL operation is achieved. Mainly, a single lobe distribution in a desirable lateral direction defined lasing of the fundamental mode. If carriers are confined to the central part of the device, this modal distribution has best overlaps with the gain radial distribution. In real device, the injected current is far from the central part of the device where the fundamental mode has appreciable amplitude. Consequently, it is necessary to funnel the current into the center of the active region. There have been several approaches to address this issue: selective oxidation [4] , proton implantation [5] or structured tunnel junction [6] . On one hand, narrowing the current aperture causes the reduction of the emitted power and on the other hand, broadening of the current aperture favors multimode operation. Hence, it is necessary to additional structuring of VCSELs for their single mode operation with broader current apertures. It can be assured by anti-resonant profile of the refractive index in the distributed Bragg reflectors (DBRs) [7] , surface etching [8] , surface grating [9] and photonic crystals (PhC) [10] . Since PhC has already verified its ability to select very narrow spectrum of allowed frequencies and discriminate all higher modes [11] , it is extremely attractive.

From one side, the PhC improves the wave guiding, which leads to better mode confinement and hence, to higher efficiency of the stimulated recombination. From another side, it destroys the reflectivity of the DBR. The accurate design can balance between these two effects and improve the device efficiency in terms of reduction of the threshold current and increase of lateral mode discrimination.

Recently, we proposed a structure for decreasing the threshold current of VCSELs [12] and showed that proposed structure decreased the threshold current about 76.52% from 2.3 mA to 0.6 mA. There have been a number of reports on the effect of the photonic crystal confined (PhC) on the modal characteristics by Czyszanowski's group [13] -[15] . However, the structures presented are different laser wavelength (1.3 μm), the etching depth and simulation method (the plane-wave admittance method).

In this paper, the impact of the photonic crystal confined and oxide confined (OC) optical confinement schemes on the threshold characteristics of single mode 1.55 μm InGaAsP VCSEL design similar to that reported in [4] is investigated and compared with previous results [12] [14] [15] . The paper is organized as follows: Section 2 briefly describes the theoretical model; Section 3 provides the details of the VCSEL structures; and Section 4 presents the obtained numerical results. Finally, in Section 5, we conclude.

THEORY

In modeling VCSEL, we must the electrical, optical and thermal interaction during VCSEL performance [16] . Thus base of simulation is to solve Poisson and continuity equations for electrons and holes. Poisson's equation is defined by [17] :

$$\nabla \cdot (\varepsilon \nabla \Psi) = \rho \tag{1}$$

where Ψ is electrostatic potential, ρ is local charge density and ε is local permittivity. The continuity equations of electron and hole are given by [17] :

$$\frac{\partial n}{\partial t} = G_n - R_n + \frac{1}{q}\nabla \cdot \vec{J_n}, \tag{2}$$

$$\frac{\partial p}{\partial t} = G_p - R_p + \frac{1}{q}\nabla \cdot \vec{J_p}. \tag{3}$$

where n and p are the electron and hole concentration, J_n and J_p are the electron and hole current densities, G_n and G_p are the generation rates for electrons and holes, R_n and R_p are the recombination rates and q is the magnitude of electron charge.

The fundamental semiconductor Equations (1)-(3) are solved self-consistently together with Helmholtz and the photon rate equations. The applied technique for solution of Helmholtz equation is based on effective frequency method [18] which shows accuracy for great portion of preliminary problems. Two-dimensional Helmholtz equation is solved to determine the transverse optical field profile and it is given by [17] :

$$\nabla^2 E(r,z,\varphi) + \frac{\omega_0}{c^2}\varepsilon(r,z,\varphi,\omega)E(r,z,\varphi) = 0 \tag{4}$$

where ω is the frequency, $\varepsilon(r,z,\varphi,\omega)$ is the complex dielectric permittivity, $E(r,z,\varphi)$ is the optical electric field and c is the speed of light in vacuum.

The light power equation relates electrical and optical models. The photon rate equation is given by [17] :

$$\frac{dS_m}{dt} = \left(\frac{c}{N_{eff}}G_m - \frac{1}{\tau_{phm}} - \frac{cL}{N_{eff}} \right)S_m + R_{spm} \tag{5}$$

where S_m is the photon number, G_m is the modal gain, R_{spm} is the modal spontaneous emission rate, L represents the losses in the laser, N_{eff} is the group effective refractive index, τ_{phm} is the modal photon lifetime and c is the speed of light in vacuum. The m refers to the modal number.

The heat flow equation has the form [17] :

$$C\frac{\partial T_L}{\partial t} = \nabla\left(\kappa\nabla T_L\right) + H$$

(6)

where C is the heat capacitance per unit volume, κ is the thermal conductivity, H is the generation, T_L is the local lattice temperature and H is the heat generation term.

The heat generation equation has the form [17] :

$$H = \left[\frac{\overline{|J_n|^2}}{q\mu_n n} + \frac{\overline{|J_p|^2}}{q\mu_p p}\right] + q(R-G)\left[\phi_p - \phi_n + T_L\left(P_p - P_n\right)\right] - T_L\left(\overrightarrow{J_n}\nabla P_n + \overrightarrow{J_p}\nabla P_p\right)$$

(7)

where: $\left[\frac{\overline{|J_n|^2}}{q\mu_n n} + \frac{\overline{|J_p|^2}}{q\mu_p p}\right]$ is   the   Joule   heating   term; $q(R-G)\left[\phi_p - \phi_n + T_L\left(P_p - P_n\right)\right] - T_L\left(\overrightarrow{J_n}\nabla P_n + \overrightarrow{J_p}\nabla P_p\right)$ is the recombination and generation heating and cooling term; and the Peltier and Thomson effects.

In addition, all important, usually non-linear, interactions between the optical, electrical, thermal and recombination phenomena are also taken into account with the aid of the self-consistent iteration algorithm, including: gain-induced wave-guiding, thermal focusing, self-focusing, temperature dependence of optical gain and absorption coefficients and the energy gaps and electrical conductivities and thermal conductivities, carrier-con- centration dependence of electrical conductivities and optical gain and absorption coefficients and the energy gaps and wavelength dependences of optical gain and absorption coefficients. Consequently, several factors are effective in determining the profiles of all model parameters within the whole VCSEL volume, including: different chemical compositions of the VCSEL layers, the distributions of the temperature, the current density, the carrier concentration and the mode radiation intensity.

Heat loss from the modeled VCSEL device was specified using thermal contacts at the top electrode, bottom electrode and the device sidewall. The thermal contacts define thermal conductivities to simulate heat loss from radiation via exposed surfaces or conduction through the semiconducting material to a heatsink.

Equation (1)-(7) provide an approach that can account for the mutual dependence of electrical, thermal, optical and elements of heat sources. In this paper, we employ numerical-based simulation software to assist in the device design and simulation [17] .

VCSEL STRUCTURE

The analyzed structure is similar to the one recently reported in [12] has been chosen as a model structure for the analysis of the 1.55 µm InGaAsP VCSEL. The active region consists of six quantum wells where the well is 5.5 nm $In_{0.76}Ga_{0.24}As_{0.82}P_{0.18}$ and the barrier is 8 nm $In_{0.48}Ga_{0.52}As_{0.82}P_{0.18}$. In both sides of this active region, there is InP and on top of it, GaAs. The top mirror is 30 layers of $GaAs/Al_{0.33}Ga_{0.67}As$ with index of refraction 3.38 and 3.05 respectively, and the bottom mirror has 28 layers of GaAs/AlAs with index of refraction 3.38 and 2.89 respectively. Layer thickness, doping, majority carrier mobility, refractive index, temperature coefficient of n, absorption coefficient, and thermal conductivity of layers are listed in Table 1. In the OC VCSEL, the incorporation of a high aluminum content layer ($Al_{0.98}Ga_{0.02}As$) in two DBR periods above the active region allows for selective oxidation [11] . Figure 1(a) shows the PhC VCSEL structure, which is similar to the above VCSEL [12] but made additionally with a single defect PhC etched throughout the whole structure. The optical confinement is achieved by means of seven air holes where the center is missed off to make the defect region, as shown in Figure 1(b).

Table 1: Parameters of the structure (N_{dop}: doping; μ: majority carrier mobility; n refractive index; dn/dT: temperature coefficient of n; α: absorption coefficient; and κ: thermal conductivity)

Parameter unit	l (µm)	N_{dop} (1/cm³)	μ (cm²/V·s)	n	dn/dT (10⁻⁴/K)	α (1/cm)	κ (W/cm·K)
Au/Ti (contact)	0.200	-	-	0.83	-	684,000	0.67
p-GaAs	0.020	2×10^{16}	-	3.38	3	500	0.44
p-GaAs	0.182	4×10^{17}	-	3.38	3	25	0.22
p-Al₀.₆₇Ga₀.₃₃As (DBR)	0.127	4×10^{17}	-	3.05	2	25	0.22
p-GaAs (DBR)	0.115	4×10^{17}	-	3.38	3	25	0.44
p-GaAs (spacer)	0.020	4×10^{17}	-	3.38	3	25	0.44
p-GaAs (spacer)	0.010	4×10^{19}	-	3.38	3	1000	0.44
p-InP (spacer)	0.178	1×10^{18}	30	3.17	2	24	0.68
p-Inp (spacer)	0.100	1×10^{16}	150	3.17	2	0.24	0.68
In₀.₇₆Ga₀.₂₄As₀.₈₂P₀.₁₈ (QW)	0.0055	-	100	3.6	2	54	0.043
In₀.₄₈Ga₀.₅₂As₀.₈₂P₀.₁₈ (barrier)	0.008	-	100	3.4	2	54	0.043
n-InP (spacer)	0.258	5×10^{18}	4600	3.15	2	8	0.68
n-GaAs (spacer)	0.050	1×10^{18}	-	3.38	3	6	0.44
n-GaAs (DBR)	0.115	1×10^{18}	-	3.38	3	6	0.22
n-AlAs (DBR)	0.134	1×10^{18}	-	2.89	1	3	0.22
n-GaAs (substrate)	450	5×10^{18}	-	3.38	3	5.8	0.44

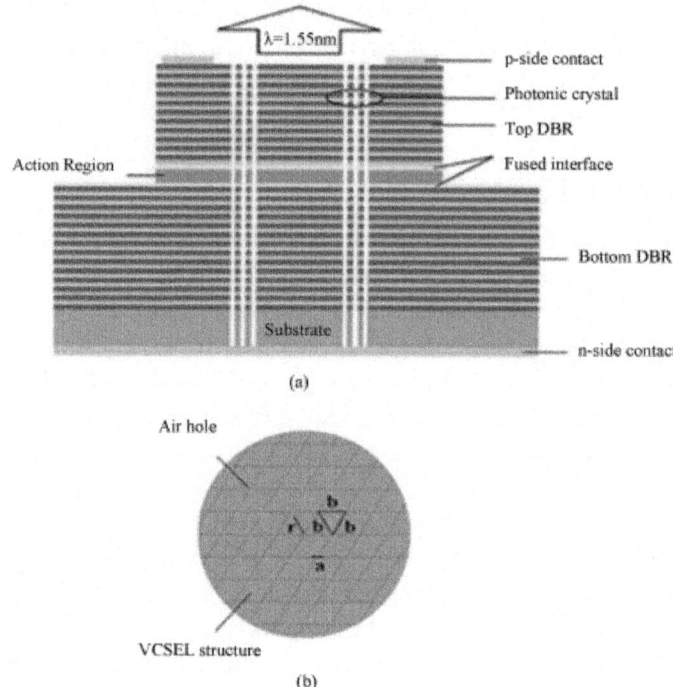

Figure 1: (a) Schematic structure of the VCSEL device and (b) top view of the triangular-lattice air holes pattern.

The crucial PhC parameters are connected by the air hole diameter (a), the pitch (b) and the optical aperture diameter (r), are defined in Figure 1(b). The transverse index guiding around the single defect region can be controlled by the air hole diameter the pitch ratio (a/b). We fixed the optical aperture diameter at 2 μm and a/b ratio is varied from 0 to 0.95 for the study.

RESULTS

The present work allows for determination of the optimum VCSEL structure that is found by a minimization of the threshold power and current. Figure 2 presents a comparison of the optical intensity distributions at the lasing threshold of OC and PhC VCSELs with different a/b values that emphasize the overlap of the holes by the modes for the cutoff condition. Fundamentally, the increase of a/b ratio and the optical aperture diameter leads to an increase of the optical confinement factor within the active region and the optical intensity distributions changes significantly. The case of a/b = 0.3 shows a star-like shape of the PhC mode that the field permeates the absorptive surrounding regions. However, at higher values of a/b ratios the mode distributions remain almost unchanged.

A comparison of the threshold characteristics of OC, PhC VCSELs and previous work [12] [14] [15] as a function of the a/b ratio is presented in Figures 3-5. As can be seen from Figure 3, the logarithmically scaled threshold power of PhC VCSELs decrease gradually with the increase in the a/b ratio. This initial decrement (from a/b = 0 to a/b = 0.9) should be mainly due to improving the optical confinement by the photonic crystal. For a/b = 0.9 the threshold power reaches a minimum value of 0.5719e−5 W and then rapidly increases for a/b = 0.95, which should be mainly due to blocking of most of the current flow by the holes. The minimal threshold power is lowered by 48.82% from 1.1176e−5 W to 0.5719e−5 W than that of the OC VCSEL, since, the electrical resistance of the etched PhC regions is increased.

Figure 4 shows a monotonic decrease in the logarithmically scaled threshold current of PhC VCSELs with the increase in the a/b ratio. This is caused by more effective optical confinement and stronger limitation of the current flow to the active region. As shown in Figure 4, photonic crystal VCSEL shows 30.86% and 57.02% lower threshold current than that of the similar oxide confined VCSEL and previous results [19] , respectively. Threshold current for two PhC VCSEL structures [14] [15] determined by r = 4 μm, a/b = 0.5 and r = 8 μm, a/b = 0.5 are 4.9 mA and 7.11 mA, respectively, which are very close to previous values of 4.89 mA and 7.12 mA, respectively [15] . We fixed the depth of the PhC holes at 7 μm for two PhC VCSEL structures.

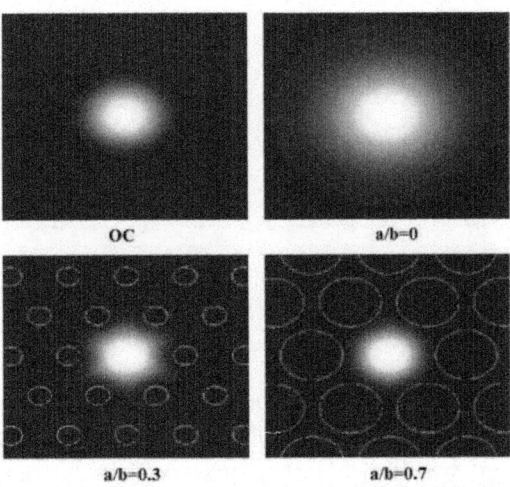

Figure 2: The optical intensity distributions at the lasing threshold of the fundamental mode within the active region in the case of the OC VCSEL and the PhC VCSEL for four different a/b ratios.

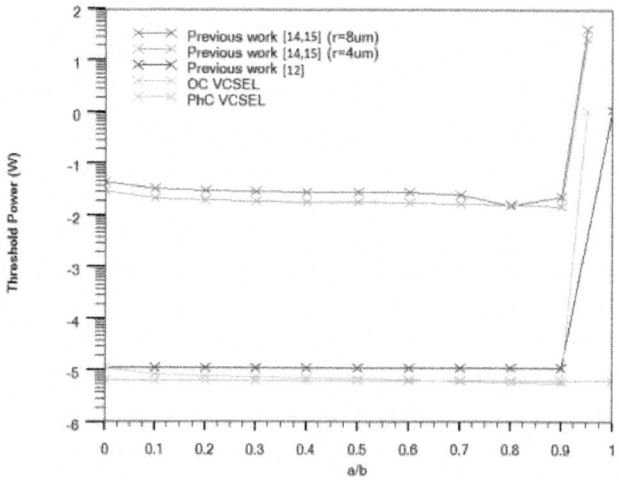

Figure 3: Threshold power of OC, PhC VCSELs and previous work [12] [14] [15] as a function of the a/b ratio.

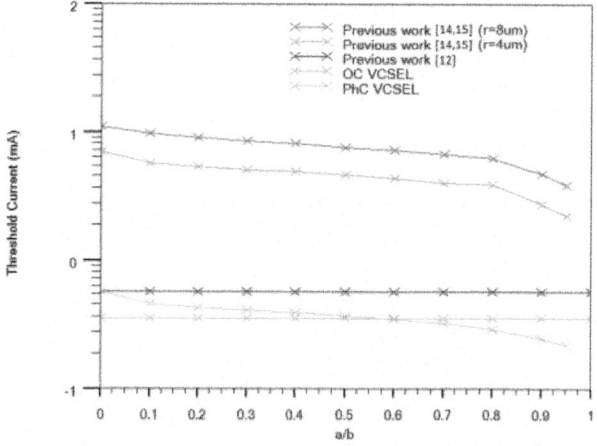

Figure 4: Threshold current of OC, PhC VCSELs and previous work [12] [14] [15] as a function of the a/b ratio.

Figure 5 shows a threshold temperature increase for a/b = 0 than that of the OC VCSEL, since, the thermal focusing is the unique wave guiding mechanism induced by the PhC structure. As can be seen from Figure 5, threshold temperature of the OC VCSEL is always smaller than that of the PhC VCSEL. This is caused by the more efficient heat sinking process in the OC VCSEL. For a/b = 0.7 the threshold temperature reaches a minimum value of 300.147 K that is the optimum configuration for over threshold operation.

CONCLUSION

Threshold characteristics of a single mode 1.55 μm InGaAsP vertical cavity surface emitting laser with two different optical confinement structures were analyzed and compared with previous work. In summary, the results indicate that lowest threshold current and threshold power is for a/b = 0.9 that shows 30.86% and 57.02%

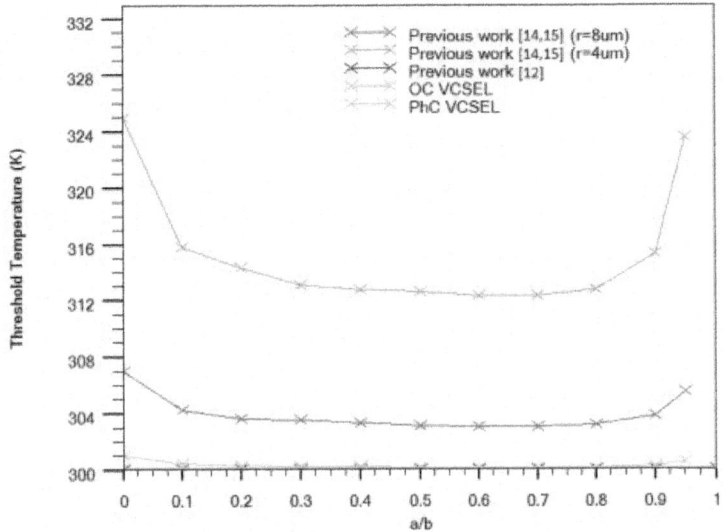

Figure 5: Threshold temperature of OC, PhC VCSELs and previous work [12] [14] [15] as a function of the a/b ratio.

lower threshold current than that of the similar oxide confined VCSEL and previous results, respectively. Our studies reveal that high-precision PhC etching is necessary in PhC VCSELs in order to optimize their threshold characteristics for light sources in high performance optical communication systems.

REFERENCES

1. Iga, M.K. (2000) Surface-Emitting Laser—Its Birth and Generation of New Optoelectronic Field. IEEE Journal of Selected Topics in Quantum Electronics, 6, 1201-1215.http://dx.doi.org/10.1109/2944.902168

2. Dems, M., Kotynski, R. and Panajotov, K. (2005) Plane Wave Admittance Method—A Novel Approach for Determining the Electromagnetic Modes in Photonic Structures. Optics Express, 13, 3196-3207. http://dx.doi.org/10.1364/OPEX.13.003196

3. Kapon, E. and Sirbu, A. (2009) Long-Wavelength VCSELS Power-Efficient Answer. Nature Photonics, 3, 27-29. http://dx.doi.org/10.1038/nphoton.2008.266

4. Jung, C., Jager, R., Grabherr, M., Schnitzer, P., Michalzik, R., Weigl, B., Muller, S. and Ebeling, K.J. (1997) 4.8 mW Single-Mode Oxide Confined Topsurface Emitting Vertical-Cavity Laser Diodes. Electronics Letters, 33, 1790-1791.http://dx.doi.org/10.1049/el:19971207

5. Morgan, R.A., Guth, G.D., Focht, M.W., Asom, M.T., Kojima, K., Rogers, L.E. and Callis, S.E. (1993) Transverse Mode Control of Vertical-Cavity Top-Surface-Emitting Lasers. IEEE Photonics Technology Letters, 4, 374-377. http://dx.doi.org/10.1109/68.212669

6. Long, C.M., Mutter, L., Dwir, B., Mereuta, A., Caliman, A., Sirbu, A., Iakovlev, V. and Kapon, E. (2014) Optical Injection Locking of Transverse Modes in 1.3-μm Wavelength Coupled-VCSEL Arrays. Optics Express, 22, 21137- 21144.http://dx.doi.org/10.1364/OE.22.021137

7. Zhou, D. and Mawst, L.J. (2002) High-Power Single-Mode Antiresonant Reflecting Optical Waveguide-Type Vertical- Cavity Surface-Emitting Lasers. IEEE Journal of Quantum Electronics, 38, 1599-1606. http://dx.doi.org/10.1109/JQE.2002.805107

8. Debernardi, P., Unold, H.J., Maehnss, J., Michalzik, R., Bava, G.P. and Ebeling, K.J. (2003) Single-Mode, Single-Po- larization VCSELs via Elliptical Surface Etching: Experiments and Theory. IEEE Journal of Selected Topics in Quan- tum Electronics, 9, 1394-1404.http://dx.doi.org/10.1109/JSTQE.2003.819487

9. Haglund, A., Gustavsson, J.S., Bengtsson, J., Jedrasik, P. and Larsson, A. (2006) Design and Evaluation of Fundamental-Mode and Polarization-Stabilized VCSELs with a Subwavelength Surface Grating. IEEE Journal of Quantum Electronics, 42, 231-240.http://dx.doi.org/10.1109/JQE.2005.863703

10. Siriani, D.F., Tan, M.P., Kasten, A.M., Lehman Harren, A.C., Leisher, P.O., Sulkin, J.D., Raftery Jr., J.J., Danner, A.J., Giannopoulos, A.V. and Choquette, K.D. (2009) Mode Control in Photonic Crystal Vertical-Cavity Surface-Emitting Lasers and Coherent Arrays. IEEE Journal of Selected Topics in Quantum Electronics, 15, 909-917.http://dx.doi.org/10.1109/JSTQE.2008.2012121

11. De La Rue, R. (2006) Photonic Crystal Components: Harnessing the Power of the Photon. Optics and Photonics News, 17, 30-35. http://dx.doi.org/10.1364/OPN.17.7.000030

12. Faez, R., Marjani, A. and Marjani, S. (2011) Design and Simulation of a High Power Single Mode 1550 nm InGaAsP VCSELs. IEICE Electronics Express, 8, 1096-1101.http://dx.doi.org/10.1587/elex.8.1096

13. Czyszanowski, T., Dems, M., Sarzala, R., Nakwaski, W. and Panajotov, K. (2011) Precise Lateral Mode Control in Photonic Crystal Vertical-Cavity Surface-Emitting Lasers. IEEE Journal of Quantum Electronics, 99, 1291-1296.http://dx.doi.org/10.1109/JQE.2011.2159363

14. Czyszanowski, T., Dems, M., Thienpont, H. and Panajotov, K. (2008) Modal Gain and Confinement Factors in Top- and Bottom-Emitting Photonic-Crystal VCSEL. Journal of Physics D: Applied Physics, 41, Article ID: 085102. http://dx.doi.org/10.1088/0022-3727/41/8/085102

15. Czyszanowski, T. (2009) Discrimination of Higher-Order Modes in Photonic-Crystal VCSEL. Proceedings of the IEEE/LEOS Winter Topicals Meeting Series, Innsbruck, 12-14 January 2009, 20-21.

16. Menon, P.S., Kumarajah, K., Ismail, M., Majlis, B.Y.M. and Shaari, S. (2010) Long-Wavelength MQW Vertical-Cav- ity Surface Emitting Laser: Effects of Lattice Temperature. Journal of Optical Communications, 31, 81-84.http://dx.doi.org/10.1515/JOC.2010.31.2.81

17. SILVACO International Incorporated (2010) ATLAS User's Manual. Version 5.12.0.R., SILVACO, Inc., USA.

18. Wenzel, H. and Wunsche, H.-J. (1997) The Effective Frequency Method in the Analysis of Vertical-Cavity Surface- Emitting Lasers. IEEE Journal of Quantum Electronics, 33, 1156-1162. http://dx.doi. org/10.1109/3.594878

19. Choquette, K.D., Geib, K.M., Ashby, C.I., Twesten, R.D., Blum, O., Hou, H.Q., Follstaedt, D.M., Hammons, B.E., Mathes, D. and Hull, R. (1997) Advances in Selective Wet Oxidation of AlGaAs Alloys. IEEE Journal of Selected Topics in Quantum Electronics, 3, 916-926. http://dx.doi. org/10.1109/2944.640645

Chapter 2

MULTI USER MIMO COMMUNICATION: BASIC ASPECTS, BENEFITS AND CHALLENGES

Ben Zid Maha[1], and Raoof Kosai[12]

[1] École Nationale Supérieure d'Ingénieurs du Mans (ENSIM), Le Mans, France

[2] Laboratoire d'Acoustique de l'Université du Maine, (LAUM), UMR CNRS 6613, Le Mans, France

INTRODUCTION

The explosive growth of Multiple Input Multiple Output (MIMO) systems has permitted for high data rate and a wide variety of applications. Some of the technologies which rely on these systems are IEEE 802.11, Third Generation (3G) and Long Term Evolution (LTE) ones. Recent advances in wireless communication systems have contributed to the design of multi-user scenarios with MIMO communication. These communication systems are referred as multi-user MIMOs. Such systems are intended for the development of new generations of wireless mobile radio systems for future cellular radio standards. This chapter provides an insight into multi-user MIMO systems. We firstly present some of the main aspects of the MIMO communication. We introduce the basic concepts of MIMO communication as well as MIMO channel modeling. Thereafter, we evaluate the MIMO system performances. Then, we concentrate our analysis on the multi-user MIMO systems and we provide the reader a conceptual understanding with the multi-user MIMO technology. To do so, we present the communication system model for such emerging technology and we give some examples which describe the recent advances for multi-user MIMO systems. Finally, we introduce linear precoding techniques which could be exploited in multi-user MIMO systems in order to suppress inter-user interference.

MIMO COMMUNICATION

An historical overview

The main historical events which make the MIMO systems [2][3] are summarized as follows:

- In 1984, Jack Winters at Bell Laboratories wrote a patent on wireless communications using multiple antennas. Jack Winters in [4] presented a study of the fundamental limits on the data rate of multiple antenna systems in a Rayleigh fading environment.

- In 1993, Arogyaswami Paulraj and Thomas Kailath proposed the concept of spatial multiplexing using MIMO.

- Several articles which focused on MIMO concept were published in the period from 1986 to 1995 [5]. This was followed by the work of Greg Raleigh and Gerard Joseph Foschini in 1996 which invented new approaches involving space time coding techniques. These approaches were proved to increase the spectral efficiency of MIMO systems.

- In 1999, Thomas L. Marzetta and Bertrand M. Hochwald published an article [6] which provides a rigorous study on the MIMO Rayleigh fading link taking into consideration information theory aspects.

- The first commercial MIMO system was developed in 2001 by Iospan Wireless Inc.

Since 2006, several companies such as Broadcom and Intel have introduced a novel communication technique based on the MIMO technology for improving the performance of wireless Local Area Network (LAN) systems. The new standard of wireless LAN systems is named IEEE 802.11n.

Nowadays, MIMO systems are implemented in many advanced technologies such as various standard proposals for the Fourth Generation (4G) of wireless communication systems and LTE. MIMO technology was shown to boost the communication system capacity and to enhance the reliability of the communication link since it uses several diversity schemes beyond the spatial diversity.

Fundamentals of MIMO system

MIMO system model is depicted in Figure 1. We present a communication system with N_T transmit antennas and N_R receive antennas.

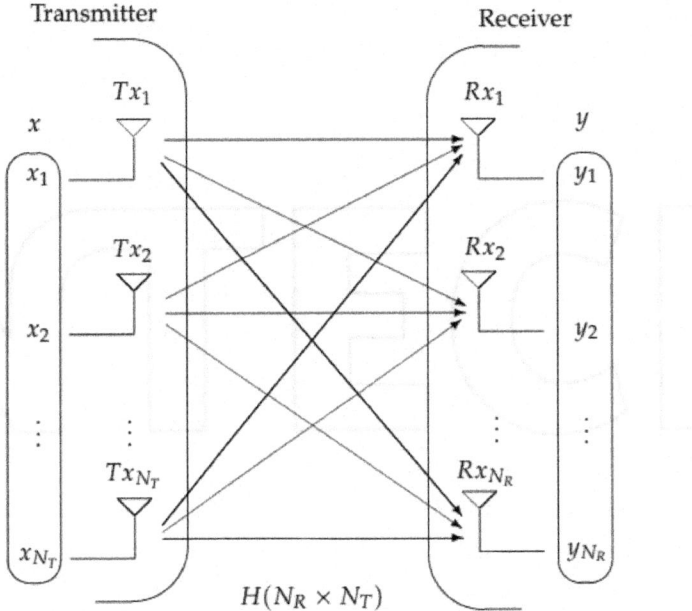

Figure 1: MIMO system model.

Antennas $Tx_1, \ldots, T_xN_T$ respectively send signals $x_1, \ldots, xN_T$ to receive antennas $Rx_1, \ldots, R_xN_R$. Each receive antenna combines the incoming signals which coherently add up. The received signals at antennas $Rx_1, \, , \ldots, RxN_R$ are respectively denoted by $y_1, \ldots, yN_R$. We express the received signal at antenna Tx_q; $q = 1, \ldots, N_R$ as:

$$y_q = \sum_{p=1}^{N_T} h_{qp} \cdot x_p + b_q \quad ; q = 1, \ldots, N_R \tag{1}$$

The flat fading MIMO channel model is described by the input-output relationship as:

$$y = H \cdot x + b \tag{2}$$

H is the $(N_R \times N_T)$ complex channel matrix given by:

$$H = \begin{pmatrix} h_{11} & h_{12} & \cdots & h_{1N_T} \\ h_{21} & h_{22} & \cdots & h_{2N_T} \\ \vdots & \vdots & \ddots & \vdots \\ h_{N_R1} & h_{N_R2} & \cdots & h_{N_RN_T} \end{pmatrix}$$

h_{qp}; $p = 1, \ldots, N_T$; $q = 1, \ldots, N_R$ is the complex channel gain which links transmit antenna Tx_p to receive antenna Rx_q.

$x = [x_1, \ldots, x_{N_T}]^T$ is the $(N_T \times 1)$ complex transmitted signal vector

$y = [y_1, \ldots, y_{N_R}]^T$ is the $(N_R \times 1)$ complex received signal vector.

$b = [b_1, \ldots, b_{N_R}]^T$ is the $(N_R \times 1)$ complex additive noise signal vector.

The continuous time delay MIMO channel model of the $(N_R \times N_T)$ MIMO channel H associated with time delay τ and noise signal b(t) is expressed as:

$$y(t) = \int_\tau H(t, \tau) x(t - \tau) d\tau + b(t)$$

(3)

- y(t) is the spatio-temporel output signal.
- x(t) is the spatio-temporel input signal.
- b(t) is the spatio-temporel noise signal.

$(\cdot)^T$ denotes the transpose operator.

MIMO channel modeling

Several MIMO channel models [7] have been proposed in literature. These models mainly fall into two categories as depicted in Figure 2.

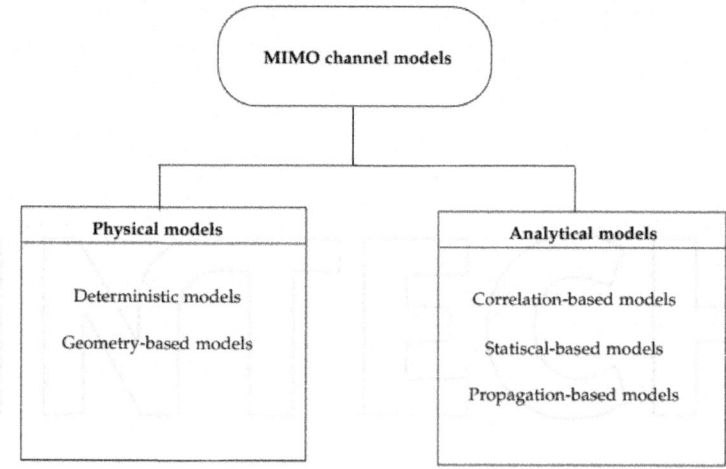

Figure 2: MIMO channel propagation models.

Physical models

MIMO channel impulse response is evaluated according to the radio wave which propagates from the transmitter to the receiver. The MIMO channel model is determined based on the experimental measurements made for extracting channel propagation parameters including antenna configuration at both the transmitter and the receiver, antenna polarization, scatterers,. . . Physical models include both deterministic models and Geometry-based stochastic channel models (GSCMs).

- Deterministic models define a channel model according to the prediction of the propagation signal.

- Geometry-based Stochastic Channel Models (GSCMs) have an immediate relation with the physical characteristics of the propagation channel.

These models suppose that clusters of scatterers are distributed around the transmitter and the receiver. The scatterers locations are defined according to a random fashion that follows a particular probability distribution. Scatterers result in discrete channel paths and can involve statistical characterizations of several propagation parameters such as delay spread, angular spread, spatial correlation and cross polarization discrimination. We distinguish two possible schemes which are the Double Bounce Geometry-based Stochastic Channel Models (DB-GSCMs) and the Single Bounce Geometry-based Stochastic Channel Models (SB-GSCMs). That is when a single bounce of scatterers is placed around the transmit antennas or receive antennas.

Analytical models

The second class of MIMO channel models includes analytical models which are based on the statistical properties obtained through measurement (Distribution of Direction of Departure (DOD), distribution of Direction of Arrival (DOA),. . .). Analytical channel models can be classified into correlation-based models (such as i.i.d model, Kronecker model, Keyhole model,. . .), statistical -based models (such as Saleh-Valenzuela model and Zwick model) and propagation-based models (such as Müller model and Finite scatterer model). We provide in [8] a detailed description of MIMO systems with geometric wide MIMO channel models where advanced polarization techniques [9][10] are exploited.

MIMO system performances

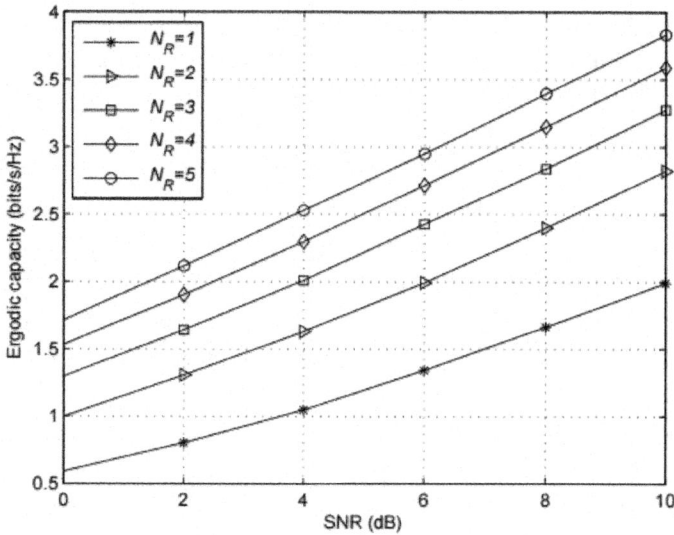

Figure 3: Ergodic capacity for MIMO systems

MIMO technology has been shown to improve the capacity of the communication link without the need to increase the transmission power. MIMO system capacity is mainly evaluated according to the following scenarios: 1. When no Channel State Information (CSI) is available at the transmitter, the power is equally split between the NT transmit antennas, the instantaneous channel capacity is then given by:

$$C(H) = \log_2 \left[\det \left(I_{N_R} + \frac{\gamma}{N_T} \cdot HH^* \right) \right] \quad bits/s/Hz$$

(4)

γ denotes the Signal to Noise Ratio (SNR). $(\cdot)^*$ stands for the conjugate transpose operator. When CSI is available at the receiver, Singular Value Decomposition (SVD) is used to derive the MIMO channel capacity which is given by:

$$C_{SVD}(H) = R \cdot \log_2 \left[\det \left(1 + \frac{\gamma}{N_T} HH^* \right) \right] \quad bits/s/Hz$$

(5)

$R = \min(N_R, N_T)$ is the rank of the channel matrix H 3. When CSI is available at both the transmitter and the receiver, the channel capacity is computed by performing the water-filling algorithm. The instantaneous channel capacity is then:

$$C_{WF}(H) = \sum_{p=1}^{R} \log_2 \left[\left(\frac{\lambda_{H,p} \cdot \mu}{\sigma_b^2} \right)^{+} \right] \quad bits/s/Hz$$

(6)

$a^{+} = \max(a, 0)$

$\lambda_{H,p}$ is the $p - th$ singular value of the channel matrix H

- μ is a constant scalar which satisfies the total power constraint
- $\sigma^2 b$ is the noise signal power

We consider the case where CSI is available at the receiver, the simulated ergodic MIMO capacity is depicted in Figure 3. For a MIMO system with two transmit antennas, numerical results show that ergodic capacity linearly increases with the number of antennas. Plotted curves are presented for different levels of the SNR. The use of additional antennas improves the performances of the communication system. Moreover, MIMO system takes advantage of multipath propagation. The performances of MIMO system are observed in the following in terms of the Bit Error Rate (BER). We consider a MIMO system with various receive antennas, the BER is evaluated for communication systems with Rayleigh fading MIMO channel and additive gaussian noise. At the receive side, the Maximum Ratio Combining (MRC) technique is performed. According to Figure 4, it is obvious that MIMO technology allows for a significant improvement of the BER. Once the MIMO technology is presented, we introduce in the following multi-user MIMO systems.

MULTI-USER MIMO SYSTEM

The growth in MIMO technology has led to the emergence of new communication systems. We are particularly interested in this chapter in multi-user MIMO (MU-MIMO) ones [11]. MU-MIMO [12] system is often considered in literature as an extension of Space-Division Multiple Access (SDMA). This technology supports multiple connections on a single conventional channel where different users are identified by spatial signatures. SDMA uses spatial multiplexing and enables for higher data rate. This could be achieved by using multiple paths as different channels for carrying data. Another benefit of using the SDMA technique in cellular networks is to mitigate the effect of interference coming from adjacent cells. Traditional communication MIMO systems are usually referred as single-user MIMO systems (SU-MIMOs) or also point-to-point MIMO. Case of MIMO systems, the access point communicates with only one mobile terminal (the user). Both the access point and the mobile terminal are equipped with multiple antennas. In contrast to the single-user case, the access point is able to communicate with several mobile terminals. SU-MIMO and MU-MIMO systems are two possible configurations for multi-user communication systems. We also find other configurations in literature such as MU-MIMO with cooperation where cooperation is established between base stations [2].

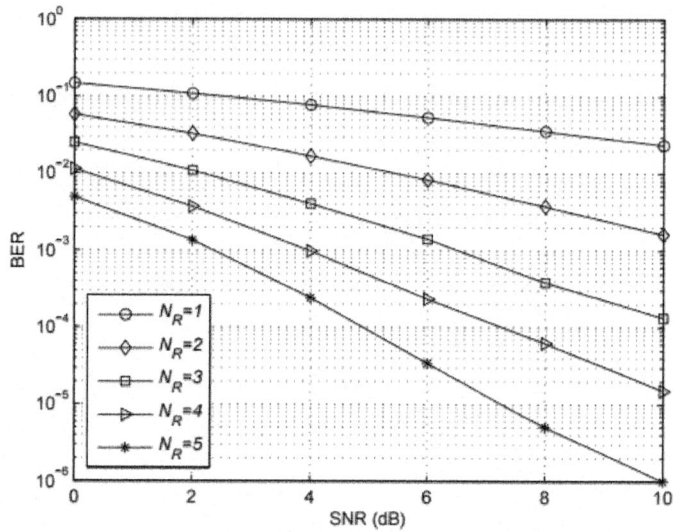

Figure 4: Improvement of the BER for MIMO (NR × 2) as a function of receive antennas number

Basic configurations of downlink multi-user MIMO systems are depicted in Figure 5. Figure 5(a) represents the SU-MIMO system where a Base Station (BS) equipped with antennas $Tx_1, \ldots, Tx_N$ communicates with user U which is equipped with M antennas $Rx_1, \ldots, R_xM$. In Figure 5(b), the presented MU-MIMO system consists of two base stations BS_1 and BS_2 each one is equipped with N antennas. Generalized MU-MIMO systems may consist of more base stations where the number of antennas could be different. At the receive side, K users $U_1, \ldots, U_K$ with respectively $M_1, \ldots, M_K$ antennas communicate with the transmit base stations. The same communication model is performed for the MU-MIMO with cooperation (Figure 5(c)) where cooperation is established between BS_1 and BS_2. Once multi-user communication systems are introduced, we explain in the following section the difference between SU-MIMO and MU-MIMO configurations.

MU-MIMO vs SU-MIMO

Table 1 summarizes the main features of both SU-MIMO and MU-MIMO systems [13]. In contrast to MU-MIMO systems where one base station could communicate with multiple users, base station only communicate with a single user in the case of SU-MIMO systems. In addition, MU-MIMO systems are intended to employ multiple receivers so that to improve the rate of communication while keeping the same level of reliability. These systems are able to achieve the overall multiplexing gain obtained as the minimum value between the number of antennas at base stations and the number of antennas at users. The fact that multiple users could simultaneously communicate over the same spectrum improves the system performance. Nevertheless, MU-MIMO networks are exposed to strong co-channel interference which is not the case for SU-MIMO ones. In order to solve the problem of interference in MU-MIMO systems, several approaches have been proposed for interference management [14][15]. Some of these approaches are based on beamforming technique [31]. Moreover, in contrast to SU-MIMO systems, MU-MIMO systems require perfect CSI in order to achieve high throughput and to improve the multiplexing gain [16]. Finally, the performances of MU-MIMO and SU-MIMO systems in terms of throughput depend on the SNR level. In fact, at low SNRs, SU-MIMO performs better. However, at high SNRs level, MU-MIMO provides better performances.

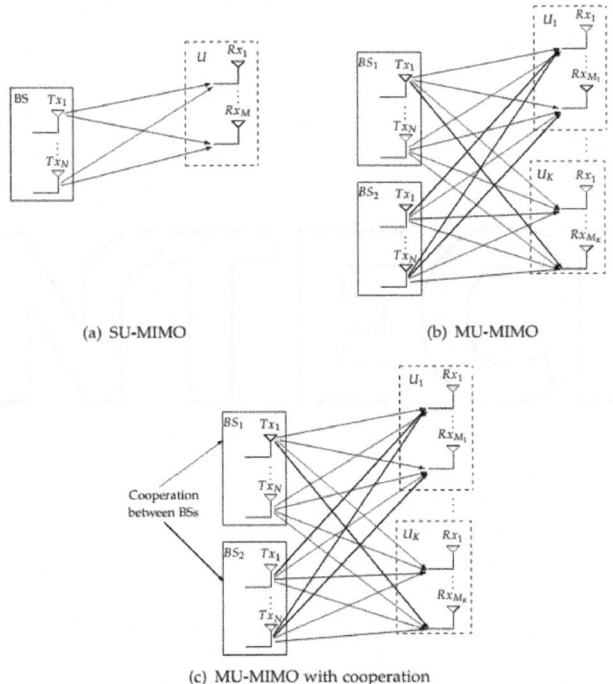

(a) SU-MIMO   (b) MU-MIMO

(c) MU-MIMO with cooperation

Figure 5: MU-MIMO configurations.

Table 1: Comparison between SU-MIMO and MU-MIMO systems

Feature	MU-MIMO	SU-MIMO
Main aspect	BS communicates with multiple users	BS communicates with a single user
Purpose	MIMO capacity gain	Data rate increasing for single user
Advantage	Multiplexing gain	No interference
CSI	Perfect CSI is required	No CSI
Throughput	Higher throughput at high SNR	Higher throughput at low SNR

COMMUNICATION SCHEMES FOR MU-MIMO SYSTEMS

Communication schemes for MU-MIMO systems include both uplink MU-MIMO (UL-MU-MIMO) and downlink MU-MIMO (DL-MU-MIMO). Case of uplink communication,users transmit signals to the base station. However, in the case of downlink communication, base station transmits signals to users. A representation of these systems is depicted in Figure 6. We assume that the base station is equipped with N antennas. Case of DL-MU-MIMO, the base station attempts to transmit signals to K users U_1 ,. . . ,U_K which are respectively equipped with antennas of numbers M_1 , . . . , M_K. For notations, if antenna k acts like a receiving antenna, it is denoted by Rx_k . Otherwise, it is denoted by Tx_k .

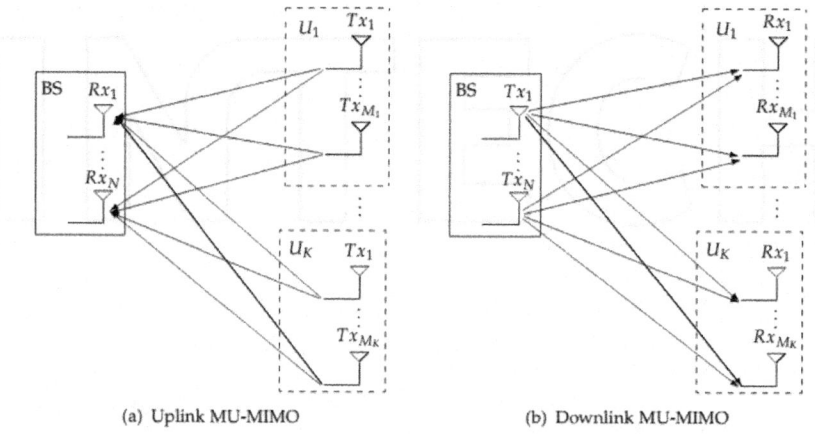

(a) Uplink MU-MIMO   (b) Downlink MU-MIMO

Figure 6: MU-MIMO communication models: UL-MU-MIMO and DL-MU-MIMO

UL-MU-MIMO

Let X_k ($M_k \times 1$), the transmit signal vector of user U_k ; k = 1 , . . . , K. We assume that data streams associated to user U_k ; k = 1 , . . . , K are zero mean white random vectors where :

$$E\{X_k X_k^*\} = I_{M_k}; \quad k = 1, \ldots, K \tag{7}$$

E denotes the expected value operator. The complex channel matrix relating user U_k; $k = 1, \ldots, K$ to the base station, H_k is of dimension ($N \times M_k$). In presence of additive noise signal $b(N \times 1)$, the received signal vector at the base station, $y(N \times 1)$ is expressed in the slow fading model by:

$$y = \sum_{k=1}^{K} H_k \cdot X_k + b \tag{8}$$

The noise signal vector is a zero mean white Gaussian variable with variance $\sigma^2 b$. The uplink scenario should satisfy two constraints:

- It should be as many receive antennas at the base station as the total number of users antennas.

Each user should have as many transmit antennas as the number of data streams. In Figure 7, the block diagram for the UL-MU-MIMO includes a joint linear precoder and decoder. Linear precoders associated to users $U_1, \ldots, U_K$ will be respectively denoted by

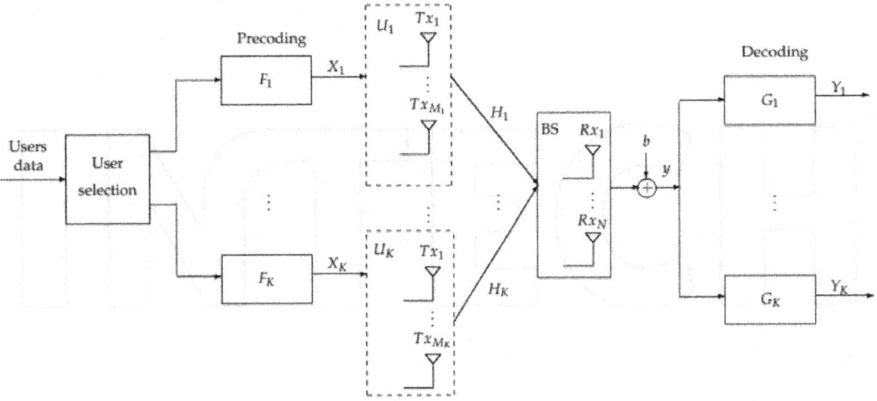

Figure 7: Block diagram for the UL-MU-MIMO with coding techniques: N antenna BS and K multiple antenna users $F_1, \ldots, F_K$. The received signal vector at the BS is then expressed as :

$$y = \sum_{k=1}^{K} H_k \cdot F_k \cdot X_k + b \tag{9}$$

An estimate of the transmitted signal vectors denoted by Y_k; $k = 1, \ldots, K$ are obtained by using the linear decoders $G_1, \ldots, G_K$. The decoding process is such that

$$Y_k = G_k \cdot y$$

DL-MU-MIMO

DL-MU-MIMO communication model assumes that K users are simultaneously receiving signals from the base station. The transmitted signal vector x(N × 1) is expressed as the sum of signals intended to users U_1 , . . . , U_K:

$$x = \sum_{k=1}^{K} X_k$$

(10)

The channel matrix between user U_k ; k = 1, . . . , K and the base station is denoted by H_k (M_k × N). At each user, received signal vector of dimension (M_k × 1); k = 1, , K is given by:

$$Y_k = H_k \cdot x + B_k \qquad ; \quad k = 1, \ldots, K$$

(11)

B_k ; k = 1, . . . , K is an additive noise signal vector of size (M_k × 1). Equation (11) could be also written:

$$Y_k = H_k \cdot x + B_k$$

(12)

$$= H_k \cdot X_k + \sum_{j \neq k}^{\wedge} (H_k \cdot X_j) + B_k \qquad ; \quad k = 1, \ldots, K$$

(13)

The second term of the sum in equation (13) represents the interference signal coming from multiple users. Processing techniques such as beamforming should be introduced in the block diagram of the MU-MIMO system for mitigating the effect of users interference and improving the performances of the communication system

FIELDS OF APPLICATION

MU-MIMO technology finds its applications in many areas and is nowadays exploited in many evolving technologies wich are described in the following. 3GPP LTE: 3rd Generation Partnership Project (3GPP) Long Term Evolution (LTE) is one of the next generation cellular networks which exploit the MU-MIMO technology. Thanks to this technology, available radio spectrum 3GPP LTE networks could achieve higher spectral efficiencies than existing 3G networks [18][19]. Release 8 of LTE: The first release of LTE (Release 8) was commercially deployed in 2009. Release 8 has introduced SU-MIMO scheme in the communication system model. This release only uses one transmission mode (Transmission mode 5) which has been defined for MU-MIMO systems. Transmission mode 5 supports rank 1 transmission for two User Equipments

(UEs). In order to achieve the performances of MU-MIMO systems, feedback parameters such as the channel Rank Indicator (RI) and the Channel Quality Indicator (CQI)/Precoding Matrix Indicator (PMI) feedback [17] are required. Release 9 of LTE: The second release of LTE (Release 9) provides enhancements to Release 8. The LTE Release 9 supports transmission mode 8 and includes both SU-MIMO and MU-MIMO schemes. LTE advanced: Other progress in LTE MIMO systems have been obtained through LTE advanced. The performed mode is the transmission mode 9. This mode allows for a possible switch between SU-MIMO and MU-MIMO. Multiple-cell networks: MU-MIMO systems have received wide spread success in wireless networks. Examples of applications include the multiple-cell networks [21] with multiple access channels where possible coordination among base stations is established. Figure 8 shows a MU-MIMO coordinated network in a cellular network. Three classes of cells are presented. These cells are referred as :

- Coordinated cells
- Central cell
- Interfering cells

The coordination between cells is performed by the Central Station (CS). The aim of this coordination is to mitigate the effect of inter-cells interference. Coding techniques should be employed in order to mitigate the effect of interfering cells.

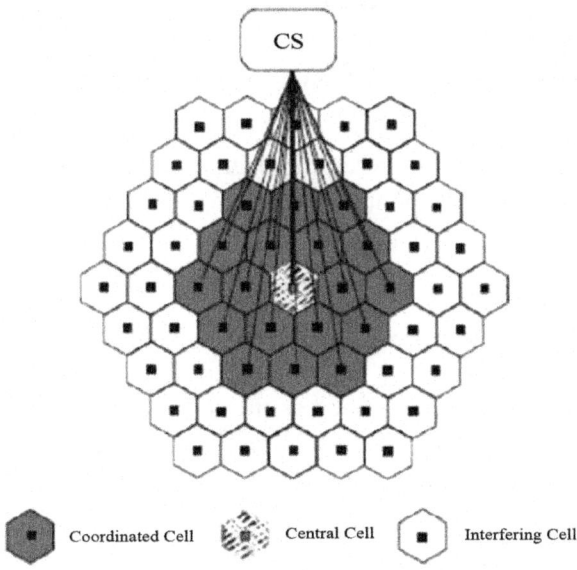

Figure 8: MU-MIMO coordinated network in cellular network [20]

Digital Subscriber Line (DSL): MU-MIMO systems are not only performed by multi-cell systems but also find their applications in other systems such as the downlink of a Digital Subscriber Line (DSL)[22][13]. The performance of MU-MIMO could be improved via the use of Orthogonal Frequency Division Multiplexing (OFDM) or Orthogonal Frequency Division Multiplexing Access (OFDMA) for multiple access scenarios in frequency selective channels. MU-MIMO systems could also improve multi-user diversity by performing High Data Rate (HDR) or Code Division Multiple Access (CDMA) techniques.

CAPACITY REGION OF MULTI-USER MIMO SYSTEM

There is no closed form for the channel capacity of multi-user MIMO systems. For this purpose, the performances of such systems will be analyzed in terms of the capacity region. This metric [23] could be defined in the usual Shannon sense as the highest rates that can be achieved with arbitrarily small error probability. Firstly, the capacity [24] needs to be evaluated for each user. Then, the capacity region is determined as the entire region for which maximum achievable rates are reached. The evaluation of the capacity region is strongly related to some constraints and should be determined according to the performed communication scenario. We address the following scenarios :

- UL-MU-MIMO with single antenna users
- UL-MU-MIMO with multiple antenna users
- DL-MU-MIMO with multiple antenna users and single antenna BS

Capacity region of UL-MU-MIMO with single antenna users

We consider the UL-MU-MIMO with N multiple antenna BS and K single antenna users. The performed communication scheme is depicted in Figure 9. The equivalent MIMO channel for the presented communication model is given by:

$$H = [H_1, \ldots, H_K]$$

(14)

$H_k(N \times 1); k = 1, \ldots, K$ represents the Single Input Multiple Output (SIMO) channel between user U_k ; k = 1, . . . , K and the BS. Case of two users (i.e. K=2), the capacity region is defined as the set of rates (R_1 , R_2) associated to users U_1 and U_2. We consider the notations:

- P_1 : average power constraint on user U_1
- P_2: average power constraint on user U_2
- N_0: noise signal power

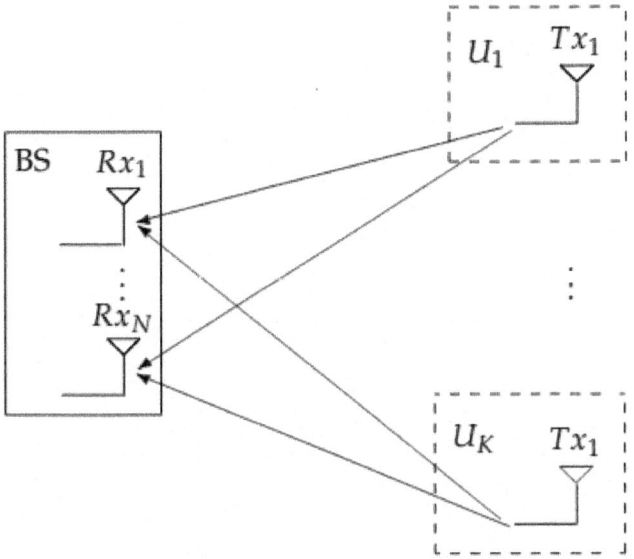

The capacity region [25] is evaluated by determining the individual rate constraint for each user. Assuming that user U_1 has the entire channel, an upper bound of the maximum achievable rate is given by

$$R_1 \leq \log_2 \left(1 + \frac{\|H_1\|^2 \cdot P_1}{N_0} \right)$$

(15)

$\| \cdot \|$ indicates the Frobenius norm. Similarly, an upper bound for the maximum achievable rate for user U_2 is:

$$R_2 \leq \log_2 \left(1 + \frac{\|H_2\|^2 \cdot P_2}{N_0} \right)$$

(16)

Finally, the sum rate constraint which is obtained when both users are acting as two transmit antennas of a single user has an upper bound expressed as :

$$R_1 + R_2 \leq \log_2 \left[\det \left(I_N + \frac{H.diag(P_1, P_2).H^*}{N_0} \right) \right]$$

(17)

The capacity region for the UL-MU-MIMO is presented in Figure 10 where two users with single antennas are considered.

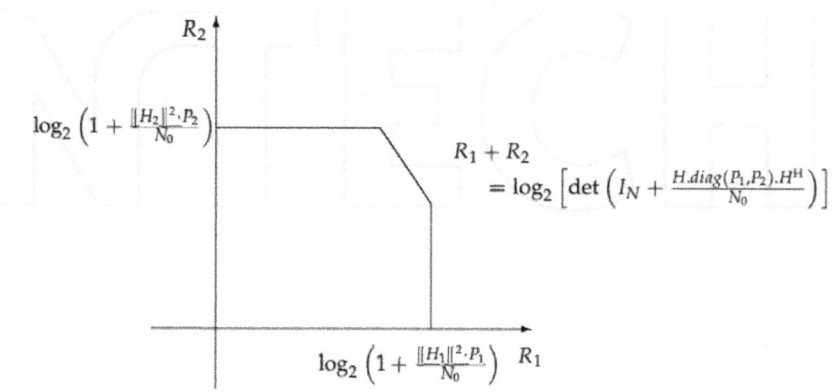

Figure 10: Capacity region of UL-MU-MIMO for two single antenna users

Case of K users, the capacity region is determined as a function of several constraints and K! corner points are determined for evaluating the boundary of the capacity region. For rates R_1 , ... , R_K respectively associated to users U_1 , ... , U_K, the sum rate is determined for an optimal receiver [25] as:

$$\sum_{k \in S} R_k \le \log_2 \left[\det \left(I_N + \frac{1}{N_0} \sum_{k \in S} P_k \cdot H_k \cdot H_k^* \right) \right] ; \quad S \subset \{1,...,K\}$$

(18)

Capacity region of UL-MU-MIMO with multiple antenna users

The capacity region could be obtained for the generalized case where the base station has N antennas and user U_k ; k = 1, ... , K is equipped with multiple antennas of number $M_k > 1$. An upper bound of the maximum achievable rate for user U_k is given by :

$$R_k \le \log_2 \left[\det \left(I_N + \frac{H_k \cdot D_k \cdot H_k^*}{N_0} \right) \right] \quad ; k = 1,...,K$$

(19)

- $H_k(N \times M_k)$ links the N antenna base station to the Mk antenna user; k = 1, ... , K.

- $D_k(M_k \times M_k)$ is a diagonal matrix formed by the power allocated at transmit antennas at user U_k.

The sum rate constraint of UL-MU-MIMO with multiple antennas users is expressed as :

$$R_1 + \ldots + R_K \leq \log_2 \left[\det \left(I_N + \frac{\sum_{k=1}^{K} H_k . D_k . H_k^*}{N_0} \right) \right]$$

$$(20)$$

DL-MU-MIMO with multiple antenna users and single antenna BS

In the case of downlink scenario, the upper bounds of the users rates are analogously determined as the uplink scenario. Nevertheless, the effect of interference could not be neglected. In fact, for the scenario with two multiple antenna users U_1 and U_2 and one antenna base station, the upper bounds of the rates achievable by users U_1 and U_2 become:

$$R_1 \leq \log_2 \left(1 + \frac{\|H_1\|^2 \cdot P_1}{N_0 + \|H_1\|^2 \cdot P_2} \right)$$

$$(21)$$

And

$$R_2 \leq \log_2 \left(1 + \frac{\|H_2\|^2 \cdot P_2}{N_0} \right)$$

$$(22)$$

Here, the signal of user U_2 is considered as interference for user U_1

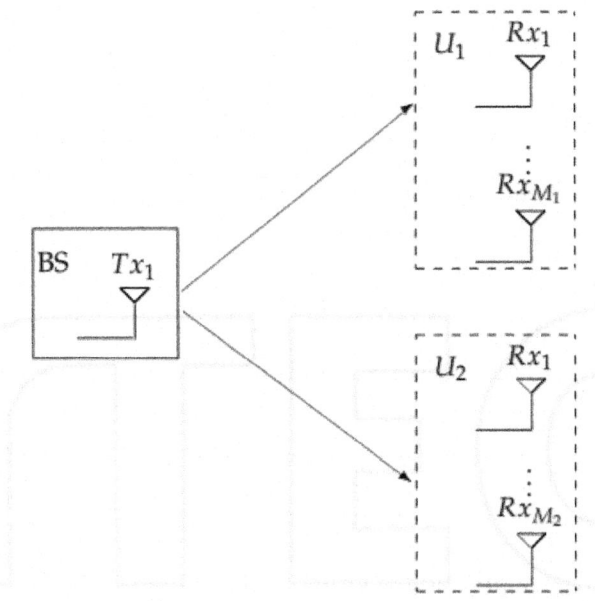

Figure 11: DL-MU-MIMO with multiple antenna users and single antenna BS

PRECODING TECHNIQUES

The DL-MU-MIMO system uses precoding techniques which are usually linear.

Zero Forcing and Block Diagonalization methods

Popular low-complexity techniques include both Zero Forcing (ZF) and Block Diagonalization (BD)[27][28] methods. Algorithms for the ZF as well as BD methods are presented in [29]. The aim of these solutions is to improve the sum rate capacity of the communication system under a given power constraint. These performances could be achieved by canceling inter-user interference. Zero Forcing Dirty Paper Coding (DPC) [30] represents a famous technique for data precoding where the channel is subject to interference which is assumed to be known at the transmitter. The precoding matrix is equal to the conjugate transpose of the upper triangular matrix obtained via the QR decomposition of the channel matrix.

MU-MIMO with Block Diagonalization precoding

We consider a communication system model with a broadcast MIMO channel where the transmitter is a base station equipped with N antennas and the receiver consists of K users U_k ; $k = 1 \ldots K$ (See figure 6(b)). The received signal at user U_k ; $k = 1 \ldots K$ with dimension ($M_k \times 1$) is expressed as

$$Y_k = H_k \cdot V_{BD}^{(k)} \cdot X_k + B_k \quad ; \quad k = 1, \ldots, K$$

(23)

$H_k(M_k \times N)$ is the channel matrix between user U_k and the base station

$V_{BD}^{(k)}(N \times M_k)$ is the BD precoding matrix for user U_k

X_k is the transmit signal for user U_k

$B_k(M_k \times 1)$ is the additive noise signal vector

We assume in the following that users $U_1, \ldots U_K$ have the same number of antennas which will be denoted by M. Block Diagonalization strategy defines a set of precoding matrices $V_{BD}^{(k)}(N \times M)$ associated to users $U_1, \ldots, U_K$. These matrices form an orthonormal basis such that:

$$[V_{BD}^{(k)}]^* \cdot V_{BD}^{(k)} = I_M \quad ; \quad k = 1 \ldots K$$

(24)

and the Block Diagonalization algorithm achieves :

$$H_k \cdot V_{BD}^{(j)} = 0 \quad ; \quad \forall \; j \neq K$$

(25)

The aim of these conditions is to eliminate multi-user interference so that to maximize the achievable throughput. The performance of downlink communication scenarios with precoding techniques depends on the SNR level. In fact, it has been shown in [27] that SU-MIMO achieves better performances than MU-MIMO at low SNRs. However, the BD MU-MIMO achieves better performances at high SNRs. As such, switching between SU-MIMO and MU-MIMO is optimal for obtaining better total rates over users.

MU-MIMO with Zero Forcing precoding

Case of Zero Forcing strategy, each transmitted symbol to the $l-$ th antenna (among M antennas of user u_k is precoded by a vector which is orthogonal to the columns of $H_j, \ j \neq k$ but not orthogonal to the $l-th$ column of H_k [26].

Beamforming for linear precoding

Beamforming paradigms represent another class of linear precoding for MU-MIMO systems. For the communication model with beamforming (Figure 12), we consider a MU-MIMO system with K multiple antenna users $U_1, \ldots, U_K$ at the receive side which are respectively equipped with $M_1, \ldots, M_K$ antennas. At the transmit side, a multiple antenna base station with N antennas transmits data signals $x_1, \ldots, x_K$ to users $U_1, \ldots, U_K$.

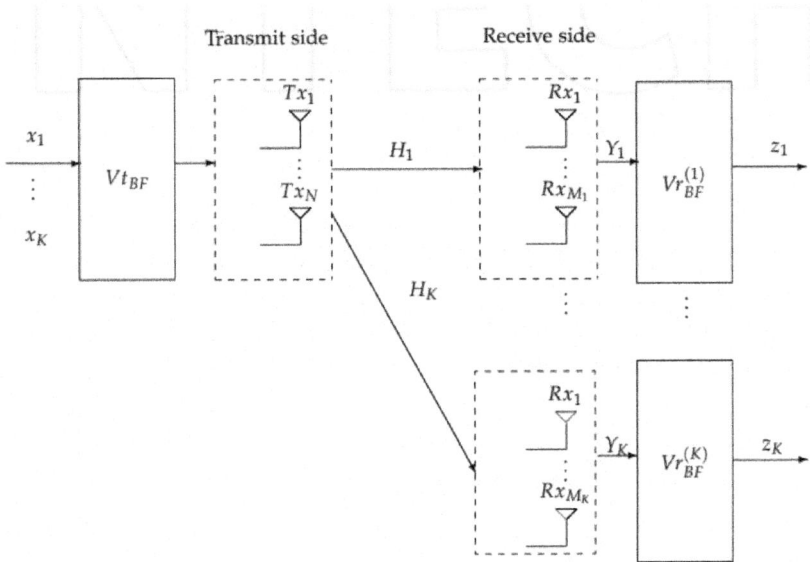

Figure 12: MU-MIMO with beamforming

The received signal vector at user U_k ; $k = 1, \ldots, K$ is expressed as :

$$Y_k = H_k \cdot Vt_{BF}^{(k)} \cdot x_k + \sum_{j=1,\, k \neq j}^{K} H_k \cdot Vt_{BF}^{(j)} \cdot x_j + B_k \quad ; \quad k = 1, \ldots, K \tag{26}$$

where:

$H_k(M_k \times N)$ is the complex channel matrix between receiver Uk and the transmit base station.

$B_k(M_k \times 1)$ is an additive noise signal vector.

$Vt_{BF}^{(k)}(N \times 1)$ is the transmit beamforming vector of index k. The transmit beamforming matrix is :

$$Vt_{BF} = [Vt_{BF}^{(1)}, \ldots, Vt_{BF}^{(k)}] \tag{27}$$

At the receive side, beamforming vectors are denoted by $Vr_{BF}^{(k)}(M_k \times 1)$.

$$Vr_{BF}^{(k)} = [vr_{BF}^{(1)}, \ldots, vr_{BF}^{(M_k)}]^{T} \tag{28}$$

The resulting signal at user U_k is:

$$z_k = Y_k^{*} \cdot Vr_{BF}^{(k)} \quad ; \quad k = 1, \ldots, K \tag{29}$$

The conjoint receive-transmit beamforming weights are obtained by maximizing the sum rate of the MU-MIMO system expressed as

$$R_{sum} = \sum_{k=1}^{K} \log_2(1 + SINR^{(k)}) \tag{30}$$

$SINR^{(k)}; \ k = 1, \ldots, K$ is the Single Interference Noise Ratio (SINR) [31] associated to user U_k . The SINR is determined as the ratio of the received strength for the desired signal to the strength of undesired signal obtained as the sum of noise and interference signal. For unit signal noise variance, the SINR for user U_k is given by

$$SINR^{(k)} = \frac{\| Vr_{BF}^{(k)} \cdot H_k^{*} \cdot Vt_{BF}^{(k)} \|^2}{\left(\sum_{k=1}^{K} \| Vr_{BF}^{(k)} \cdot H_k^{*} \cdot Vt_{BF}^{(k)} \|^2 \right) + 1} \tag{31}$$

Beamforming weights at the receiver are determined so that to suppress inter-user interference such as [32]:

$$Vr_{BF}^{(k)} = \frac{[C^{(k)}]^{-1} \cdot H_k \cdot Vr_{BF}^{(k)}}{\|[C^{(k)}]^{-1} \cdot H_k \cdot Vt_{BF}^{(k)}\|_2}$$

(32)

$C^{(k)}$ is the covariance matrix of H_k

$\|\cdot\|_2$ stands for the 2-norm operator

CONCLUSION

This chapter presents a basic introduction to the fundamentals of multi-user MIMO communication. MU-MIMO is considered as an enhanced form of MIMO technology. Such technology has been a topic of extensive research since the last three decades. The attractive features of MIMO systems have shown that the use of multiple antennas at both the ends of the communication link significantly improves the spectral efficiency of the communication system as well as the reliability of the communication link. In multiuser channels and cellular systems, MIMO is offered for MU-MIMO communication to allow for spatial sharing of the channel by several users. Nowadays, it has been a great deal with MU-MIMO systems. Several approaches are adopted and different scenarios may be considered. Throughout this chapter, we have presented possible configurations associated with MU-MIMO with particular emphasis on the fundamental differences between SU-MIMO and MU-MIMO. Some scenarios have been considered for performance evaluation of MU-MIMO communication in terms of the capacity region metric. MU-MIMO scenarios follow into UL-MU-MIMO for Multiple Access Channel (MAC) and the DL-MU-MIMO for Broadcast Channel (BC). The DL-MU-MIMO is the more challenging scenario since optimum strategies for interference cancelation are required. Througout this chapter, we have presented precoding techniques used within MU-MIMO systems for efficient transmission and interference cancelation. Among the existing techniques, we have introduced ZF and BD methods. Of particular interest, we have described the linear beamforming algorithms. The design of multi-user MIMO systems is attractive for the research field as well for the industrial one and the field of application is extensively growing

REFERENCES

1.   Ben Zid, M., Raoof, K. and Bouallègue, A. (2012). MIMO spectral efficiency over energy consumption requirements: Application to WSNs, Int'l J. of Communications, Network and System Sciences Vol. 5(No. 2): 121–129.

2.   Ben Zid, M., Raoof, K. and Bouallègue, A. (2012). MIMO Systems and

Cooperative Networks Performances, in: Advanced Cognitive Radio Network, Scientific Research Publishing, ISBN: 193–506–874–1.

3. Raoof, K. and Zhou, H. (2009). Advanced MIMO systems, Scientific Research Publishing, ISBN:978–1–935068–02–0.

4. Winters, J.H. (1987). On the capacity of radio communication systems with diversity in a Rayleigh fading environment, IEEE Journal on Selected Areas in Communications Vol. 5(No. 5): 871–878.

5. Telatar, I. E.(1995). Capacity of multi-antenna Gaussian channels, European Transactions on Telecommunications Vol. 10(No. 6): 585–595.

6. Marzetta, T. L. and Hochwald, B. M. (1999). Capacity of a mobile multiple-antenna communication link in Rayleigh flat fading, IEEE Transactions on Information Theory Vol. 45(No. 1): 139–157.

7. Almers, P., Bonek, E., Burr, A., Czink, N.and Debbah, M.(2007). Survey of channel and radio propagation models for wireless MIMO systems, EURASIP Journal on Wireless Communications and Networking Vol. 2007(No. 1): 1–19.

8. Ben Zid, M.(2012). Emploi de techniques de traitement de signal MIMO pour des applications dédiées réseaux de capteurs sans fil,Thesis dissertation-UJF-Grenoble I. http://tel.archives-ouvertes.fr/tel-00745006.

9. Prayongpun, N. (2009). Modélisation et étude de la capacité du canal pour un système multi-antennes avancé exploitant la diversité de polarisation, Thesis dissertation-UJF-Grenoble I http://tel.archives-ouvertes.fr/tel-00396666/en/

10. Dao, M.T., Nguyen, V.A., Im, Y.T., Park, S.O. & Yoon,G.(2011). 3D Polarized channel modeling and performance comparison of MIMO antenna configurations with different polarizations, IEEE Transactions on Antennas and Propagation Vol. 59(No. 7): 2672–2682.

11. Soysal, A. and Ulukus, S. (2007). Optimum power Allocation for Single-User MIMO and Multi-User MIMO-MAC with Partial CSI, IEEE Transactions on Selected Areas in Communications Vol. 25(No. 7): 1402–1412.

12. Serbetli, S. and Yener, A. (2004). Transceiver Optimization for Multiuser MIMO Systems, IEEE Transactions on Signal Processing Vol. 52(No. 1): 214–226.

13. Bengtsson, M. (2004). From single link MIMO to multi-user MIMO, Proceedings of IEEE International Conference on Acoustics, Speech, and Signal Processing (ICASSP '04), pp. iv-697–iv-700.

14. Codreanu, M., Tolli, A., Juntti, M. & Latva-aho, M.(2007). Joint Design

of Tx-Rx Beamformers in MIMO Downlink Channel, IEEE Transactions on Signal Processing Vol. 55(No. 9): 4639–4655.

15. Gomadam, K., Cadambe, V. R., & Jafar, S. A.(2011).A Distributed Numerical Approach to Interference Alignment and Applications to Wireless Interference Networks, IEEE Transactions on Information Theory Vol. 57(No. 6): 3309–3322.

16. Hassibi, B. and Sharif, M. (2007). Fundamental limits in MIMO broadcast channels, IEEE Journal on Selected Areas in Communications Vol. 25(No. 7): 1333–1344.

17. Duplicy, J., Badic, B. and Balraj, R. (2011). MU-MIMO in 4G systems, Eurasip Journal on Wireless Communications and Networking Vol. 2011: 1–12.

18. Frank, P., Müller, A. and Speidel, J. (2010). Performance of CSI-based multi-user MIMO for the LTE downlink, Proceedings of 6th International Wireless Communications and Mobile Computing Conference (IWCMC), USA, pp. 1086–1090.

19. Frank, P., Müller, A., Droste, H. and Speidel, J. (2010). Cooperative interference-aware joint scheduling for the 3GPP LTE uplink, Proceedings of IEEE 21st International Symposium on Personal Indoor and Mobile Radio Communications (PIMRC), pp. 2216–2221.

20. Benjebbour, A., Shirakabe, M., Ohwatari, Y., Hagiwara, J. and Ohya, T. (2008). Evaluation of user throughput for MU-MIMO coordinated wireless networks, Proceedings of IEEE 19th International Symposium on Personal, Indoor and Mobile Radio Communications (PIMRC), pp. 1–5.

21. Gesbert, D., Hanly, S., Huang, H., Shamai,S. and Yu, W.(2010). Multi-Cell MIMO Cooperative Networks: A New Look at Interference, IEEE Journal on Selected Areas in Communications Vol. 28(No. 9): 1380–1408.

22. Spencer, Q. H. and Peel, C. B. and Swindlehurst, A. L. and Haardt, M. (2004). An introduction to the multi-user MIMO downlink, IEEE Communications Magazine Vol. 42(No. 10): 60–67.

23. Weingarten, H., Steinberg, Y. and Shamai, S. (2006). The Capacity Region of the Gaussian MIMO Broadcast channel, IEEE Journal on Selected Areas in Communications Vol. 52(No. 9): 3936–3964.

24. Hemrungrote, S., Hori, T., Fujimoto, M. and Nishimori, K. (2010). Channel Capacity Characteristics of Multi-User MIMO Systems in Urban Area, Proceedings of 2010 IEEE Antennas and Propagation Society International Symposium (APSURSI), pp. 1–4.

25. Tse, D. and Viswanath, P. (2005). Fundamentals of wireless communication, Cambridge University Press, ISBN: 0–521–84527–0.

26. Ravindran, N. and Jindal, N. (2007). Limited feedback-based block diagonalization for the MIMO broadcast channel, IEEE Journal on Selected Areas in CommunicationsVol. 26(No.8): 1473–1482.

27. Zhang, J. and Andrews, J. G., and Heath, R. W. (2009) . Block Diagonalization in the MIMO Broadcast Channel with Delayed CSIT, Proceedings of IEEE Global Telecommunications Conference (GLOBECOM), Hawaii, pp. 1–6.

28. Zhang ,R. (2010). Cooperative Multi-Cell Block Diagonalization with Per-Base-Station Power Constraints, Proceedings of IEEE Wireless Communications and Networking Conference (WCNC), pp. 1–6.

29. Spencer, Q.H., Swindlehurst, A.L.and Haardt, M. (2004). Zero-forcing methods for downlink spatial multiplexing in multiuser MIMO channels, IEEE Transactions on Signal ProcessingVol. 52(No.2): 461–471.

30. Lee, J. and Jindal, N. (2007). High SNR Analysis for MIMO Broadcast Channels: Dirty Paper Coding versus Linear Precoding, IEEE Transactions on Information Theory Vol. 53(No.12): 4787–4792.

31. Tarighat, A., Sadek, M. and Sayed, A.H. (2005). A multi user beamforming scheme for downlink MIMO channels based on maximizing signal-to-leakage ratios, Proceedings of IEEE International Conference on Acoustics, Speech, and Signal Processing (ICASSP), Philadelphia, pp. iii/1129–iii/1132.

32. Mundarath, J.C. and Kotecha, J.H. (2008). Multi-User Multi-Input Multi-Output (MU-MIMO) Downlink Beamforming Systems with Limited Feedback, Proceedings of IEEE Global Telecommunications Conference (GLOBECOM), pp. 1–6.

Chapter 3

PHYSICAL LAYER SECURITY FOR MULTIUSER MIMO COMMUNICATIONS

Giovanni Geraci[1,2], and Jinhong Yuan[1]

[1] School of Electrical Eng. & Telecommunications, The University of New South Wales, Australia

[2] Wireless and Networking Technologies Laboratory, CSIRO ICT Centre, Sydney, Australia

INTRODUCTION

Wireless multi-user MIMO communications are used more and more often to exchange sensitive data. Because of the broadcast nature of the physical medium, unauthorized receivers located within the transmission range can observe the signals sent by the transmitter to a legitimate receiver and eavesdrop them. Therefore, security has become an extremely important issue to deal with. Multiuser MIMO communications are particularly sensitive to the problem of security, because each confidential message must be kept secret not only from external nodes, but also from all the users other than the intended one. Traditionally, wireless security is ensured by network-layer cryptography techniques. However, these techniques may not be suitable in the case of large dynamic wireless networks, since they raise issues like key distribution and management (for symmetric cryptosystems), and high computational complexity (for asymmetric cryptosystems).

Moreover, these schemes are potentially vulnerable, since they rely on the limited resources of the eavesdropper and on the unproven assumption that certain encryption algorithms are hard to invert. Methods exploiting the randomness inherent in noisy channels, known as physical layer security, have been proposed to enhance the protection of transmitted data and achieve perfect secrecy [1, 2]. Physical-layer security allows secret communications over a wireless channel without requiring an encryption key, and it works by limiting the amount of information that can be extracted at the physical level by an unintended receiver. This is performed by designing appropriate coding and precoding schemes, and by exploiting the channel state information available at the network nodes [3]. Physical layer security for communications was proposed in the 1970's by Wyner [4], who studied a three-terminal network consisting of a transmitter, an intended user and an eavesdropper, known as the wiretap channel. For this network, the secrecy capacity was defined as the maximum rate at which a message can be transmitted reliably to the intended user while the rate of information leakage to the eavesdropper vanishes asymptotically with the code length. For the case when the eavesdropper's channel is a degraded version of the intended user's channel, Wyner showed that it is possible to have secret communication without using an encryption key. This can be achieved by a randomized coding scheme where the information is hidden in the additional noise seen by the eavesdropper. Each message is mapped to many codewords, thus inducing maximal equivocation at the eavesdropper. Csizar and Korner generalized Wyner's work by considering a nondegraded version of the wiretap channel [5].

Physical layer security was then applied to Gaussian channels [6], and it was observed that a secret transmission can be achieved only if the channel at the eavesdropper is noisier than the channel at the intended user. The presence of slow fading was shown to significantly change the situation, since it allows the transmitter to employ a variable-rate transmission, thus achieving secrecy even when the eavesdropper's channel is better than the intended receiver's channel on average [7]. Also the use of multiple antennas can enhance the secrecy capability, because it enables the transmitter to beamform in a direction as orthogonal to the eavesdropper and as close to the intended user as possible [8–10]. Even when the channel at the eavesdropper is unknown by the transmitter, artificial noise can be transmitted to degrade the eavesdropper's channel and thus reduce its signal-to-noise ratio, while being harmless to the intended receiver [11–13].

More recently, physical layer security has also been extended to multiuser MIMO channels. In this chapter, we will survey the research in the field of physical layer security for multiuser MIMO communications, especially focusing on the case when multiple malicious users are present in the network, and they can eavesdrop on each other. For these complex scenarios, we will present some suboptimal low-complexity transmission schemes, discuss their performance and quantify the sum-rate penalties imposed by the secrecy requirements and by the presence of multiple users. We will discuss the challenges that arise in networks with a large number of malicious receivers, we will identify potential ways to deal with these challenges, and present an outlook on future directions for research.

Since the transmitter cannot always predict the behavior of the users, the multiuser MIMO channel with malicious users is now regarded with large interest. This is also denoted as the broadcast channel with confidential messages (BCC). Consider a broadcast channel with two independent confidential messages sent to two receivers, where each receiver acts as an eavesdropper for the other one. In other words, the first message is intended for the first receiver but needs to be kept secret from the second receiver, and viceversa. This scenario was studied in [15] for the multiple-input single-output (MISO) Gaussian case and in [16] for general MIMO Gaussian case. In this case it was shown that both confidential messages can be simultaneously transmitted at their respective maximum secrecy rates, and the achievability was obtained using the dirty-paper coding.

Let us cosider now a larger multiuser network with more than two malicious users. For this network, it is required that the base station (BS) securely transmits each confidential message, ensuring that none of the other unintended users receive any information. Since in general the behavior of the users cannot be determined by the transmitter, a conservative worst-case scenario can be assumed for each user, where all the remaining users can cooperate to jointly eavesdrop. In this case, for each user, the alliance of the cooperating eavesdropper is equivalent to a single multi-antenna eavesdropper. The MISO BCC with a generic number of malicious receivers was studied in [17, 18], and it consists of a BS with M antennas that simultaneously transmit independent confidential messages to K spatially dispersed single-antenna users, which can cooperate and eavesdrop on each other. Although determining the secrecy capacity region for the generic MISO BCC is still an open problem, suboptimal transmission schemes have been proposed to achieve high secrecy sum-rates by controlling the amount of crosstalk between the users [19]. These schemes are based on linear precoding, and unlike dirty-paper coding, their low complexity makes them suitable for practical implementation. In the following sections, we present some new results on the secrecy sum-rates achieved by multiuser MIMO linear precoding.

PHYSICAL LAYER SECURITY WITH MULTI-USER MIMO LINEAR PRECODING

Although suboptimal, linear precoding schemes are of particular interest because of their low-complexity implementations and because they can control the amount of crosstalk between the users to maintain a high sum-rate in the broadcast channel [20–27]. In the MISO BCC, linear precoding can be employed to control the amount of interference and information leakage to the unintended receivers introduced by the transmission of each confidential message [17–19]. Let the transmitted signal be denoted by x, then the received signal is given by

$$y = Hx + n \tag{1}$$

where $H = [h_1,\ldots,h_K]$ is the $K \times M$ channel matrix, h_k is the k-th column of H and it represents the channel between the BS and the k-th user, and n is complex Gaussian noise. In linear precoding, the transmitted vector x is derived from the vector containing the confidential messages $u = [u_1,\ldots,u_K]^T$ through a deterministic linear transformation (precoding) [22–25]. Let

$W = [w_1,\ldots,w_K]$ be the $M \times K$ precoding matrix, where w_k is the k-th column of W. Then the transmitted signal is

$$x = Wu = \sum_{k=1}^{K} w_k u_k. \tag{2}$$

Achievable secrecy sum-rates with linear precoding

The secrecy sum-rates achievable by linear precoding were obtained in [18] by considering the worst-case scenario, where for each intended receiver k the remaining K − 1 users can form an alliance $\tilde{k}$, and cooperate to jointly eavesdrop on the message u_k. By noting that each user k, along with its own eavesdropper ˜k and the transmitter, forms an equivalent multi-input, single-output, multi-eavesdropper (MISOME) wiretap channel [10], an achievable secrecy sum-rate Rs is given by

$$R_s = \sum_{k=1}^{K} \max \left\{ \log_2 \left(1 + \text{SINR}_k\right) - \log_2 \left(1 + \text{SINR}_{\tilde{k}}\right), 0 \right\}, \tag{3}$$

where SINR_k and $\text{SINR}_{\tilde{k}}$ are the signal-to-interference-plus-noise ratios for the message u_k at the intended receiver k and the eavesdropper ˜k, respectively, given by

$$\text{SINR}_k = \frac{\rho \left|h_k^H w_k\right|^2}{1 + \rho \sum_{j\neq k} \left|h_k^H w_j\right|^2} \tag{4}$$

And

$$\text{SINR}_{\tilde{k}} = \rho \left\|H_k w_k\right\|^2, \tag{5}$$

and where ρ is the transmit SNR, and H_k is the matrix obtained from H by removing the k-th row

Particular attention was given to the Regularized Channel Inversion (RCI) precoder, because it achieves better performance than the plain Channel Inversion precoder, especially at low SNR [24, 25]. A linear precoder based on RCI was proposed for the MISO BCC in [19]. The RCI precoding matrix is given by

$$\mathbf{W} = \frac{1}{\sqrt{\gamma}}\mathbf{H}^H \left(\mathbf{H}\mathbf{H}^H + M\xi\mathbf{I}_K\right)^{-1},$$

(6)

where γ is a long-term power normalization constant, given by

$$\gamma = \mathrm{tr}\left\{\mathbf{H}^H\mathbf{H}(\mathbf{H}^H\mathbf{H} + M\xi\mathbf{I}_M)^{-2}\right\}.$$

(7)

For each message, the function of the regularization parameter ξ is to achieve a tradeoff between maximizing the signal power at the intended user and minimizing the interference and information leakage at the other unintended users. In [19], the regularization parameter is optimized to maximize the secrecy sum-rate.

Large-system results

The secrecy sum-rate achievable by the RCI precoder in the MISO BCC was obtained in [19] by large-system analysis, where both the number of receivers K and the number of transmit antennas M approach infinity, with their ratio β = K/M being held constant. Unless otherwise stated, the results presented in the following refer to the large-system regime. We note that these results are accurate even when applied to small systems with a finite number of users. An expression for the secrecy sum-rate R_s° in the large-system regime is given by [19]

$$R_s^\circ = \max\left\{K\log_2 \frac{1 + g\,(\beta,\xi)\,\frac{\rho + \frac{\alpha_s^2}{\beta}[1+g(\beta,\xi)]^2}{\rho + [1+g(\beta,\xi)]^2}}{1 + \frac{\rho}{(1+g(\beta,\xi))^2}}, 0\right\}$$

(8)

With

$$g\,(\beta,\xi) = \frac{1}{2}\left[\mathrm{sgn}(\xi)\cdot\sqrt{\frac{(1-\beta)^2}{\xi^2} + \frac{2\,(1+\beta)}{\xi} + 1} + \frac{1-\beta}{\xi} - 1\right].$$

(9)

In [19], a closed form expression was also derived for the optimal regularization parameter $\xi^{\star\circ}$, given by

$$\xi^{\star\circ} = \frac{-2\rho^2\,(1-\beta)^2 + 6\rho\beta + 2\beta^2 - 2\,[\beta\,(\rho+1) - \rho]\cdot\sqrt{\beta^2\,[\rho^2 + \rho + 1] - \beta\,[2\rho\,(\rho-1)] + \rho^2}}{6\rho^2\,(\beta+2) + 6\rho\beta}.$$

(10)

For the specific case $\beta = 1$, i.e. $M = K$, the value of $\zeta^{\star\circ}$ reduces to [18]

$$\zeta^{\star\circ} = \frac{1}{3\rho + 1 + \sqrt{3\rho + 1}}, \quad \text{for } \beta = 1.$$

(11)

We note that the value of the regularization parameter $\xi^{\star\circ}$ that maximizes the secrecy sum-rate differs from the value $\zeta^{\star\circ}_{ns} = \beta/\rho$ that maximizes the sum-rate without secrecy requirements [28].

By substituting the optimal value of the regularization parameter (10) in (8), it is possible to obtain the optimal secrecy sum-rate $R_s^{\star\circ}$ achievable by RCI precoding in the large-system regime. The secrecy sum-rate $R_s^{\star\circ}$ is a function of K, β and ρ. When $\beta = 1$, the optimal secrecy sum-rate $R_s^{\star\circ}$ has a simple expression, given by

$$R_s^{\star\circ} = K \log_2 \frac{9\rho + 2 + (6\rho + 2)\sqrt{3\rho + 1}}{4(4\rho + 1)}, \quad \text{for } \beta = 1.$$

(12)

Although the optimal value of the regularization parameter $\xi^{\star\circ}$ in (10) was derived in [19] in the large-system regime, using $\xi^{\star\circ}$ in a finite-size system does not cause a significant loss in the secrecy sum-rate compared to using a regularization parameter ξfs(H) optimized for every channel realization.

Fig. 1 shows the complementary cumulative distribution function (CCDF) of the normalized secrecy sum-rate difference between using $\xi^{\star\circ}$ and ξfs(H) as the regularization parameter of the RCI precoder for K = 4, 8, 16, 32 users, for $\beta = 1$ and at an SNR of 10dB. The difference is normalized by dividing by the secrecy sum-rate of the precoder that uses ξ_{fs}(H). We observe that the average normalized secrecy sum-rate difference is less than 2.4 percent for all values of K. As a result, the large-system regularization parameter $\xi^{\star\circ}$ may be used instead of the finite-system regularization parameter with only a small loss of performance. Moreover, the value of $\xi^{\star\circ}$ does not need to be calculated for each channel realization.

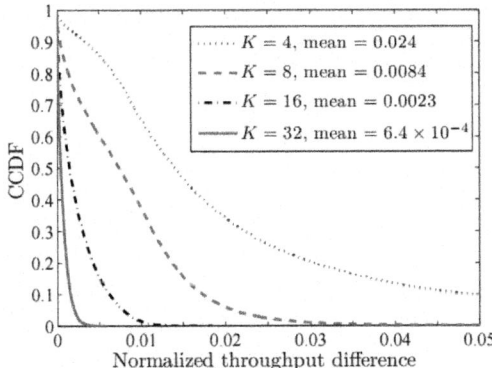

Figure 1: Complementary cumulative distribution function (CCDF) of the normalized secrecy sum-rate difference between using $\zeta_{fs}(\mathbf{H})$ and $\zeta^{\star\circ}$, with $\beta = 1$ and $\rho = 10$dB.

Fig. 2 compares the secrecy sum-rate $R_s^{\star\circ}$ of the RCI precoder from the large-system analysis to the simulated ergodic secrecy sum-rate Rs with a finite number of users, for different values of β. We observe that as M increases, the simulated rates approach the curves from large-system analysis. For $\beta \leq 1$, $R_s^{\star\circ}$ is always positive and monotonically increasing with the SNR ρ. However when $\beta > 1$, the secrecy sum-rate does not monotonically increase with ρ. There is an optimal value of the SNR beyond which the achievable secrecy sum-rate $R_s^{\star\circ}$ starts decreasing, until it becomes zero for large SNR. When $\beta \geq 2$ no positive secrecy sum-rate is achievable at all [19].

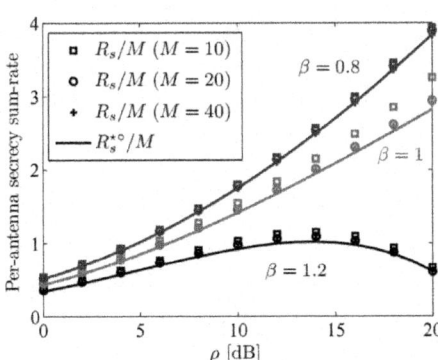

Figure 2: Comparison between the secrecy sum-rate with RCI precoding in the large-system regime (8) and the simulated ergodic secrecy sum-rate for finite M. Three sets of curves are shown, each one corresponds to a different value of β.

The expression of the secrecy sum-rate $R_s^{\star o}$ becomes simpler in the limit of large SNR. In fact, it can be approximated by

$$R_s^{\star o} \approx \begin{cases} K \log_2 \frac{1-\beta}{\beta} + K \log_2 \rho & \text{for } \beta < 1 \\ \frac{K}{2} \log_2 \frac{27}{64} + \frac{K}{2} \log_2 \rho & \text{for } \beta = 1 \\ \max \left\{ 3K \log_2 \frac{\beta}{\beta-1} - K \log_2 \rho, 0 \right\} & \text{for } \beta > 1 \end{cases} , \quad \text{as } \rho \to \infty.$$

$$(13)$$

We note from (13) that for high SNR, the behavior of the secrecy sum-rate significanly depends on the ratio β between the number of users K and the number of transmit antennas M. When K < M, the secrecy sum-rate scales linearly with the factor K. If K = M, the scaling factor reduces to K/2. When the number of users K exceeds the number of antennas M, then the secrecy sum-rate decreases with the SNR ρ, and there is a value of ρ beyond which the achievable secrecy sum-rate becomes zero.

Effect of the network load

Fig. 3 depicts the asymptotic secrecy sum-rate per transmit antenna as a function of β, for several values of the SNR. We denote by β_{opt} the value of the ratio β that maximizes the secrecy sum-rate per transmit antenna $R_s^{\star o}$ /M. It is possible to see from Fig. 3 that β_{opt} is an increasing function of the SNR. The value of β_{opt} falls between 0 and 1, and it tends to 1 in the limit of large SNR.

We denote by β_{max} the maximum value of β allowed for non-zero secrecy sum-rates. The value of β_{max} represents the maximum number of users per transmit antenna that can be

served with non-zero secrecy sum-rate. Fig. 3 shows that β_{max} is a decreasing function of the SNR. The value of β_{max} can be found by solving the following cubic equation [19]

$$(\rho + 1) \beta_{max}^3 - (3\rho + 2) \beta_{max}^2 + 3\rho \beta_{max} - \rho = 0. \qquad (14)$$

The value of β_{max} falls between 1 and 2. This means that if $K \geq 2M$, i.e. if $\beta \geq 2$,, then the secrecy sum-rate is zero for all SNRs. In the limit of large SNR, equation (14) reduces to

$$(\beta_{max} - 1)^3 = 0, \qquad (15)$$

yielding to β_{max} = 1. These results can be explained as follows. In the worst-case scenario, the alliance of cooperating eavesdroppers can cancel the interference, and its received SINR is the ratio between the signal leakage and the thermal noise. In the limit of large SNR, the thermal noise vanishes, and the only means for the transmitter to limit the eavesdropper's SINR is by reducing the signal leakage to zero by inverting the channel matrix. This can only be accomplished when the number of transmit antennas is larger than or equal to the number of users, hence only if $\beta \leq 1$. When $\beta > 1$ this is not possible, and no positive secrecy sum-rate can be achieved. When $\beta \geq 2$, the eavesdroppers are able to drive the secrecy sum-rate to zero irrespective of ρ. This is consistent with the results presented in [10] for a single-user system.

THE COST OF PHYSICAL LAYER SECURITY IN MULTI-USER MIMO

Guaranteeing secrecy and serving multiple (and potentially malicious) users at the same time both come at a cost in terms of the per-user transmission rate. In this section, we discuss the cost of achieving physical layer security in multiuser MIMO communications. 4.1. Secrecy loss The cost due to the secrecy requirements, which we denote by secrecy loss, can be obtained by comparing the secrecy sum-rate $R_s^{\star\circ}$ s achieved by the RCI precoder to the sum-rate $R_s^{\star\circ}$ achieved by an optimized RCI precoder without secrecy requirements. The gap between $R_s^{\star\circ}$ and R $^{\star\circ}$ represents how much guaranteeing secrecy costs in terms of the achievable sum-rate. The optimal sum-rate $R_s^{\star\circ}$ without secrecy requirements is obtained by using the precoder in (6), and it is given by [29]

$$R^{\star\circ} = K \log_2 \left[1 + g\left(\beta, \zeta_{ns}^{z\star\circ}\right) \right], \tag{16}$$

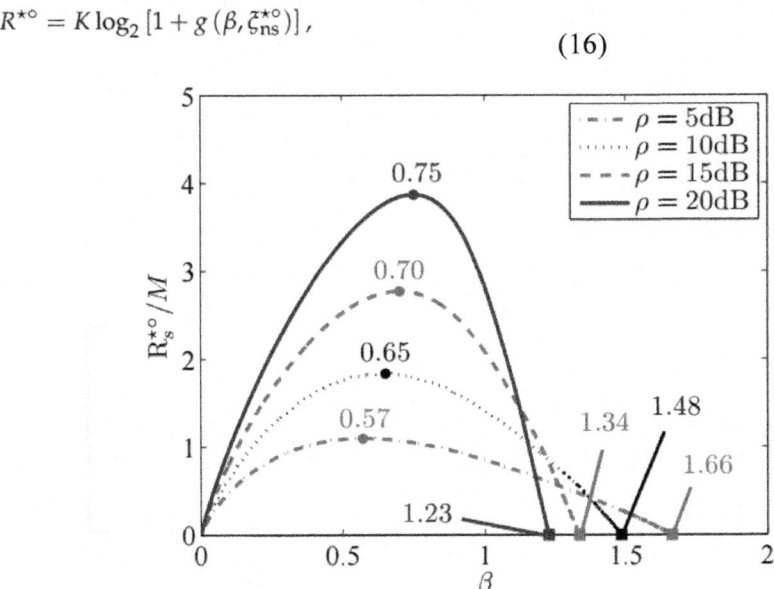

Figure 3: Asymptotic secrecy sum-rate per transmit antenna with RCI as a function of β. Circles denote β_{opt}, squares denote β_{max}.

With $\xi_{ns}^{\star\circ} = \beta/\rho$. It is easy to show that $R^{\star\circ} \geq 0$ for all values of β and ρ, with equality only for $\rho = 0$, and that R $^{\star\circ}$ tends to zero as $\beta \to \infty$. Hence, there is no limit to the number of users per transmit antenna β that the system can accommodate with a non-zero sum-rate. However if we impose the secrecy requirements, the secrecy sum-rate R $_s^{\star\circ}$ is zero for $\beta \geq \beta_{max}$, with β_{max} given by (14). Therefore, introducing the secrecy requirements will limit to βmax the number of users per transmit antenna that can be served with a non-zero sum-rate.

We now compare the secrecy sum-rate $R_s^{\star\circ}$ to the sum-rate R $^{\star\circ}$ in the limit of large SNR. Again by using the regularization parameter $\xi_{ns}^{\star\circ} = \beta/\rho$ we obtain the following large-SNR approximation for the secrecy sum-rate without secrecy requirements [19]

$$R^{\star\circ} \approx \begin{cases} K\log_2 \frac{1-\beta}{\beta} + K\log_2 \rho & \text{for } \beta < 1 \\ \frac{K}{2}\log_2 \rho & \text{for } \beta = 1 \ , \quad \text{as } \rho \to \infty. \\ K\log_2 \frac{\beta}{\beta-1} & \text{for } \beta > 1 \end{cases}$$

(17)

By comparing (17) to (13), we can draw the following conclusions regarding the large-SNR regime. If the number of transmit antennas M is larger than the number of users K, then $\beta < 1$, $R_s^{\star\circ} = R^{\star\circ}$, , and the secrecy requirements do not decrease the sum-rate of the network. Therefore, by using $\xi \star\circ$ from (10) one can achieve secrecy while maintaining the same sum-rate, i.e. there is no secrecy loss. If M = K, then $\beta = 1$, the secrecy requirements only reduce the sum-rate by a constant value, and the scaling factor K/2 remains unchanged. Alternatively, one can achieve secrecy while maintaining the same sum-rate, by increasing the transmitted power by a factor $64/27 \approx 3.75$dB. If M < K, i.e. $\beta > 1$, then the secrecy requirements result in a value of $R_s^{\star\circ}$ that decreases with the SNR, as opposed to a constant sum-rate $R_s^{\star\circ}$ without secrecy. Therefore if the SNR is too large, then the secrecy sum-rate becomes zero

Multiuser Loss

The cost due the interference caused by the presence of multiple users in the system, which we denote by multiuser loss, is given by the gap between the per-user secrecy rate $R_s^{\star\circ}/K$ and the secrecy capacity $C_s S_U$ of the single-user MISOME wiretap channel, where one user is served at a time and the remaining users can eavesdrop. The value of $C_s S_U$ was obtained in [10], and for large SNR it can be approximated by

$$C_{s,SU} \approx \begin{cases} \log_2 \rho & \text{for } \beta < 1 \\ \frac{1}{2} \log_2 \rho & \text{for } \beta = 1 \ , \quad \text{as } \rho \to \infty. \\ \max\left\{\log_2 \frac{1}{(\beta-1)}, 0\right\} & \text{for } \beta > 1 \end{cases}$$

$$\text{(18)}$$

We compare $R_s^{\star\circ}/K$ to $C_{s,SU}$ in the large-SNR regime. We note that in $C_{s,SU}$ from [10] a single-user system is considered. Therefore, only one message is transmitted to one legitimate user, and the user does not experience any interference. By comparing (18) to $R^{\star\circ} s/K$, we can conclude that for $\beta \le 1$, the RCI precoder achieves a per-user secrecy rate which has the same linear scaling factor as the secrecy capacity of a single-user system with no interference. When $1 < \beta < 2$, the presence of interference results in a value of $R_s^{\star\circ}/K$, that decreases with the SNR, as opposed to a constant value for $C_{s,SU}$. When $\beta \ge 2$, the secrecy rate is zero irrespective of the presence of interference.

Power allocation

In some cases, the rate loss generated by the secrecy requirements and by the interference due to the presence of multiple users can be compensated by a power allocation scheme. In [18], an iterative power allocation algorithm was proposed to obtain the maximum secrecy sum-rate for a fixed regularization parameter ξ. The algorithm was also extended to maximize the secrecy sum-rate by jointly optimizing the regularization parameter ξ and the power allocation vector. However, in many cases there is a negligible performance difference between the joint and the separate optimization. As a result, a low-complexity, near-optimal RCI precoder may be implemented by using $\xi^{\star\circ}$ from (10) and optimizing the power vector separately [18]. The RCI precoder with optimal power allocation (RCI-PA) outperforms the RCI precoder with equal power (RCI-EP), and the gain does not vanish at high SNR. The RCI-PA precoder thus reduces the rate loss due to secrecy requirements and interference, and in some cases it achieves a per-user rate which is as high as the rate achieved by the optimal RCI-EP precoder without secrecy requirements, and as high as the secrecy capacity of a single-user system [18].

Numerical results

Fig. 4 compares the simulated ergodic sum-rates Rs and R of the RCI precoder with and without secrecy requirements, respectively. These were obtained by using the regularization parameters $\xi^{\star\circ}$ and $\xi_{ns}^{\star\circ}$, respectively. For $\beta < 1$, the difference between R and Rs becomes negligible at large SNR, and secrecy can be achieved without additional costs. For $\beta = 1$, the two curves tend to have same slope at large SNR, but there is a residual gap between them. Therefore, secrecy can be achieved at a lower sum-rate. We note that in order to achieve secrecy without decreasing the sum-rate, the required additional power is less than 4dB at all SNRs. For $\beta > 1$, the sum-rate tends to saturate for large SNR, whereas the secrecy sum-rate starts decreasing. If the SNR is too large, then the secrecy requirements force the sum-rate to zero.

Fig. 4 also shows the simulated secrecy capacity $C_{s,SU}$ of the MISOME wiretap channel. For $\beta \leq 1$, the RCI precoder achieves a per-user secrecy rate which has the same linear scaling factor as $C_{s,SU}$. When $1 < \beta < 2$, $C_{s,SU}$ saturates at high SNR, while the secrecy sum-rate decreases. All these numerical results confirm the ones obtained from the large-system analysis.

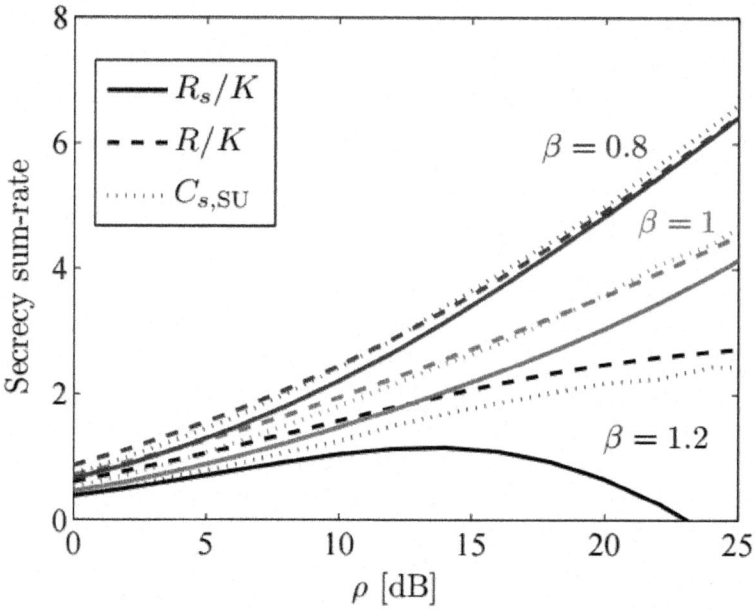

Figure 4: Comparison between the simulated ergodic per-user secrecy rate with RCI (solid) and the two upper bounds: (i) per-user rate without secrecy requirements (dashed) and (ii) MISOME secrecy capacity (dotted), for $K = 12$ users. Three values of β are considered: 0.8, 1, and 1.2, corresponding to $M = 15$, 12 and 10 antennas.

Fig. 5 shows the simulated per-user secrecy rate of the RCI-PA precoder from [18], with optimal power allocation. This is compared to the RCI-EP precoder. Fig. 5 also shows that the power allocation scheme reduces the sum-rate loss due to the secrecy requirements. For $\rho \geq 15$dB, RCI with power allocation achieves a per-user secrecy rate which is even higher than the per-user rate achieved by the optimal RCI-EP without secrecy requirements. Furthermore, Fig. 5 shows the simulated secrecy capacity $C_{s,SU}$ of a MISOME channel with the same per-message transmit power. Although $C_{s,SU}$ is obtained in a single-user and interference-free system [10], at high SNR, RCI with power allocation achieves a per-user secrecy rate as large as $C_{s,SU}$.

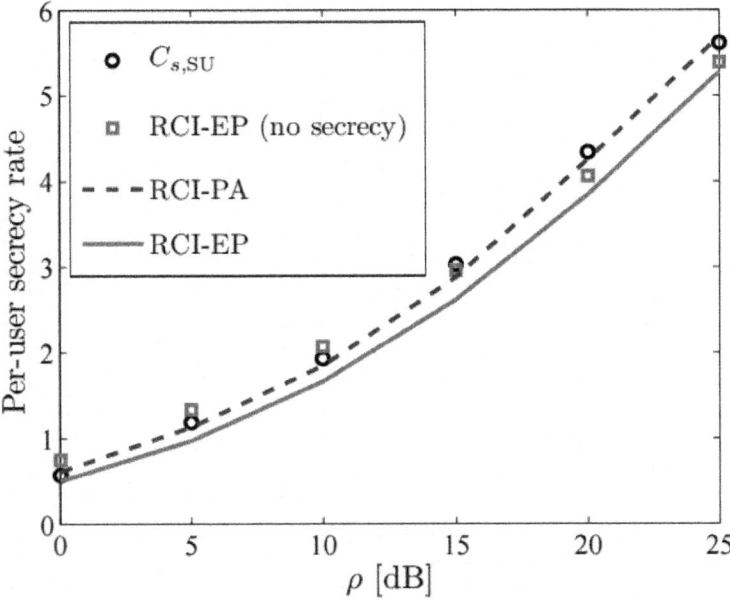

Figure 5: Per-user secrecy rate vs. ρ for $\beta = 1$ and K = 4 users: with equal power allocation (solid) and with optimal power allocation (dashed). The rate of the optimal RCI precoder without secrecy requirements (squares) and the secrecy capacity of the MISOME channel (circles) are also plotted.

CURRENT RESEARCH ON MULTIUSER

MIMO physical layer security Before concluding this chapter, we briefly discuss current research topics on physical layer security for multiuser MIMO communications, and we mention possible extensions of the results presented.

Power reduction strategy

Since for $\beta > 1$ the RCI precoder performs poorly in the high-SNR regime, a linear precoder based on RCI and power reduction could significantly increase the high-SNR secrecy sum-rate. In fact, we can observe from Fig. 2 that when $\beta > 1$ there is an optimal value of the SNR beyond which the achievable secrecy sum-rate R_s^{*o} starts decreasing.

A power reduction strategy would prevent the secrecy sum-rate from decreasing at high SNR by reducing the transmit power, and therefore reducing the SNR to the value that maximizes the secrecy sum-rate. For $1 < \beta < 2$ and large SNR, the RCI precoder with power reduction would thus achieve a constant nonnegative secrecy sum-rate. However, this strategy would not be effective for $\beta \geq 2$, since in this case the secrecy sum-rate is zero irrespective of the SNR.

Secrecy sum-rates in the presence of channel estimation error

In Sections 3 and 4, we discussed the secrecy rate performance of multi-user MIMO linear precoding for the case when perfect channel state information (CSI) is available at the transmitter. However, a more realistic scenario is the one where only an estimation of the channel is available at the transmitter. The relation between the true channel $\hat{H}$ and the estimated channel H^ is usually modeled as

$$H = \hat{H} + E \qquad\qquad (19)$$

where the matrix E represents the channel estimation error, and it is independent from H^ . The knowledge of $\hat{H}$ is used by the transmitter to obtain the RCI precoding matrix. The entries of $\hat{H}$ and E are i.i.d. complex Gaussian random variables with zero mean and variances $1 - \tau^2$ and τ^2, respectively. The value of $\tau \in [0, 1]$ depends on the quality and technique used for channel estimation. When $\tau = 0$ the CSI is perfectly known, whereas $\tau = 1$ corresponds to the case when no CSI is available at all.

Future research could analyze the performance of linear precoding in the presence of imperfect CSI, deriving the achievable secrecy sum-rate as a function of the channel estimation error variance τ^2. This would allow to study how the CSI estimation error must scale with the SNR, in order to maintain a given high-SNR rate gap to the case with perfect CSI, so that the multiplexing gain is not affected. More specifically, the case of frequency division duplex (FDD) systems could be studied. Assuming that users quantize their channel directions by using B bits and employing random vector quantization (RVQ), and that they feed the quantization index back to the transmitter [30, 31], it would be interesting to determine how many feedback bits are required by each user in order to maintain a constant gap to the case with perfect CSI.

CONCLUSIONS

Throughout this chapter, we presented an up-to-date summary of the research in the field of physical layer security for multiuser MIMO communications. Unlike classical cryptography, physical layer security does not require key distribution and management, it does not rely on the limited computational power of the eavesdroppers, and it does not employ complex encryption algorithms. For these reasons, it is suitable for large dynamic wireless networks, and it has been proposed to enhance the protection of confidential messages transmitted over wireless channels. In this chapter, we especially focused on the problem of secret communication in a multiuser MIMO system. We considered the general case where a multiantenna base station transmits independent confidential messages to a generic number of users. We assumed that the users can potentially act maliciously and eavesdrop on each other. For this system set-up, we presented some transmission schemes based on linear precoding. We discussed the performance of these schemes as well as the cost of simultaneously guaranteeing secrecy to multiple users. It has been recently shown that, in the large SNR regime, a linear precoding scheme based on regularized channel inversion can achieve secrecy without reducing the sum-rate at no additional cost when the number of transmit antennas M is larger than the number of users K. If K = M, secrecy can be achieved with a small rate loss or, alternatively, without reducing the sum-rate at a cost of less than 4dB in terms of the power transmitted. However, the secrecy requirements limit the maximum number of users that can be served with a non-zero rate. When K > M, there is an optimal value of the SNR beyond which the achievable rate starts decreasing, and at large SNR the secrecy sum-rate achievable by RCI precoding is poor. The base station could prevent the secrecy sum-rate from decreasing by reducing the transmit power, and therefore the SNR, to the value that maximizes the secrecy sum-rate. This would result in a constant nonnegative high-SNR secrecy sum-rate. However, this strategy would not be effective if $K \geq 2M$.

ACKNOWLEDGEMENTS

The work of G. Geraci was supported in part by the Australian Government under International Postgraduate Research Scholarship, and in part by the Wireless Technologies Laboratory, CSIRO ICT Centre, Sydney, Australia. The work of J. Yuan was supported in part by the Australian Research Council Discovery Project (Grant DP120102607).

REFERENCES

1.   Y. Liang, H. V. Poor, and S. Shamai (Shitz). Information Theoretic Security. Dordrecht, The Netherlands: Now Publisher, 2009.

2.   R. Liu and W. Trappe. Eds, Securing Wireless Communications at the Physical Layer. New York: Springer Verlag, 2010.

3.   A. Mukherjee, S. A. A. Fakoorian, J. Huang, and A. L. Swindlehurst. Principles of physical-layer security in multiuser wireless networks:

Survey. 2010. arXiv:1011.3754.

4.   A. D. Wyner. The wire-tap channel. Bell System Tech. J., 54:1355–1387, 1975.

5.   I. Csiszár and J. Körner. Broadcast channels with confidential messages. IEEE Trans. Inf. Theory, 24(3):339–348, May 1978.

6.   S. Leung-Yan-Cheong and M. Hellman. The Gaussian wire-tap channel. IEEE Trans. Inf. Theory, 24(4):451–456, July 1978.

7.   J. Barros and M.R.D. Rodrigues. Secrecy capacity of wireless channels. In Proc. IEEE Int. Symp. on Inform. Theory (ISIT), pages 356–360, july 2006.

8.   Z. Li, W. Trappe, and R. Yates. Secret communication via multi-antenna transmission. In Proc. CISS, March 2007.

9.   A. Khisti, G. Wornell, A. Wiesel, and Y. Eldar. On the Gaussian MIMO wiretap channel. In Proc. IEEE Int. Symp. on Inform. Theory (ISIT), pages 2471–2475, 2007.

10.   A. Khisti and G.W. Wornell. Secure transmission with multiple antennas I: The MISOME wiretap channel. IEEE Trans. Inf. Theory, 56(7):3088–3104, July 2010.

11.   S. Goel and R. Negi. Guaranteeing secrecy using artificial noise. IEEE Trans. Wireless Commun., 7(6):2180–2189, 2008.

12.   X. Zhou and M.R. McKay. Secure transmission with artificial noise over fading channels: Achievable rate and optimal power allocation. IEEE Trans. Veh. Technol., 59(8):3831–3842, Oct. 2010.

13.   A. Mukherjee and A.L. Swindlehurst. Robust beamforming for security in MIMO wiretap channels with imperfect CSI. IEEE Trans. Signal Process., 59(1):351–361, Jan. 2011.

14.   E. Ekrem and S. Ulukus. The secrecy capacity region of the Gaussian MIMO multi-receiver wiretap channel. IEEE Trans. Inf. Theory, 57(4):2083–2114, April 2011.

15.   R. Liu and H.V. Poor. Secrecy capacity region of a multiple-antenna Gaussian broadcast channel with confidential messages. IEEE Trans. Inf. Theory, 55(3):1235–1249, 2009.

16.   R. Liu, T. Liu, H.V. Poor, and S. Shamai. Multiple-input multiple-output Gaussian broadcast channels with confidential messages. IEEE Trans. Inf. Theory, 56(9):4215–4227, 2010.

17.   G. Geraci, J. Yuan, A. Razi, and I. B. Collings. Secrecy sum-rates for multi-user MIMO linear precoding. In Proc. IEEE Int. Symp. on Wireless Commun. Systems (ISWCS), Nov. 2011.

18. G. Geraci, M. Egan, J. Yuan, A. Razi, and I. B. Collings. Secrecy sum-rates for multi-user MIMO regularized channel inversion precoding. IEEE Trans. Commun., 2012. to appear. Available: http://arxiv.org/abs/1207.5063.

19. G. Geraci, J. Yuan, and I. B. Collings. Large system analysis of the secrecy sum-rates with regularized channel inversion precoding. In Proc. IEEE Wireless Commun. Networking Conference (WCNC), Apr. 2012.

20. Q.H. Spencer, C.B. Peel, A.L. Swindlehurst, and M. Haardt. An introduction to the multi-user MIMO downlink. IEEE Comms. Mag., 42(10):60–67, Oct. 2004.

21. Qinghua Li, Guangjie Li, Wookbong Lee, Moon Lee, D. Mazzarese, B. Clerckx, and Zexian Li. MIMO techniques in WiMAX and LTE: a feature overview. IEEE Comms. Mag., 48(5):86–92, May 2010.

22. Q.H. Spencer, A.L. Swindlehurst, and M. Haardt. Zero-forcing methods for downlink spatial multiplexing in multiuser MIMO channels. IEEE Trans. Signal Process., 52(2):461–471, 2004.

23. T. Yoo and A. Goldsmith. On the optimality of multiantenna broadcast scheduling using zero-forcing beamforming. IEEE J. Sel. Areas Commun., 24(3):528–541, March 2006.

24. C. B. Peel, B. M. Hochwald, and A. L. Swindlehurst. A vector-perturbation technique for near-capacity multiantenna multiuser communication - Part I: Channel inversion and regularization. IEEE Trans. Commun., 53(1):195–202, January 2005.

25. M. Joham, W. Utschick, and J.A. Nossek. Linear transmit processing in MIMO communications systems. IEEE Trans. Signal Process., 53(8):2700–2712, August 2005.

26. L. Sun and M.R. McKay. Eigen-based transceivers for the MIMO broadcast channel with semi-orthogonal user selection. IEEE Trans. Signal Process., 58(10):5246–5261, Oct. 2010.

27. S. Jin, M. R. McKay, X. Gao, and I. B. Collings. MIMO multichannel beamforming: SER and outage using new eigenvalue distributions of complex noncentral wishart matrices. IEEE Trans. Commun., 56(3):424–434, 2008.

28. V.K. Nguyen and J.S. Evans. Multiuser transmit beamforming via regularized channel inversion: A large system analysis. In Proc. IEEE Global Commun. Conf. (GLOBECOM), pages 1–4, Dec. 2008.

29. V. K. Nguyen, R. Muharar, and J. S. Evans. Multiuser transmit beamforming via regularized channel inversion: A large system analysis.

Technical Report, Nov. 2009. Available: http://cubinlab.ee.unimelb.edu. au/ rmuharar/doc/manuscriptrevRusdha22- 11.pdf.

30. N. Jindal. MIMO broadcast channels with finite-rate feedback. IEEE Trans. Inf. Theory, 52(11):5045–5060, November 2006.

31. D. Ryan, I. B. Collings, I. V. L. Clarkson, and R. W. Heath Jr. Performance of vector perturbation multiuser MIMO systems with limited feedback. IEEE Trans. Commun., 57(9):2633–2644, 2008.

Chapter 4

AN OVERVIEW OF VISIBLE LIGHT COMMUNICATION SYSTEMS

Taner Cevik[1] and Serdar Yilmaz[2]

[1]Department of Computer Engineering, Fatih University Istanbul, Turkey

[2]Department of Electrical and Electronics Engineering, Fatih University Istanbul, Turkey

ABSTRACT

Visible Light Communication (VLC) has gained great interest in the last decade due to the rapid developments in Light Emitting Diodes (LEDs) fabrication. Efficiency, durability and long life span of LEDs make them a promising residential lighting equipment as well as an alternative cheap and fast data transfer equipment. Appliance of visual light in data communication by means of LEDs has been densely searched in academia. In this paper, we explore the fundamentals and challenges of indoor VLC systems. Basics of optical transmission such as transmitter, receiver, and links are investigated. Moreover, characteristics of channel models in indoor VLC systems are identified and theoretical details about channel modelling are presented in detail.

INTRODUCTION

Depending on the technological developments, the variety and quality of the communication devices and applications running on these devices have increased dramatically. These high quality applications require excessive data transfer capacity and speed. Much of the internet transmission at the backbone is handled by the Optical Fiber Infrastructure that can achieve data speeds on the order of Tb/s. On the other hand, these high data rates at the backbone part cannot be perceived by the end users. Nevertheless, it is not always beneficial and conceivable to deploy a cable infrastructure to every point of a site. Therefore, the importance of wireless communication increases day by day and is being widely used in the last-meter such as home, office and campus environments. Even though wireless communication is favorable in terms of cost, practicality and ease of operation, it brings about the bottleneck problem. RF waves that fall beneath the 10 GHz frequency portion of the electromagnetic spectrum have been widely used in wireless communication. However, since the existing bandwidth cannot satisfy the required capacity and speed demands, as well as multiple technologies contemporaneously share the same bandwidth (Wi-fi, bluetooth, cellular phone network, cordless phones), scientists and professionals have focused on new research areas in wireless communications. An alternative solution proposed for this first-meter bottleneck problem is shifting the working frequency interval to the unlicensed 60 GHz band. By this way, it is desired to widen the bandwith and achieve higher data rates [1]. Given the name WiGig, and standardized by Wireless Gigabit Alliance [2], it has become possible to reach about 6-7 Gb/s data rates with this technology [3]

However, shifting towards the right side of the frequency spectrum, reduces the wavelength of the electromagnetic waves. The propogation range of short wavelength signals is very limited. As the signal spreads over longer distances, the error rate increases due to the weakening of the energy [4]. Therefore, WiGig technology is intended to be used for data communication at high speeds in more enclosed areas. Regarding to these quests, it is desired to utilize the mm-length electromagnetic waves ($\lambda <= 1$mm, $f > 100$ GHz) with the aim of enabling supplementary communication channels. Commuication with the mm wavelength on the right side of the spectrum is called Optical Wireless Communication (OWC). Data transfer on the infra-red band is already provided. Around 100 million electronic devices per year take place on the shelves adopted with infra-red technology. Moreover, the next generation wireless communication technology 4G and the follower are not built on a single technology. These Technologies are desired as an integrated topone system that will compromise multiple technologies working in harmony. The OWC technology is expected to be an important figure of 4G and 5G systems especially in the section that the end users are connected to the internet [5]. The outstanding advantages of OWC when compared with Radio Frequency Communication (RF) can be listed as follows [6-10]:

- Unregulated 200 THz bandwith in the range of 155-700nm wavelengths.
- No licensing fee requirement
- Optical signals can not pass through walls like radio waves penetrate.

Therefore, the signals emitted in a room provides significant benefits in terms of security by staying in that room. For long-distance communications Line-of-Sight (LoS) is essential, that is, the sender and the receiver must see each other directly. Any intervening situation or barrier can be easily recognized. Thus, OWC is significantly preferred in the military and state mechanisms that require high information privacy and security.

- Stay of signals in the room or office, elliminates the possibility of any interference in adjacent rooms or offices. By this way, each room will constitute a cell and the capacity prodcutivity will rise to the top levels.

- The equipment used is cheaper when compared with RF devices.

- Optical signals are not as detrimental as RF signals to the human health.

- OWC requires lower energy consumption than RF systems. Data transfer by using the infra-red portion of the spectrum is already provided.

Latest research activities have been focused on achieving data transfer simultaneously with enlightenment by means of using LED lighting equipment. These energy stingy and cost effective LED devices are desired to be used for data transfer without using RF signals, especially in short ranges. By using visible light, it is intended to achieve wireless communication in the environments and situations such as airplanes, hospitals etc, where it is not convenient to use RF waves. The idea of illumination and data communication simultaneously by using the same physical carrier is firslty suggested by Nakagawa et al. in 2003 (Nakagawa Laboratory). Their studies [11- 15] pioneereed many following research activities. Later on, the Nakagawa Laboratory went into coperation with the famous Japan technology firms and they established the Visual Light Communication Consortium (VLCC). Followingly, many research activities have been done that the most outstanding is the European OMEGA Project. Eventually, in 2011, IEEE completed the release and visual light wireless communication gained a global standard with the name 802.15.7- 2011 [16]- IEEE Standard for Local and Metropolitan Area Networks--Part 15.7: Short-Range Wireless Optical Communication Using Visible LightUsing Visible Light [17]. Though a standard of visual light communication has been released in 2011, prevalent usage of this technology will take further time.

In this paper, we investigate the fundamentals and challenges of indoor VLC systems. In this paper, we explore the fundamentals and challenges of indoor VLC systems. Basics of optical transmission such as transmitter, receiver, and links are investigated. Moreover, characteristics of channel models in indoor VLC systems are identified and theoretical details about channel modelling are presented in detail. The remainder of the paper is organized as follows. In section 2, we describe the fundamentals of VLC. Section 3 discusses the physical design of VLC links. Followingly, section 4 gives details about the channel modeling issues in VLC systems. Lastly, we give the concluding remarks and the future directions in section 5.

BASICS OF VLC

In recent years, one of the ideas put forward for wireless optical communication is the visible light communication method. The signals in the 380-780 nm wavelength interval of the electromagnetic spectrum are the light signals that can be detected by the human eye. It is possible to achieve illumination and data transfer simultaneously by means of LEDs that is the prominent lighting equipment lately. By this way, both interior lighting of a room and data transfer will be achieved without the need of an additional communication system. This technology is given the name of Visual Light Communication. Basic configuration of a VLC communication system is given in Figure 1.

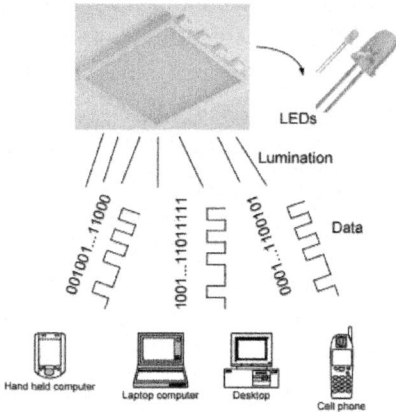

Figure 1: Basic VLC configuration.

Fundamental entries in a VLC system are the transmitter (LEDs), receivers (photodetectors), modulation of data to optics and the optical communication channel as shown in Figure 2. We will discuss these main figures of a VLC system in the following section in detail.

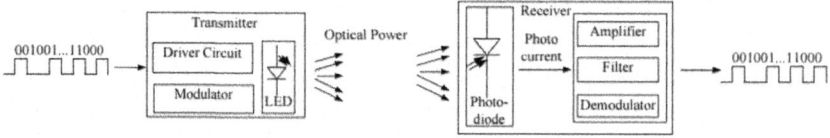

Figure 2: Block diagram of a VLC architecture.

International Journal of Computer Networks & Communications (IJCNC)
Vol.7, No.6, November 2015

Transmitter

There are different possible light sources used for illumination. However, Laser Diodes (LD) and LEDs are the two most popular ones among these especially preferred for optical data communication. Since the purpose of this study is about VLC, that is the concept of maintaining illumination and data transfer simultaneously, discussion of LDs is out of the scope. Thus, we will give details only about LEDs. The major difference between a LD and a LED is that, the former is a coherent light source and other is an incoherent source. That is, in the LED structure photons are emitted spontaneously with different phases. However, with LDs, a photon stimulates another photon that is radiated with a phase correlated with the previous which is called coherent radiation [10, 18].

Basic Principles Of Leds

The idea behind the emission of light with a p-n type LED is that, a bias voltage is applied to the p-n junction and by this way, holes in the p-junction move towards the opposite side. Also, the electrons residing in the n-junction are induced towards the p-junction. These minority carriers recombine in the depletion layer which is also called band-gap. In order for a high-energy level electron to recombine with a lower-energy proton, it should release the excessive energy. That is, the electron passes from the conduction band to the valence band. This excessive energy is released as a photon and approximately equal to the band-gap energy. The magnitude of the energy of the photon determines its wavelength which can be adjusted by the type of the semiconductor material. Relation between the band-gap energy and the wavelength of the emitted photon is given as follows [19]:

$$E_g = h * f \tag{1}$$

where h denotes the Planck's constant with the value equal to 6.626×10^{-34} Js and f expresses the frequency of the radiated photon.

$$h * f = h * (c / \lambda) \tag{2}$$

where c denotes the speed of light $= 3 * 10^8$ m/s and λ is the wavelength of the emitted photon.

Efficiencies Of Leds

An ideal LED should emit a photon per an injected electron. The ratio of the number of photons emitted to the number of electrons injected is defined as the internal quantum efficiency and represented in Eq. (1) [10, 20, 21].

$$\eta_{int} = n_{p\text{-}emt} / n_{e\text{-}inj}$$

(3)

where η_{int}, $n_{p\text{-}emt}$ and $n_{p\text{-}emt}$ denote the internal quantum efficiency, number of protons radiated from the active region and number of electrons injected respectively. In an ideal LED, all of the photons in the active region, should leave te diode, which is called the unit extraction efficiency. However, not all of the photons emitted in the active region, leave the diode due to several reasons such as absorption, etc. So-called the extraction efficiency is another important parameter defining the efficiency of a LED. That is the ratio of the number of photons radiated into the free space per second to the number photons emitted in the active region per second and represented in Eq.(4) [20].

$$\eta_{extract} = n_{p\text{-}emt\text{-}air} / n_{p\text{-}emt}$$

(4)

where $\eta_{extract}$, $n_{p\text{-}emt\text{-}air}$ and $n_{p\text{-}emt}$ represent the extraction efficiency, number of photons emitted into the air and number of photons radiated in the active region respectively. Consequently, we can derive the external quantum efficieny of a LED by combining the Equations (3-4) [22]:

Consequently, we can derive the external quantum efficieny of a LED by combining the Equations (3-4) [22]:

$$\eta_{ext} = n_{p\text{-}emt\text{-}air} / n_{e\text{-}inj}$$

(5)

Ultimately, the power efficiency of a device is the ratio of the output power to the ratio of the input power, that is the electrical power applied to the LED $(P_{inj\_elec} = V * I)$ to the emitted photon energy (P_{photon}).

$$\eta_{pow} = P_{photon} / (V * I)$$

(6)

Radiation Pattern Of Leds A L

Comprises of a semiconductor light source and a surrounding material with different refracting indexes respectively. The most popular LEDs are the ones with planar surfaces of which the emission pattern is modeled with Lambertian Law [23]. Although alternative surface shapes such as hemsipehere or parabolic are possible, fabrication of these LEDs are much more complicated. Light emerging from the semiconductor light source, faces with the surrounding material and refracts into the air with an angle different from the incoming one which can be explained by the Snell Law (Figure 3):

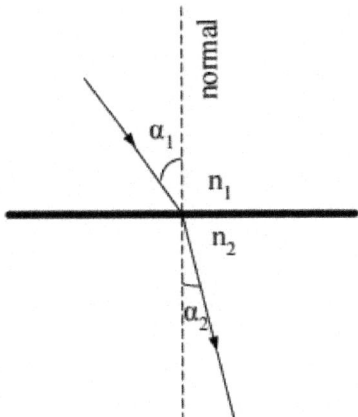

Figure 3: Refraction of a signal during the transition between two different materials with different refraction indexes

This situation decreases the external power efficiency of the device. If the signal incomes with an angle larger than the threshold, most of the signal will be reflected which reduces the external power efficiency to a fw percent [24]. Deriving from the Snell Law, a point directly on the direction of the light source with an angle of 0, gets the maximum intensity of the flux (I0). However, at the point with an angle of θ, the luminance intensity (Iθ) is calculated as:

$$n_1 * sin\alpha_1 = n_2 * sin\alpha_2$$
(8)

$$I_\theta = (P_{source} / (4\pi r^2)) * cos\theta$$
(9)

Since the air and semiconductor have different refractive indexes, the light intensity in air is derived as follows [20]:

$$I_{air} = (P_{source} / (4\pi r^2)) * cos\theta ((n_1^2) / (n_2^2))$$
(10)

The above equations are valid for the assumption that the surrounding materila is hemisphere shaped and Iair is the light intensity of the single point on the surrounding hemisphere. Hence, the total power power emitted into the air can be calculated as integrating the total intensities on the surface of the hemispehere as follows [20]:

$$P_{air} = \int_{\theta=0°}^{90°} I_{air} * 2\pi r * sin\theta * r * d\theta$$

(11)

RECEIVER

Photodetectors are the receiver entity of an OWC system that absorbs the photons impinging on its frontend surface and overagainst generates an electrical signal. The conversion of photonic energy to the electrical energy can be achieved in alternative way. For example, in vacuum photodiodes or photomultipliers, the absorption of photons created photoelectric effects and free electrons emerge as a result that are used as carriers. Another way is that, by the falling of the photons into the junction area of a semiconductor photodiode such as p or pin diodes, an electron and hole pair is released.

Followingly, these released carriers move to the corresponding regions such as conductance and valance bands in order to release their excessive energies [10, 25]. There are many types of photodetectors exist such as photomultipliers, photoconductors, phototransistors, and photodiodes that owning specific qualities. However, photodiodes are the most preferred devices as a photodetector due to their small size, high sensivity and fast response. P-I-N (PIN) and Avalanche Photo-Diode (APD) are the favored types photodiodes as a photodetector [26]. There are some important requirements that an ideal photodetector should cover:

- sensitive to the wavelength interval associated
- long operational life
- minimally affected from the temperature fluctuations
- efficient accomplishment of noise such ambient, dark, etc.
- noiseless physical structure
- small in size
- reliable
- cost effective

Quantum Efficiency

Different definitons are given for Quantum Efficiency. One of the definitions is that it is the probability of a single inicident photon to generate an electron-hole pair that contribute to the detection of electric current. Another definition for quantum efficiency is that it is the ratio of electron-hole pair generated that contribute to the detector current, to the incident photon flux and calculated as follows [27]:

$$\bar{\eta}_{quantum\text{-}eff} = (1-R) * \zeta * (1-e^{-ad})$$

$$(12)$$

where R denotes the reflectance of the surface and that can be reduced by non-reflective coatings. ζ represents the fraction of electron-hole pairs that contribute to the photocurrent. a and d exzpress the absorption coefficient (per cm^2) and the photodetector depth respectively.

Responsivity

The responsivity of a photodetector is the ratio of the output current to the input optical power and the relation of it to the quantum efficiency is as follows [28]:

International Journal of Computer Networks & Communications (IJCNC) Vol.7, No.6, November 2015

$$R = \frac{e*\eta}{h*v}$$

$$(13)$$

where e denotes the electron charge, h is the Planck's constant and v represents the light frequency.

PHYSICAL DESIGN OF VLC LINKS

One of the main challenges to be carefully considered during the design and modelling stages of an VLC system is the localization status of the transmitter and receiver pair which mainly defines how the signal is transmitted. Design of a VLC link can be classified in two ways as depicted in Figure 4. The first method of the categorization can be made whether the transmitter and receiver is directed or not to a specific point or coordinate. Under this category, three different options are possible regarding to direcitonality of the transmitter and receiver. The first option is that, both the transmitter and receiver are directed to a specific point. This type of configuration enhances power efficiency as well as immunity to the environmental distorting effects such as ambient and artificial light sources. The second category under the directionality classification is the nondirectional configurations in which the transmitter and receiver are not particularly focused to a specific direction or point. In order to achieve signal transmission, wide beam transmitters and wide FOV receivers are required. The main drawbacks of this configuration is the need for high power levels to combat with the high optical loss and the multipath-induced distortions. Although, multipath fading is overcomed by means of the immense ratio of the detector size and signal wavelength. The other link configuration of this type of classification is the hybrid design method in which the tranmitter and receiver can have different levels of directionalities, such as a narrow beam transmitter directed to a specific point and a wide FOV receiver which are not aligned to a particular direction [10, 29]. The second type of design choice is the existence of a LOS path between the transmitter-receiver pair. There are two options in this category of configuration. First is the Line of Sight (LOS) configuration in which no interruption or obstacles are present between the transmitter and receiver. Thus, no reflection consideration is considered that simplies the path loss calculation. Besides, high power efficiency is achievable. In contrast, in the Non-LOS architectures, signals emerge from the source do not directly arrive at the receiver. They are refleected from different surfaces or objects and arrive in different time intervals to the receiver. This causes multipath distortions and maket he estimation of path loss much more difficult. The Non-LOS architecture with nondirected transmitter-receiver pair is called diffuse systems which is the most robust system and easy to implement for especially the mobile communication scenarios [10, 29, 30].

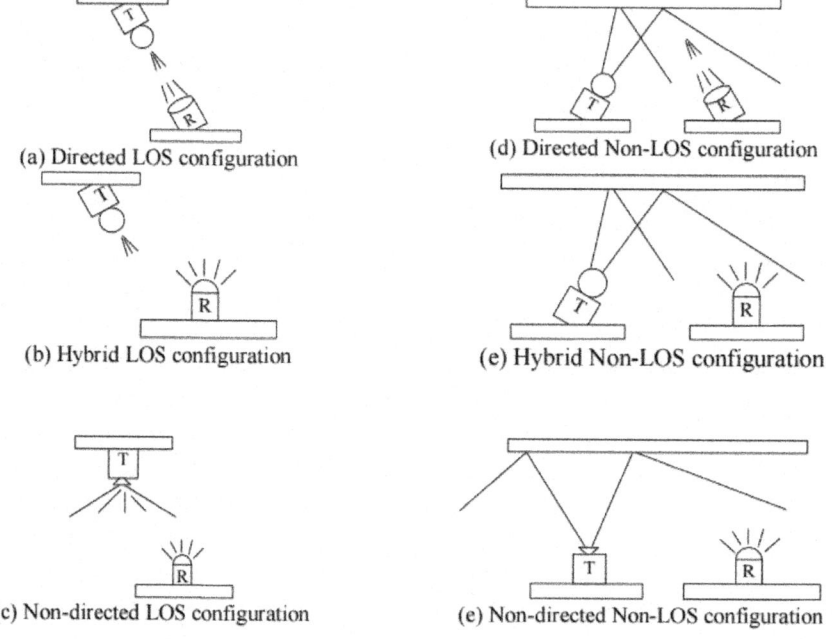

(a) Directed LOS configuration

(b) Hybrid LOS configuration

(c) Non-directed LOS configuration

(d) Directed Non-LOS configuration

(e) Hybrid Non-LOS configuration

(e) Non-directed Non-LOS configuration

Figure 4: Channel configuration models

CHANNEL MODELING

The most practical, cost effective and easy modulatin technique in VLC systems is the Intensity Modulation (IM) / Direct Detection (DD) [29, 31] . Unlike the coherent transmission techniques IM/DD does not concern with the nature of the signal such as phase or frequency. In contrast, in coherent transmission technology, information is coded on the optical beam by means of phase, frequency or amplitude modulation techniques. The receiver side owns the staff called downconverter which is comprised of an oscillator and a mixer. Mixer combines both the arriving optical beam and the one generated by the local oscillator. Then the combined optical signal falls onto the photodetector and depending on the similarity of the oscillator and incoming signal frequencies, either demodulation operation is applied or not. If the frequencies of the incoming signal and the oscillator are different, then a modulation operation is performed which is called Heterodyne Detection. Otherwise, the combined signal is directly downconverted to the baseband signal that is called Homodyne Detection [32]. For optical wireless systems, in which the system cost and complexities are desired to be low, IM/DD modulation technique takes over as a prominent method. In this technology, the desired waveform is modulated onto the instantaneous power of the carrier. At the receiver side, the detector uses the down-conversion technique DD, during which a photocurrent is produced directly proportional with the incoming photonic power [33]. Since the photodetector area is larger in the order of magnitudes when compared with the signal wavelength, VLC links do not suffer the impacts of multipath fading. However, since it is possible for signals reflected from ceilings, walls or other intervening reflective objects also arrive at the receiver by travelling in a dispersive way that results with the event multipath distorsion called InterSymbol Interference (ISI). For indoor VLC systems, this dispersion with ISI effect can be modeled as baseband linear impulse response (h(t)). This multipath distortion effect is mostly concerned especially for non-directional and Non-LOS channel models. Indoor VLC channels are generally assumed as quasi-static. In doing this, the positions of the tranmitter, receiver and the intervening objects are assumed to be static or moving with very low speeds and bit rate is very high. Thus, channel variations occur in the order of many bits periods [10, 29, 34]. The channel model for indoor VLC systems is based on the Eq. (14):

where R denotes the responsivity of the photodetector and formulated as in Eq. (13). n(t) represents the gaussian modeled noise comprising ambient and preamplifier receiver noises. $\otimes$ expresses the convolution operation and x(t) denotes the instantaneous input power which can not be negative (x(t)>0). Calculation of the average power received by the photodetector is as follows [29]:

$$P_r = H(0)P_t \tag{15}$$

International Journal of Computer Networks & Communications (IJCNC) Vol.7, No.6, November 2015 147 where p_t denotes the average trasmitted power and calculated as in Eq. (16):

$$lim_{T \to \infty} \left(\frac{1}{2T} \int_{-T}^{T} x(t) \, dt \right)$$

$$(16)$$

and H_0 represents the channel dc gain that is defined as follows:

$$H(0) = \int_{-\infty}^{+\infty} h(t) dt$$

$$(17)$$

The channel modelling of indoor VLC systems differs according to the LOS presence between the transmitter and receiver. Following sections give details about the different models applied for the LOS and Non-LOS indoor VLC channels.

Los Channel Model

In an indoor VLC system, channel dc gain for LOS propogation model as depicted in Figure 5, with a source assumued as radiating in Lambertian model, concentrator with a gain of $g(\psi)$ at the receiver and an optical bandpass filter with the function $Ts(\psi)$ is given as follows:

$$H_{LOS(0)} = \begin{cases} A_r(m_1+1) \, (2\pi d^2) \, (cos^{ml}\phi)(\\ \quad T_s(\psi))(g(\psi))(cos(\psi)), \\ 0, \end{cases} \quad \begin{array}{l} 0 \leq \psi \leq \psi_c \\ otherwise \end{array}$$

$$(18)$$

where A_r is the effective area of the detector collecting the incoming optical signals with the arrival angle (ψ) smaller than the FOV angle that is expressed with ψ_c. m_1 is related with the directivity of the source beam and denotes the Lambert's mode number. In order to increase the efficiency of the system, transmitted optical power level should be increased. However, that is not possible because of the eye-safety and power efficiency reasons. An alternative solution can come into mind to employ larger dtector areas in the receiver. Although it can be thought as a promising solution, due to reasons of increeased cost, complexity, capacitance and reduced bandwith, this choice is unfavourable. The way of increasing the efficiency is using a concentrator at the frontend as mentioned before. $g(\psi)$ denotes the gain of the concentrator used at the front-end of the detector and calculated as follows [10, 29, 35]

$$(12)$$

Non-Los Channel Model

Channel modelling varies depending on the presence and degree of directivity of the transmitter and receiver. Althougy there are three types of Non-LOS configurations, there are two model generalizations covering the all. A single model can be used for the directed and hybrid Non-LOS configurations (Figure 6) as follows:

International Journal of Computer Networks & Communications (IJCNC) Vol.7, No.6, November 2015

$$H_{Non\text{-}LOS(0)} = \begin{cases} (\rho A h\, T_s(\psi) g(\psi) \cos(\psi))/(\pi(h^2 + d_{sr}^2)^{3/2}, & 0 \le \psi \le \psi_c \\ 0, & \theta > \psi_c \end{cases}$$

$$(19)$$

where the parameter ρ denotes the Lambertian reflectivity index of the ceiling. As clarified in Eq. (19), the best way of increasing the dc channel gain can be achieved by increasing the concentrator gain of the receiver by means of increasing its refractive gain and decreasing FOV angle [29].

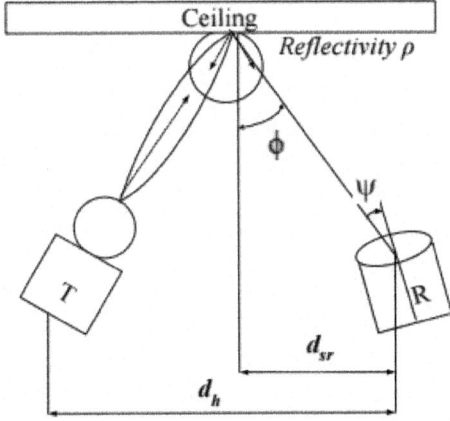

Figure 6: Directed or Hybrid Non-LOS Channel Configuration.

In non-directed non-LOS link configurations, high order reflections of the light emerging from the transmitter must be considered. Especially for the indoor environments such as with the higher than 5m in height, it is strongly advised to consider the reflections up to fifth order [36]. However, especially for small offices, the approach of considering the one-bounce reflection model estimates the dc channel gain with an error of a few decibels. First oder calculation of dc channel gain for the link configuration as depicted in Figure 7 is given in Eq. (20).

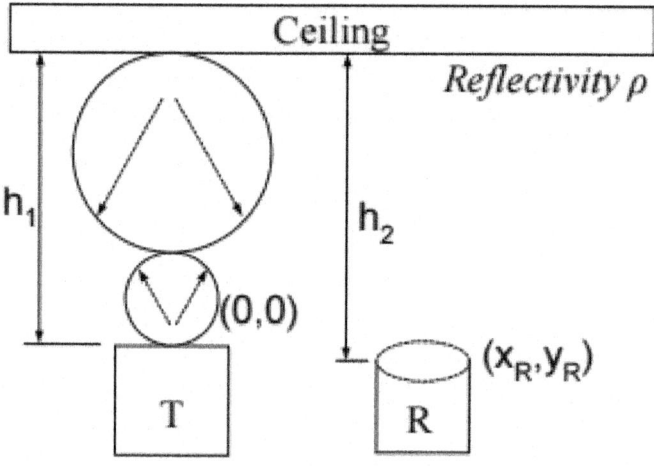

Figure 6: Directed or Hybrid Non-LOS Channel Configuration.

$$H_{Non\text{-}LOS(0)} = ((\rho T_s g A h_1^2 h_2^2)/\pi^2) \iint (h_1^2 + x^2 + y^2)^2 (h_2^2 + (x-x_R)^2 + (y-y_R)^2)^2$$

(20)

where $Ts \approx T_s(\psi)$ since the optical filter is assumed as if it is omnidirectional and the concentrator is also omnidirectional with a gain $g \approx g(\psi)$, FOV $\psi_c \approx \pi/2$. Obviously, dc channel gain $H_{(0)}$ is inversely proportional with dh^{-4} [29].

CONCLUSION

Much of the internet transmission at the backbone is handled by the Optical Fiber Infrastructure that can achieve data speeds on the order of Tb/s. On the other hand, these high data rates at the backbone part cannot be perceived by the end users. Since the existing bandwidth cannot satisfy the required capacity and speed demands, as well as multiple technologies contemporaneously share the same bandwidth (Wi-fi, bluetooth, cellular phone network, cordless phones), scientists and professionals have focused on new research areas in wireless communications. An alternative solution proposed for this first-meter bottleneck problem is shifting the working frequency interval to the unlicensed 60 GHz band. By this way, it is desired to widen the bandwith and achieve higher data rates. Regarding to these quests, it is desired to utilize the mm-length electromagnetic waves ($\lambda \leq$ 1mm, f > 100 GHz) with the aim of enabling supplementary communication channels. In recent years, one of the ideas put forward for wireless optical communication is the visible light communication method. It is possible to achieve illumination and data transfer simultaneously by means of LEDs that is the prominent lighting equipment lately. By this way, both interior lighting of a room and data transfer will be achieved without the need of an additional communication system. This paper explores the fundamental issues and concepts of VLC systems by supporting theoratical details.

REFERENCES

1. http://www.wi-fi.org/news-events/newsroom/wi-fi-alliance-and-wireless-gigabit-alliance-to-unify

2. http://www.wi−fi.org/download.php?file=/sitesi/default/files/private/Wigig_White_Paper_20130909.pdf

3. https://gigaom.com/2009/05/06/wigig-alliance-to-push-6-gbps-wireless-in-the-home/

4. Stallings, W., "Wireless Communications and Networks", NJ: Pearson Prentice Hall, 2005.

5. O'Brien, D. C., Katz, M., "Short-Range Optical Wireless Communications", Wireless World Research Forum (WWRF) White Papers, 2005.

6. Barry, R., "Wireless Infrared Communications", Boston: Kluwer Academic Publishers, 1994.

7. Ramirez-Iniguez R., Idrus, S. M., Sun, Z., "Optical Wireless Communications: IR for Wireless Connectivity", Boca Raton: CRC Press, 2008.

8. Ciaramella, E., Arimoto, Y., Contestabile, G., Presi, M., D'Errico, A., Guarino, V., Matsumoto, M., "128 terabit/s (32 × 40 Gbit/s) WDM transmission system for free space optical communications", IEEE Journal on Selected Areas in Communications, 27, 1639–1645, 2009.

9. Rajagopal, S., Roberts, R. D., "IEEE 802.15.7 Visible Light Communication: Modulation Schemes and Dimming Support", IEEE Comnmunications Magazine, 72-83, 2012.

10. Ghassemblooy, Z., Popoola, W., Rajbhandari, S., "Optical Wireless Communications System and Channel Modelling with MATLAB", CRC Press Taylor&Francis Group, 2013.

11. Komine, T., Tanaka, Y., Haruyama, S., Nakagawa, M., "Basic study on Visible-Light Communication using Light Emitting Diode Illumination", Proceedings of the 8 th International Symposium on Microwave and Optical Technology, Canada, pp. 45-48, 2011.

12. Tanaka, Y., Haruyama, S., Nakagawa, M., "Wireless optical transmission with the White colored LED fort he wireless home links", Proceedings of the 11th International Symposium on Personal, Indoor and Mobile Radio Communications, London, UK, pp. 1325-1329, 2000.

13. Komine, T., Nakagawa, M., "Integrated System of White LED Visible-Light Communication and Power-Line Communication", IEEE Transactions on Consumer Electronics, vol. 49, no. 1, pp. 71-79, 2003.

14. Tanaka, Y., Komine, T., Haruyama, S., Nakagawa, M, "Indoor Visible Light Transmission System Utilizing White LED Lights", IEICE Transactions on Communications, vol. E86-B, no. 8, pp. 2440- 2454, 2003.

15. Komine, Toshihiko, Nakagawa, M., Fundamental Analysis for Visible-Light Communication System using LED Lights", IEEE Transactions on Consumer Electronics, Vol. 50, No. 1, 2004.

16. Sklavos, N., Hübner, M., Goehringer, Kitsos, P., "System-Level Design Methodologies for Telecommunication", Springer, 2013.

17. S M Sze and K K Ng, Physics of Semiconductor Devices, 3rd ed Hoboken, New Jersey: John Wiley & Sons Inc., 2007.

18. J. M. Senior, "Optical Fiber Communications Principles and Practice", 3rd ed Essex: Pearson Education Limited, 2009.

19. E. Fred Schubert, "Light –Emitting Diodes", 2nd Edition, Cambridge University Press, 2006.

20. Hongping Zhao, Guangyu Liu, Jing Zhang, Ronald A. Arif, Nelson Tansu, "Analysis of Internal Quantum Efficiency and Current Injection Efficiency in III-Nitride Light-Emitting Diodes", Journal of Display Technology, Vol. 9, No. 4, pp. 212-225, 2013. International Journal of Computer Networks & Communications (IJCNC) Vol.7, No.6, November 2015 150

21. Han-Youl Ryu, Guen-Hwan Ryu, Sang-Ho Lee, Hyun Joong KIM, "Evaluation of the Inter Quantum Efficiency in Blue and Green Lighyt-emtting Diodes Using the Rate Equation Model", Journal of the Korean

Physical Society, Vol. 63, No. 2, pp. 180-184, 2013.

22.   Alex Ryer, "Light Measurement Handbook", www.intl-light.com/customer/handbook/handbook.

23.   John M. Senior, "Optical Fiber Communications Principles and Practice", 3rd edition, Pearson Prentice Hall, 2009.

24.   C C Davis, Lasers and Electro-Optics: Fundamentals and Engineering, Cambridge, UK: Cambridge University Press, 1996.

25.   G Keiser, Optical Fiber Communications, 3rd ed New York: McGraw-Hill, 2000.

26.   Bahaa E. A. Saleh, Malvin Carl Teich, "Fundamentals of Photonics", John Wiley & Sons, Inc., 1991.

27.   Sasa Radovanovic, Anne-Johan Annema, Bram Nauta, "High-Speed Photodiodes in Standard CMOS Technology", Springer, 2006.

28.   Joseph M. Kahn, John R. Barry, "Wireless Infrared Communications", Proceedings of the IEEE, Vol. 85, No. 2, pp. 265-298, 1997.

29.   Joseph M. Kahn, William J. Krause, Jeffrey B. Carruthers, "Experimental Characterization of NonDirected Indoor Infrared Channels", IEEE Transactions on Communications, Vol. 43, No.2/3/4, pp. 1613- 1623, 1995.

30.   Raed Mesleh, Rashid Mehmood, Hany Elgala, Harald Haas, "Indoor MIMO Optical Wireless Communication Using Spatial Modulation", IEEE International Conference on Communications, pp. 1-5, Cape Town, 2010.

31.   http://www.invocom.et.put.poznan.pl/~invocom/C/P1-9/swiatlowody_en/p1-1_8_2.htm

32.   Raed Mesleh, Hany Elgala, Harald Haas, "Optical Spatial Modulation", ournal of Optical Communications and Networking, vol. 3, no. 3, pp. 234-244, 2011.

33.   A. M. Street, P. N. Stavrinou, D. C. Obrien and D. J. Edwards, "Indoor optical wireless systems—A review", Optical and Quantum Electronics, Vol. 29, pp. 349–378, 1997.

34.   X. Ning, R. Winston, J. O. Gallagher, "Dielectric totally internally reflecting concentrators", Applied optics, Vol. 26, No. 2, pp. 300-305, 1987.

35.   F. R. Gfeller, P. Bernasconi, W. Hirt, C. Elisii, B. Weiss, "Dynamic cell planning for wireless infrared in-house data transmission", Proceedings of the International Zurich Seminar on Digital Communications: Mobile Communications: Advanced Systems and Components, pp. 261-272, 1994.

Chapter 5

EXPLORATION OF SPATIAL DIVERSITY IN MULTI-ANTENNA WIRELESS COMMUNICATION SYSTEMS

INTRODUCTION

Background

Since the invention of the radio telegraph by Marconi in 1895, wireless communication has attracted great interest and is now one of the most rapidly developing technologies. From narrow-band voice communications to broadband multimedia communications, the data rate of the wireless communications has been increased dramatically, from kilobits per second to megabits per second. However, with increasing demand on wireless internet and personal multimedia, the data rate of wireless communications needs be further expanded. Future wireless networks face challenges of supporting data rates higher than one gigabits per second [1]. Among numerous factors that limit the data rate of wireless communications, multipath propagation plays an important role [2]. In wireless communications, the radio signals may arrive at the receiver through multiple paths because of reflection, diffraction, and scattering. This phenomena is called multipath propagation, which causes constructive and destructive effects due to signal phase shifting. Channels with multipath fading fluctuate randomly, resulting in significant degradation of signal quality. When the bandwidth of the signal is greater than the coherence bandwidth of the fading channel, different frequency components of the signals experience different fading. This frequency-selective fading may further limit the data rate of wireless communications. To combat multipath fading, code division multiple access (CDMA) and orthogonal frequency-division multiplexing (OFDM) were developed [3,4]. As a spread spectrum modulation, CDMA overcomes multipath fading by transmitting signals which occupy a wider bandwidth.

Even though a small portion of this wideband channel undergoes deep fading, the overall channel could be in good condition. The loss of the signals can be recovered by using the Rake receiver and/or maximum-ratio combining [5]. On the other hand, the OFDM scheme splits the channel into many small bandwidth carriers, each of which occupies a narrowband channel. The information in the carriers under deep fading can be recovered if forward error correction is applied [6]. Although CDMA and OFDM are effective in combating against multipath fading, they cannot provide a higher data rate compared to other techniques. In multiple access systems, all users have to share the available bandwidth. Users in a CDMA system share the same bandwidth by using different spreading codes, while in an OFDM system, each user is assigned to a subset of the carriers, a frequency division multiple access system (FDMA) in essence.

In general, these traditional wireless communication schemes employ time, frequency, and code dimensions for multiple access. Given a bandwidth and a number of users, other dimensions of freedom are expected to increase data rates. During the past decade, multi-antenna systems, which are also referred to as multipleinput multiple-output (MIMO) systems, have attracted significant interest. The benefits of multiple antennas arise from the use of extra spatial dimension. By employing multiple antennas at the transmitter and/or receiver in a wireless system, the rich scattering channel can be exploited to create a multiplicity of parallel links over the same radio band. This novel property provides MIMO with several advantages, including array gain, spatial diversity gain, and spatial multiplexing gain [7]. Array gain refers to the average increase in the signal-to-noise ration (SNR) that results from a coherent combining of signals from multiple transmit-receive antenna pairs. The coherent combining may be achieved at the receiver by beamforming; if the channel state information (CSI) is known to the transmitter, the coherent combining may also be realized at the transmitter by weighting the transmission signals into multiple antennas, known as transmit beamforming. Array gain improves the system robustness to the noise, thereby improving the coverage of the system.

By providing the receiver with multiple copies of the transmitted signal in space, MIMO systems achieve spatial diversity and effectively mitigate multipath fading, thereby improves the quality and reliability of the reception. The achieved spatial diversity order depends on the channel environment, specifically, the channel coherence. In a rich scattering environment, a transmitter with an antenna array may transmit multiple independent data streams within the bandwidth of operation, and the receiver with an antenna array can successfully separate the data streams. MIMO systems, therefore, offer an increase in data rate through spatial multiplexing.

Multipath scattering is commonly seen as detrimental to wireless communications. However, with the emergence of MIMO systems, multipaths have been effectively converted into a benefit for wireless communications. Due to this advantage, MIMO technology is considered key to future gigabit wireless communications [1]. MIMO can also be integrated with various modulation schemes to enhance the system performance. For example, MIMO has been employed in wideband CDMA (WCDMA) systems and the long term evaluation (LTE) to enhance CDMA and OFDM systems, respectively [8, 9].

Motivation and Dissertation

Outline Although multi-antennas have presented various advantages and have been utilized for more than ten years, a theoretical analysis of some aspects in multi-antenna systems and the potential applications of MIMO technology remain unexplored. Studies of MIMO are still ongoing in both academia and industry. New concepts and methodologies such as distributed MIMO, cooperative MIMO, network MIMO, coordinated multi-point, and physical layer security have been proposed in recent years [10–13]. These studies will undoubtedly enhance wireless communications in the near future. This dissertation explores the spatial diversities provided by multi-antenna transmission. Solutions to two existing problems in MIMO-OFDM systems are presented, a spread spectrum system is enhanced by using the multi-antenna technique, and a novel application of the multi-antenna system is explored for securing wireless communications. The rest of the dissertation is organized as follows: Chapter 2 introduces the fundamentals of multi-antenna wireless communications. Chapter 3 investigates the BER performance of the limited feedback precoded spatial multiplexing MIMO-OFDM system with channel prediction at the receiver. A virtual channel based approach is proposed to analyze the I/Q imbalances in MIMO-OFDM systems in Chapter 4. In Chapter 5, a detailed investigation of the recent advances in distributed MIMO technologies in cooperative wireless networks is presented. In Chapter 6, the multi-antenna technique is incorporated into self-encoded spread spectrum (SESS) systems to improve the system performance. Chapter 7 proposes a novel physical layer method to secure wireless communications by taking advantages of the extra dimensions of freedom provided by multi-antenna transmission. Finally, Chapter 8 concludes this dissertation.

Notations

In this dissertation, the following notations are used. A small (capital) letter represents a variable in time (frequency) domain, and a bold small (capital) letter represents a vector (matrix). The superscripts $*$, T, H, and -1 represent the conjugate, the transpose, the Hermitian transpose, and the matrix inverse operations, respectively. $| \cdot |$ and $\| \cdot \|$ kdenote the absolute value of a scalar and the Euclidean norm of a vector. $C^{M \times N}$ denotes complexvalued $M \times N$ matrices. The I_k and 0_k represent a $k \times k$ identity matrix and a vector with k zero elements, respectively. The vec$\{\cdot\}$ and $E\{\cdot\}$ represent the vectorization operation and the expectation operation, respectively. The symbol $\perp$ denotes the orthogonality between two vectors. Operation $(a)+$ is to find the maximum value between the real number a and zero. The minimum singular value of matrix A is denoted by $\lambda\min\{A\}$. diag$(\cdot)$ denotes the diagonal vector of a matrix. $\otimes$ denotes convolution. The Dirac delta function, Q-function, gamma function, and incomplete gamma function are denoted by $\delta\{\cdot\}$, $\mathcal{Q}(\cdot)$, $\Gamma(a)$, and $\Gamma(a, x)$, respectively.

FUNDAMENTALS OF MULTI-ANTENNA WIRELESS COMMUNICATIONS

This chapter introduces basic concepts in multi-antenna wireless systems that are related to this dissertation research. Radio propagation presents some unique characteristics, including large-scale pass loss and small-scale fading. The study of a point-to-point wireless link helps understand the performance of radio systems. When multiple antennas are employed, space-time channel models should be constructed, upon which the multi-antenna system can be modeled. The three basic architectures of the multi-antenna systems are beamforming, spatial multiplexing, and space-time coding. Multi-user MIMO and distributed MIMO are the extended schemes. When the channel experiences frequency-selective fading, OFDM modulation can be used in a multi-antenna system to mitigate the fading effects.

Radio Propagation

Unlike wired channels, the radio channels are extremely random because of the reflection, diffraction, and scattering of the electromagnetic wave propagation [2]. In order to analyze wireless communication systems, it is important to study the statistical characteristics of radio channels. Large-scale path loss and small scale fading are two important characteristics of the radio channel. The commonly used methodology to study these characteristics is to create propagation models.

Large-Scale Path Loss

Radio wave power decays along the propagation path. A large-scale model determines the average received signal strength at a long distance from the transmitter. It is useful in estimating the radio coverage area of a transmitter. Free-space propagation model is an important theoretical large-scale model [2]. According to this model, the power received by a receiver antenna at a distance d from the transmitter is given by

$$P_r(d) = \frac{P_t G_t G_r \lambda^2}{(4\pi)^2 d^2 L},$$

(2.1)

where P_t is the transmitted power, $P_r(d)$ is the received power, G_t is the transmit antenna gain, Gr is the receive antenna gain, λ is the wavelength in meters, and L is the system loss factor. The difference in d_B between the transmitted power and the received power is defined as path loss. For the free-space model, the path loss is given by

$$PL(dB) = 10 \log \frac{P_t}{P_r} = -10 \log \left[\frac{P_t G_t G_r \lambda^2}{(4\pi)^2 d^2 L} \right].$$

(2.2)

In practice, the free-space model may not be accurate because of the presence of trees, buildings, and other obstacles. Different terrain profiles may cause very different propagation models. The most widely used propagation models in practice include the Okumura model and the Hata model. The Okumura model is used for urban cellular communications with base station antenna heights ranging from 30 m to 1000 m, cell radius of 1 km to 100 km, and application frequencies from 150 MHz to 1920 Mhz. This model is expressed as

$$L_{50}(dB) = L_F + A_{mu}(f, d) - G(h_{te}) - G(gre) - G_{AREA},$$

(2.3)

where L_{50} is the 50th percentile value of propagation path loss, LF is the free space propagation loss, Amu is the median attenuation relative to free space, f is the carrier frequency, $G(h_{te})$ is the base station antenna height gain factor, $G(h_{re})$ is the mobile antenna height gain factor, and GAREA is the gain due to the type of environment. The Hata model is an empirical formulation of the graphical path loss data provided by Okumura, and is valid from 150 MHz to 1500 MHz. According to this model, the median path loss in urban areas is given by

$$L_{50}(urban)(dB) = 69.55 + 26.16 \log f_c - 13.82 \log h_{te} - a(h_{re}) + (44.9 - 6.55 \log h_{te}) \log d,$$

(2.4)

where f_c is the frequency in MHz and $a(h_{re})$ is the correction factor for effective mobile antenna height, which is a function of the size of the coverage area.

Small-Scale Fading

Small-scale fading describes the rapid fluctuations of the amplitude and phases of a radio signal over a short period of time or propagation distance. This rapid signal fluctuation is caused by the constructive and destructive combination of the signals arriving at the receiver through multiple paths with different delays. In other words, multipath in the radio channel creates small-scale fading effects. The time-dispersive properties of multipath channels are quantified by multipath delay spread. If the delay spread is small, the radio channel has a constant gain and linear phase response over the bandwidth of the transmitted signal. In this situation, the received signal undergoes flat fading. Otherwise, the received signal may undergo frequency-selective fading. In wireless communications, the statistical time-varying nature of the received envelope of a flat fading signal is commonly modeled by the Rayleigh distribution [14], whose probability density function (pdf) is given by

$$
p(r) = \begin{cases} \frac{r}{\sigma^2} \exp\left(-\frac{r^2}{2\sigma^2}\right), & (0 \le r \le \infty) \\ 0, & (r < 0) \end{cases},
$$

$$(2.5)$$

where σ is the rms value of the received voltage signal before envelope detection.

Space-Time Channel Models

Consider a MIMO system with N_t transmit antennas and Nr receiver antennas. The spacetime channel is given by the $N_r \times N_t$ matrix

$$
\mathbf{H}(\tau,t) = \begin{bmatrix} h_{1,1}(\tau,t) & h_{1,2}(\tau,t) & \cdots & h_{1,N_t}(\tau,t) \\ h_{2,1}(\tau,t) & h_{2,2}(\tau,t) & \cdots & h_{2,N_t}(\tau,t) \\ \vdots & \vdots & \ddots & \vdots \\ h_{N_r,1}(\tau,t) & h_{N_r,2}(\tau,t) & \cdots & h_{N_r,N_t}(\tau,t) \end{bmatrix},
$$

$$(2.6)$$

where $h_{i,j}$ denotes the impulse response between the jth transmit antenna and the ith receive antenna from time t to time $t - \tau$. If the transmitted signal from the jth transmit antenna is denoted by $s_j(t)$, the signal received at the ith receive antenna is given by

$$
y_i(t) = \sum_{j=1}^{N_r} h_{i,j} \times s_j(t), \; i = 1, 2, \cdots N_r.
$$

$$(2.7)$$

When the channel is flat fading, the delay τ can be dropped from the channel matrix expression. When the channel is frequency-selective fading, the channel may be expressed in the frequency domain via a Fourier transform. By applying orthogonal frequency-division multiplexing (OFDM), the frequency-selective channel can be divided into a number of narrow-band subchannels that are considered as flat fading. Due to this reason, the channels used in this dissertation are considered as flat-fading channels unless noted otherwise.

I.i.d. Model

In the classic independent identically distributed (i.i.d.) channel model, the elements of H are independent zero mean circularly symmetric complex Gaussian (ZMCSCG) random variables with unit variance. The i.i.d. channel matrix is denoted by H_w. Some properties of H_w are summarized below:

$$\mathcal{E}\left\{[\mathbf{H}_w]_{i,j}\right\} = 0,$$

(2.8)

$$\mathcal{E}\left\{\left|[\mathbf{H}_w]_{i,j}\right|^2\right\} = 1,$$

(2.9)

$$\mathcal{E}\left\{[\mathbf{H}_w]_{i,j}[\mathbf{H}_w]_{m,n}^*\right\} = 0, \ if \ i \neq m \ or \ j \neq n.$$

(2.10)

GPP Spatial Channel Model

In 2003, the 3rd Generation Partnership Project (3GPP) standards body developed a spatial channel model (SCM) [15]. SCM is a type of statistics channel model. It adopts a sum-ofsinusoids method to describe channel spatial characteristics. Figure 2.1 illustrates the SCM model. There are M=20 MPCs (also called sub-paths) that are grouped into one cluster with certain powers, delays, angle of arrivals (AoAs) and angle of departures (AoDs), which are randomly selected according to specific probability density functions (PDF) and their cross-correlations. All sub-paths in one cluster have identical powers and delays but different angle offsets for both AoA and AoD. These offsets are chosen from fixed tables in order to meet the desired per-path angle spread [15]. A fixed number of N = 6 clusters is defined in this model. Scatters related to the clusters are placed randomly and are determined by the AoAs and AoDs of the sub-paths. SCM is designed for a bandwidth up to 5 MHz, which is not sufficient for high-speed data service requirements. By introducing intra-cluster delay spread, SCME supports higher bandwidth. The targeted bandwidth of SCME is 100 MHz. Because both SCM and SCME were intended for cellular environments [16], only three scenarios are defined, including suburban macro, urban macro, and urban micro. This drawback limits their applications in outdoor-to-indoor environments. In [17], SCM is extended into a three-dimensional channel which is suitable for outdoor-to-indoor environments.

Multi-Antenna System Model

According to the number of the antennas used at the transmitter and the receiver, multiantenna systems are divided into four schemes: single-input single-output (SISO), singleinput multiple-output (SIMO), multiple-input single-output (MISO), and MIMO [7]. SISO, SIMO, and MISO can be treated as special cases of MIMO. Figure 2.2 shows a typical block diagram of the MIMO system with N_t transmit antennas and N_r receive antennas.

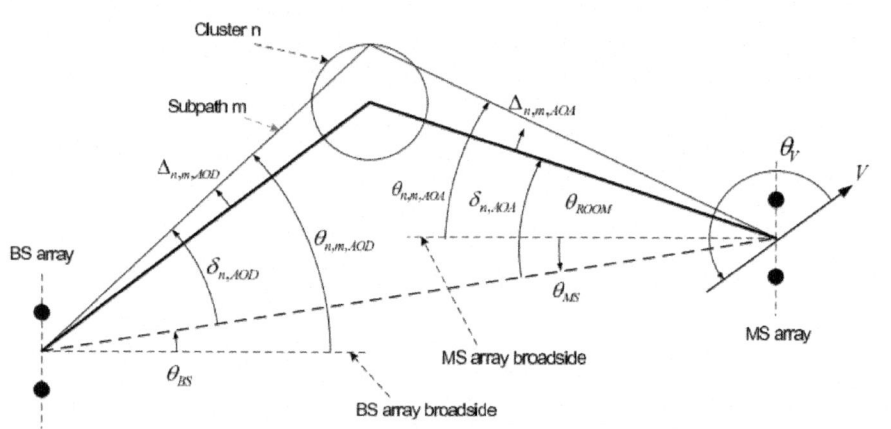

Figure 2.1: 3GPP spatial channel model

The input information stream s is assumed to be symbols that have been coded and mapped onto constellations. The input symbols are first divided into N_s data streams denoted by s, and then are pre-processed in space (and in time) domain into $N_t \times K$ blocks X, which are transmitted via the radio channel. Here K is the length of the space-time codewords. If no space-time coding is applied, $K = 1$. The transmitted signal can be expressed as

$$\mathbf{X} = f(\mathbf{s}),$$

$$(2.11$$

where f denotes the operation of the space(-time) pre-processing. Under the narrow-band flat fading assumption, the channel is represented by a $N_r \times N_t$ matrix H, where H_{ij} denotes the channel coefficient between the ith receive antenna and the jth transmit antenna. Thus, the received signals are given as

$$\mathbf{Y} = \mathbf{HX} + \mathbf{n},$$

$$(2.12)$$

where n is the complex white Gaussian noise vector. The received signals are then postprocessed in space(-time) domain to obtain the estimated signal.

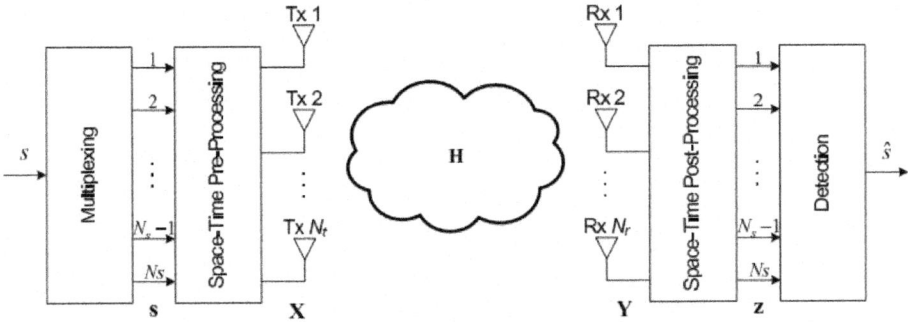

Figure 2.2: Block diagram of a MIMO system.

The data rate of the system can be calculated as

$$R = \frac{N_s}{K},$$

(2.13)

which is a function of N_s and K

Basic Multi-Antenna Schemes

The choice of N_s and K not only determines the data rate of the system, but also implies the architecture of the MIMO systems. Based on different configurations, there are three basic multi-antenna schemes: beamforming, spatial multiplexing (SM), and space-time coding (STC).

Beamforming

Beamforming is a powerful technique which can increase the link signal-to-noise ratio (SNR) by focusing the energy into desired directions. In this scheme, $N_s = 1$ and $K = 1$, resulting in the data rate $R = 1$. To enable beamforming, the number of transmit antennas and/or the number of the receive antennas should be greater than one.

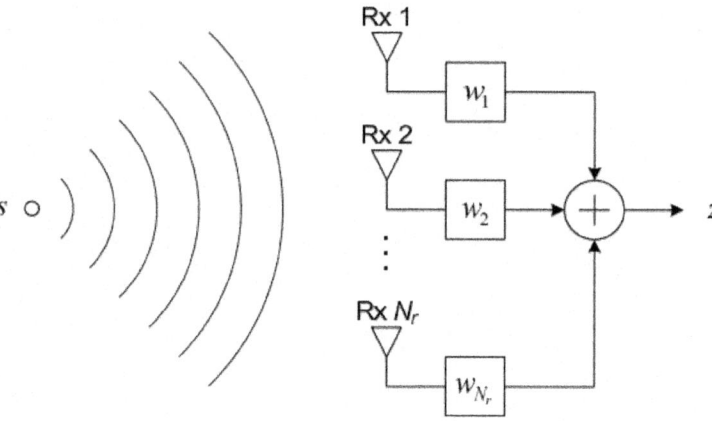

Figure 2.3: Illustration of a receive beamforming system.

Receive Beamforming

With a single transmit antenna ($N_t = 1$) and multiple receive antennas ($N_r > 1$), beamforming can be realized at the receiver. Figure 2.3 illustrates the receive beamforming scheme. The source symbol s arrives at the antenna array at the receiver side through the wireless channel. The received signals from the multiple antennas are first weighted by a beamforming vector w and then are combined. After the receiver has acquired a channel estimate, it can set the beamforming vector w to its optimal value to maximize the received SNR. This is done by aligning the beamforming vector with the channel via the so-called maximum ratio combining (MRC) $w = H^H$, which can be viewed as a spatial version of the well-known matched filter [18]. Another method for selecting the beamforming vector is to maximize the receive signal energy by steering the radiation pattern of the antenna array to the direction of arrival (DOA). This technique is known as smart antenna [19]. DOA can be estimated by using MUSIC (multiple signal classification) [20] and ESPRIT (estimation of signals parameters via rotational invariance technique) [21] algorithms.

Although the matched beamforming maximizes the SNR in AWGN channels, it is not optimal when there is co-channel interference. In this case, null-steering beamformer can be used to cancel the interference from other interference sources. By selecting the beamforming vector to be orthogonal to the channel from the interference sources, cancelation of up to N_{r-1} interfering signals is theoretically feasible.

Transmit Beamforming

Beamforming can also be realized at the transmitter [18]. Similar to the receive beamforming scheme, the transmitting signal is precoded by a beamforming vector and then transmitted via multiple transmit antennas. The beamforming vector is selected based on the channel state information (CSI) to maximize the receive SNR of the received signal. To enable the vector selection, the CSI must be known at the transmitter. This requires CSI feedback from the receiver, where the CSI is estimated, to the transmitter. A transmit beamforming system is used in Section 7 to enable physical layer security.

Spatial Multiplexing

With beamforming, a multi-antenna system can only transmit one symbol at a time. However, the multiple transmit-receive paths provide extra degrees of freedom. More than one symbol could be transmitted via the MIMO channel simultaneously. Spatial dimension may be utilized to increase the system throughput. With $N_t > 1$ transmit antennas and $N_r > 1$ receive antennas, up to $min(N_t, N_r)$ streams can be multiplexed in space dimension. Therefore, in a spatial multiplexing multi-antenna system, $N_s > 1$ and $K = 1$, leading to the data rate $R > 1$. In spatial multiplexing, a precoding matrix W, also called precoder, is used to precode the symbols in the vector s as

$$\mathbf{X} = \mathbf{W}\mathbf{s}.$$

(2.14)

The precoder has two effects: decoupling the input signal into orthogonal spatial modes in the form of eigen-beams, and allocating power over these beams, based on the CSI at the transmitter. If the precoded orthogonal spatial modes match the channel eigen-directions, there will be no interference among signals sent on these modes, creating parallel channels and allowing transmission of independent signal streams. Moreover, the precoder allocates power on these beams. For orthogonal eigen-beams, if the beam powers are different, the overall transmit radiation pattern will have a specific shape. The precoder effectively creates a radiation shape matching to the channel based on the CSI, so that more power is sent in the directions where the channel is strong and less or no power where it is weak. More transmit antennas will increase the ability to finely shape the radiation pattern and therefore are likely to deliver more precoding gain. At the receiver, the signals can be recovered by linear detection algorithms.

Space-Time Coding

In beamforming and spatial multiplexing schemes, K is chosen to be 1, which means no modulation or coding are involved in time domain. However, joint coding in space and time domains could lead to full spatial diversity with no CSI required at the transmitter. This technique is called space-time coding (STC) [22]. In a space-time coding system, the Ns information symbols are coded into a $N_t \times K$ block, where K is the length of the codewords. To successfully decode the symbols at the receiver, it is usually required that K be greater than or equal to the number of symbols N_s. Hence, the data rate of a STC system is R <= 1. When R = 1, the system is called a full-rate system. The most common used STC is space-time block codes (STBC). Alamouti first proposed the code for two transmit antennas [23]. In this scheme, the input symbols are grouped into blocks. Each block consists of two consecutive symbols, denoted by s = $[s_1 s_2]$. The blocks are coded into 2 × 2 blocks by using the codeword

$$
\mathcal{G}_2 = \begin{pmatrix} s_1 & s_2 \\ -s_2^* & s_1^* \end{pmatrix}
$$

(2.15)

This 2 × 2 block information is transmitted through two antennas and in two time slots: in the first time slot, s_1 and $-s_2^*$ are transmitted, and in the second time slot, s_2 and s_1^* are transmitted. Alamouti code is the only STBC with full rate. For more than two transmit antennas, no full-rate STBC exists, but it has been proven that for any number of transmit antennas, orthogonal STBCs with data rate less than one can be found [24].

Single-User vs Multiple-User MIMO

In the aforementioned three basic schemes, the information transmitted via the multiantenna channels is limited to one user. These schemes are called single-user MIMO (SUMIMO). Because the multiple antennas provide extra spatial dimension, it is possible to transmit information of multiple users simultaneously in the same bandwidth, while the users are differentiated in the spatial domain. The technology that implement this scheme is called multi-user MIMO (MU-MIMO) [25]. It is an attractive approach to increase spectral efficiency in wireless links. MU-MIMO is an extended concept of space-division multiple access (SDMA), which allows a terminal to transmit (or receive) signal to (or from) multiple users in the same band simultaneously. MU-MIMO can also be thought of as an extension of MIMO applied in various ways as a multiple access strategy. The performance of MU-MIMO relies on precoding capability. Its performance advantages are not achievable if the transmitter does not use precoding. The precoding matrix is generated from the CSI, which means that the feedback of CSI from the receiver to the transmitter is inevitable.

Per-User Unitary Rate Control (PU^{2}RC) is a practical MU-MIMO scheme [26]. It utilizes the concept of both pre-coding matrices and scheduling to enhance the system performance of multi-antenna wireless networks. Recently, PU2RC has been adopted in the IEEE 802.16m system [27] and the concept of this scheme was included in the 3GPP LTE standard [9].

Distributed MIMO

Traditionally, spatial diversity is achieved by using multiple antennas at the transmitter and/or receiver, where the antennas are packed together with spacing of the order of wavelength, referred to as co-located MIMO. However, the benefits of the co-located MIMO technique are limited in practical systems. The reasons for this limitation are two-fold. First, spatial correlation causes performance degradation. In a co-located MIMO system, antennas at each node have to be placed close to each other. Thus, radio signals at the co-located antennas experience a similar scattering environment, and the channels may be correlated, especially when a line-of-sight (LOS) channel between the transmitter and receiver dominates. The channel matrices could be ill-conditioned, resulting in significant capacity decrease. Second, due to the terminal size limitation, the node cannot be equipped with many antennas. Since the diversity gain is proportional to the number of antennas, a co-located MIMO system with few antennas cannot produce the expected performance. To mitigate the aforementioned drawbacks in co-located MIMO systems, a new technique named "distributed MIMO" was proposed and has attracted much attention [28]. The major difference between the distributed MIMO and the co-located MIMO is that multiple antennas at the front-end of wireless networks are distributed among widely-separated radio nodes. In a distributed MIMO system, each node may be only equipped with one antenna. Many nodes at different locations transmit the same information to the receiver. In this manner, multiple nodes form a virtual antenna array that achieves higher spatial diversity gain. This kind of spatial diversity is referred to as user cooperation diversity [28], or simply cooperative diversity [29]. Distributed MIMO can provide full user cooperation diversity, and the data rate in the cooperative networks can be significantly increased by using distributed space-time coding [10].

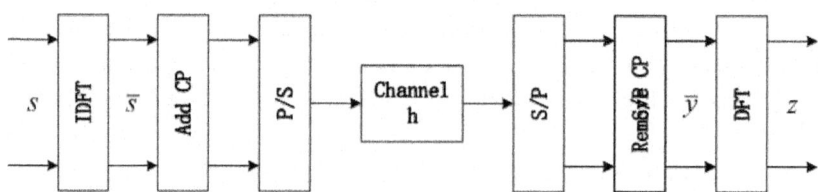

Figure 2.4: Block diagram of an OFDM system

OFDM and Broadband Communications OFDM is an attractive modulation scheme used in broadband wireless systems with large delay spread. By converting the frequency-selective fading channel into a parallel collection of flat-fading subchannels, OFDM can mitigate multi-path dispersion and provide high data-rate transmissions [30]. The high spectral efficiency and the efficient implementation by using fast Fourier transform (FFT) also make OFDM an popular modulation scheme. OFDM-based physical layers have been selected for a number of wireless applications including digital broadcast systems [31], wireless local area networks (LANs) [32] [33], and broadband wireless access systems [34]. A block diagram of an OFDM system is shown in Figure 2.4. The transmitted block of the modulated symbols is denoted by

$$\mathbf{s} = [\mathbf{s}(1)\,\mathbf{s}(2)\,\cdots\,\mathbf{s}(N)]^T$$

$$(2.16)$$

where N is the number of the sub-carriers. Each block is transformed to the time domain by the inverse discrete Fourier transform (IDFT):

$$\bar{\mathbf{s}} = \mathbf{F}^*\mathbf{s}$$

$$(2.17)$$

where F is the unitary discrete Fourier transform (DFT) matrix of size N defined by

$$\mathbf{F}_{ik} = \frac{1}{\sqrt{N}} e^{-\frac{j2\pi ik}{N}}, \quad i,k = 0,1,\ldots,N-1$$

$$(2.18)$$

A cyclic prefix (CP) of length P is added to the head of each transformed block. The time-domain blocks are then serially transmitted over the time-varying Rayleigh fading channel. A finite impulse response (FIR) model with L+ 1 taps is assumed for the channel, i.e.,

$$\mathbf{h} = [h_0\,h_1\,\cdots\,h_L]^T$$

$$(2.19)$$

with L ≤ P in order to preserve the orthogonality between tones. At the receiver, the signal corresponding to the transmitted block ⁻s is sampled into a vector. After discarding the CP, the received block of data can be written as [35]

$$\bar{\mathbf{y}} = \mathbf{H}\bar{\mathbf{s}} + \bar{\mathbf{v}} \tag{2.20}$$

Where

$$\mathbf{H} = \begin{bmatrix} h_0 & h_1 & \cdots & h_{L-1} & h_L & & & & \\ & h_0 & h_1 & \cdots & h_{L-1} & h_L & & & \\ & & \ddots & \ddots & & \ddots & \ddots & & \\ & & & h_0 & h_1 & \cdots & h_{L-1} & h_L \\ h_L & & & & h_0 & h_1 & \cdots & h_{L-1} \\ \vdots & \ddots & & & & \ddots & \ddots & \vdots \\ h_2 & \cdots & h_L & & & & h_0 & h_1 \\ h_1 & h_2 & \cdots & h_L & & & & h_0 \end{bmatrix} \tag{2.21}$$

is an N × N matrix, and $\mathbf{v}^-$ is additive white noise at the receiver. H presents the form of a circulant matrix due to the existence of the CP. It is well known that H can be diagonalized by the DFT matrix as

$$\mathbf{H} = \mathbf{F}^* diag\{\lambda\}\mathbf{F} \tag{2.22}$$

where the vector λ is related to h via

$$\lambda = \sqrt{N}\mathbf{F}^* \begin{bmatrix} \mathbf{h} \\ \mathbf{0}_{N-(L+1))\times 1} \end{bmatrix} \tag{2.23}$$

and $diag\{\lambda\}$ is an $N \times N$ matrix with λ as its main diagonal.

Then, substituting (2.22) into (2.20), the received block of data can be expressed as

$$\bar{\mathbf{y}} = \mathbf{F}^* diag\{\lambda\}\mathbf{F}\bar{\mathbf{s}} + \bar{\mathbf{v}} \tag{2.24}$$

Applying DFT, the received frequency-domain data is given by

$$\mathbf{z} = \mathbf{F}\bar{\mathbf{y}} \tag{2.25}$$

Substituting (2.24) into (2.25), the received data is given by

$$\mathbf{z} = diag\{\lambda\}\mathbf{s} + \mathbf{v} \tag{2.26}$$

where v is a transformed version of the original noise vector $\mathbf{v}^-$. Although OFDM is a type of advanced modulation technique, it suffers from degradations resulting from the imbalances between the In-phase and the Quadrature-phase (IQ) branches. IQ imbalance is caused by the analog front-end imperfections. It may lead to inter-carrier interference in an OFDM system, resulting in limited operating SNR.

LONG-RANGE CHANNEL PREDICTION IN SPATIAL MULTIPLEXING MIMO-OFDM SYSTEMS

With ever increasing demands for multimedia services and web-related contents, high data throughput is becoming one of the major features in the next generation of wireless communication systems. Among the existing techniques, spatial multiplexing (SM) MIMO-OFDM is considered as one of the most promising physical-layer architectures to provide high-speed communications [36]. In a less-severe scattering environment, the channel matrix could become ill-conditioned, resulting in the performance degradation of a spatial multiplexing MIMO-OFDM system. However, a precoding technique can be used to improve the robustness of spatial multiplexing to the rank deficiencies of the MIMO channel matrix [37,38]. According to this method, the transmitted data streams are customized to the eigenstructure of the channel matrix by being multiplied with a precoding matrix, which is generated based on the channel state information (CSI). This implies that the transmitter requires some sort of knowledge of the CSI. In most communication systems, the CSI at the transmitter is acquired from a low data-rate feedback channel where the receiver provides information about the forward link condition to the transmitter [26,39].

In an OFDM system, the CSI of all subchannels should be fed back. However, it is usually unrealistic to feed back very much information to the transmitter due to the bandwidth efficiency. Recently, an alternative way was proposed for precoded systems to select the codewords for each subchannel at the receiver and feed the indices back to the transmitter [40–43]. With this method, the total feedback bits for each channel realization can be significantly reduced. Another concern in the feedback systems is feedback delay. Due to the reverse propagation delay and the processing and queuing time at the transceiver, the feedback delay may last for several milliseconds, or tens of the OFDM symbol time [44, 45]. The feedback CSI at the transmitter may become outdated and hence cause a significant performance degradation. As an effective countermeasure against feedback delay, channel prediction has been used in many systems. An adaptive channel predictor based on linear minimum mean square error (MMSE) estimation was proposed in a single antenna single carrier system [46]. By using this adaptive algorithm, this predictor can successfully predict the channel variation within a longer range. Linear channel predictors were also used in transmit-beamforming systems

[43, 47–49]. So far, only Kalman filter-based methods have been employed to predict the fading channel for spatial multiplexing MIMO systems [41, 42].

However, the pilotbased linear channel estimates cannot be utilized to improve the adaptive performance of the Kalman filter, resulting in a relatively short prediction period which cannot cover the long-range delay. In this chapter, a multi-block autoregressive (AR) model-based linear MMSE predictor is proposed to predict the time-varying fading channel in the limited feedback precoded SMMIMO-OFDM systems [50]. Instead of using an adaptive algorithm, multi-block prediction is utilized to iteratively predict the long-range channel variation. Codewords for precoding the future blocks are selected at the receiver based on the predicted CSI, and the indices of the codewords are fed back to the transmitter. The bit-error rate (BER) performance of the system is investigated. Simulation results show that the proposed approach can accurately predict long-range channel variation and almost completely compensate for the error performance degradation.

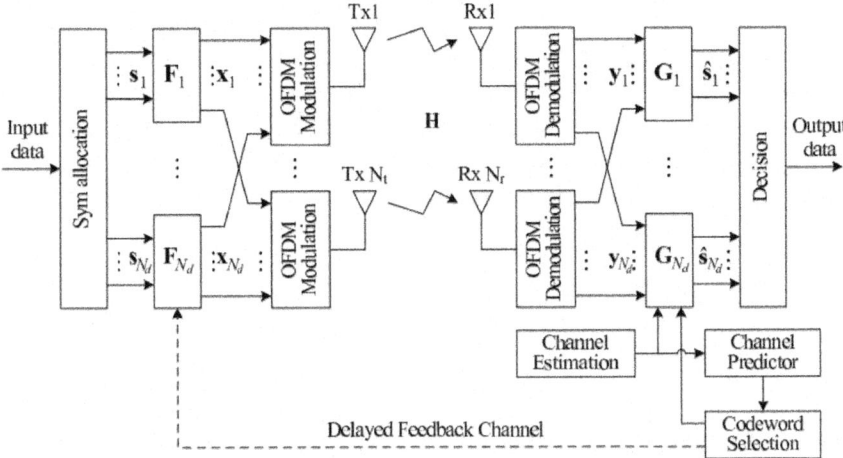

Figure 3.1: Block diagram of a precoded spatial multiplexing MIMO-OFDM system.

System Model

The block diagram of a limited feedback precoded spatial multiplexing MIMO-OFDM system with channel prediction is depicted in Figure 3.1, where N_t transmit antennas and N_r receive antennas are deployed for MIMO architecture. OFDM utilizes N_f subcarriers, wherein N_d and N_p subcarriers are allocated for user data and pilot symbols, respectively. In Figure 3.1, the input data are first divided into N_d groups of N_s substreams $\mathcal{C}^{N_s \times 1}$, where $k \in \{1, 2, \cdots, N_d\}$ is the index of the subcarriers and n is the index of the OFDM symbols. It is assumed that the input data is chosen independently from the CSI and $E\{s_k(n)s_k^H(n)\} = (\mathcal{E}_s/N_s)I_{N_s}$, where $\mathcal{E}_s$, where Es is the total transmit power. Each group of the Ns substreams is then multiplied by a precoding matrix $F_k(n) \in \mathcal{C}^{N_t \times N_s}$ which is selected from a codebook F = $\{F_1, F_2, \cdots_{F2} N_b\}$ according to the N_b-bit indices fed back from the receiver over an delayed error-free feedback channel. The precoded data $x_k(n) \in \mathcal{C}^{N_t \times 1}$

$$x_k(n) = F_k(n)s_k(n).$$

(3.1)

These coded data are then modulated after OFDM modulation, including pilot insertion, inverse discrete Fourier transform (IDFT), adding cyclic prefix (CP), and serial-parallel conversion. Finally, the OFDM modulated signal is transmitted over a time-varying fading channel. It is assumed that the channel length N_l is shorter than the CP length N_{cp}, leading to a flat fading over each subcarrier. At the receiver, the received signal is demodulated in term of OFDM demodulation, including CP remove, discrete Fourier transform (DFT), and separation between user data and pilot symbols. According to [51], the received symbol vector can be written as

$$\begin{aligned} y_k(n) &= H_k(n)x_k(n) + w_k(n) \\ &= H_k(n)F_k(n)s_k(n) + w_k(n), \end{aligned}$$

(3.2)

where $H_k(n) \in \mathcal{C}^{N_r \times N_t}$ is the channel matrix of the kth subchannel during the nth OFDM symbol, and $w_k(n) \in \mathcal{C}^{N_r \times 1}$ is a noise vector whose entries consist of the independent and identically distributed (i.i.d.) complex Gaussian distribution with zero mean and variance N_0.

Channel Estimation

After OFDM demodulation, the pilot symbols are extracted from the received data and are used to estimate the current CSI at the pilot subcarriers. The CSI at all subcarriers is obtained by linear interpolation at the receiver [52]. It is assumed that the CSI is correctly estimated and is denoted by $\hat{H}_k(n)$.

Linear Receiver

Although the maximum likelihood (ML) receiver outperforms with linear receiver, its computation complexity increases exponentially with the number of substreams. A linear

MMSE receiver may also provide good performance. The received data are estimated by using linear decoding matrices, $\mathbf{G}_k(n) \in \mathcal{C}^{N_s \times N_r}$, defined as follows:

$$\hat{\mathbf{s}}_k(n) = \mathbf{G}_k(n)\mathbf{y}_k(n), \tag{3.3}$$

where the decoding matrix $G_k(n)$ for the MMSE criterion is given as [51, 53]

$$\mathbf{G}_k(n) =$$
$$\left[\frac{N_s N_0 I_{N_s}}{\mathcal{E}_s} + \mathbf{F}_k^H(n)\hat{\mathbf{H}}_k^H(n)\hat{\mathbf{H}}_k(n)\mathbf{F}_k(n) \right]^{-1} \mathbf{F}_k^H(n)\hat{\mathbf{H}}_k^H(n). \tag{3.4}$$

Codeword Selection Criteria

Many criteria have been used to select the optimal precoding matrix from a given codebook for linear receivers, such as minimum singular value selection criterion (MSV-SC), mean square error selection criterion (MSE-SC), and capacity selection criterion (Capacity-SC) [54]. MSV-SC is chosen as the selection criterion because it provides a close approximation to maximize the minimum signal-to-noise ratio (SNR) for dense constellations. Based on MSV-SC, the optimal codeword F_{opt} is selected from the codebook F given a channel realization H as follows [54]

$$\mathbf{F}_{opt} = \operatorname*{argmax}_{\mathbf{F}_i \in \mathcal{F}} \lambda_{\min}\{\mathbf{HF}_i\}. \tag{3.5}$$

Codebook Design

Codebook design has attracted much attention recently. An excellent overview of the development of codebook design is found in [26]. Generally, the codebook should be designed according to the codeword selection criteria. For MSV-SC, the codebook is designed by maximizing the minimum projection norm-2 distance between any pair of codeword matrix column spaces, which is known as Grassmannian subspace packing [54]. This designed codebook is also known as the best codebook [55].

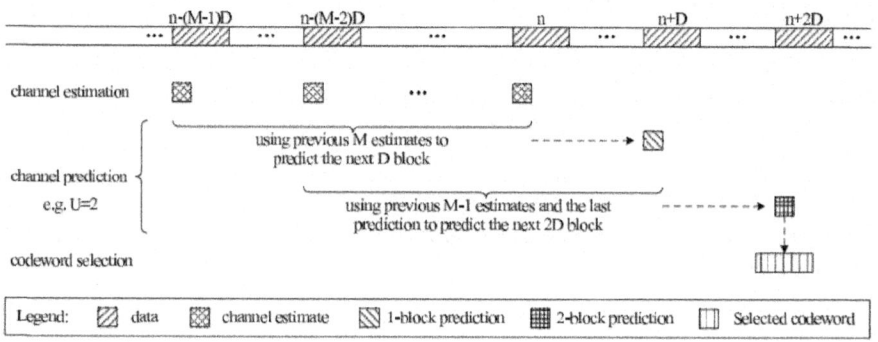

Figure 3.2: Illustration of the multi-block prediction procedure.

Proposed Multi-Block Linear Predictor

Wireless channels usually demonstrate time-varying characteristics. In some environments, the channel is fast fading, e.g., the terminal user is moving around quickly. According to the Jakes model [56], the autocorrelation of a fading channel is given as

$$r(t) = J_0(2\pi f_d t),\tag{3.6}$$

where $J_0(\cdot)$ is the zero-order Bessel function of the first kind, and f_d is the maximum Doppler shift, which is given by the terminal's moving speed v as

$$f_d = \frac{v}{c}f_c,\tag{3.7}$$

where c and f_c are the light speed and the carrier frequency, respectively. The autoregressive (AR) model-based linear predictor is the most commonly used method to predict time-varying fading channels. Usually, the channel prediction is given by a AR(M) model as

$$\bar{\mathbf{h}}_k(n+1) = \sum_{m=1}^{M} a_m \hat{\mathbf{h}}_k(n-m+1),\tag{3.8}$$

where M is the order of the AR model, am is the mth AR model coefficient, and $\hat{\mathbf{h}}_k(n) = \text{vec}\{\hat{\mathbf{H}}_k(n)\}$, where vec$\{\cdot\}$ means of stacking all columns of $\hat{\mathbf{H}}$ into one column vector. This

implies that the predictor can only predict one OFDM symbol ahead. However, the feedback delay may last for several milliseconds, or tens of time slots of the OFDM symbol [44, 45]. To enable a long-range prediction, a multi-block linear predictor is proposed, shown in Figure 3.2, where a 2-block prediction procedure is illustrated. The proposed predictor first utilizes M previous channel estimates, which are separated from each other by one block of D OFDM symbols, to predict the Dth next channel matrix as follows:

$$\tilde{h}_k(n+D) = \sum_{m=1}^{M} a_m \hat{h}_k(n - mD + D).$$

$$(3.9)$$

The AR model coefficients are determined by the MMSE criterion as [46]:

$$\mathbf{a}_k = \mathbf{R}^{-1}\mathbf{r},$$

$$(3.10)$$

Where $\mathbf{a}_k = [a_1 \; a_2 \; \cdots \; a_M]^T$, $\mathbf{R} \in \mathcal{C}^{M \times M}$ is the autocorrelation matrix with elements $\mathbf{R}_{ij}(n) = r(D|i-j|)$, and $\mathbf{r} \in \mathcal{C}^{M \times 1}$ is the autocorrelation vector with elements $r_i(n) = r(D_i)$.

It should be noted that the autocorrelation matrix and vector, and thereby the AR model coefficients, are slow fading. They can be calculated by using the measured Doppler frequency. As a matter of convention, these parameters are assumed to be time invariant and their time indices are dropped. Using Equation (3.9), the prediction range is the period of D OFDM symbols. If the OFDM symbol time is T, the prediction range is DT. The intuitive sense is that this method can predict infinity delay by increasing the block size D. However, the channel estimates used in (3.9) must be sampled at least at the Nyquist rate given by twice of the maximum Doppler shift f_d. This requirement limits the prediction range. For example, if the maximum Doppler shift is 200 Hz, the channel matrix has to be taken every 2.5 ms, i.e., the largest prediction delay must be limited within 2.5 ms. In practice, the channel sampling frequency has to be set higher than the Nyquist rate in order to achieve an accurate estimation. As shown in Section 3.4, when the maximum Doppler shift is 200

Hz, the channel estimates should be selected every 1 ms to guarantee the accuracy of the channel prediction. In order to increase the prediction range, an iterative method is further utilized to predict a few more blocks ahead. For a Q-block predictor, the QDth next channel prediction is given as

$$\tilde{h}_k(n + QD) = \sum_{q=1}^{Q-1} a_q \tilde{h}_k(n + QD - qD)$$
$$+ \sum_{m=Q}^{M} a_m \hat{h}_k(n - mD + QD).$$

$$(3.11)$$

The multi-block prediction exploits the previous channel estimate and previously predicted channel matrices. Although the iterative procedure introduces error propagation, this multiblock predictor can effectively predict a few blocks of data stream ahead of the channel variation. The performance improvement can be validated by the numerical simulation in the next section.

Simulation Results

In this section, we present the Monte Carlo simulation results of the BER performance in a limited feedback precoded spatial multiplexing MIMO-OFDM system with multi-block channel prediction. The system parameters are listed in Table 3.1. A 4 × 4 MIMO architecture is employed and 512 subcarriers are allocated for OFDM scheme. No pilot symbols are used * . Two substreams are transmitted over each subchannel. ITU Vehicular A channel model [57] is adopted to simulate the multipath channel, and the channel time-varying characteristic is modeled as a Jakes model [56]. It is assumed that the terminal user moves with a fixed speed of 108 km/h, leading to a fixed maximum Doppler shift of 200 Hz. As the sampling frequency is set as 5 MHz, one OFDM symbol time is 108.8 μs. If the block size is set to 10, the channel matrix used in the linear predictor is taken as a sampling rate 919 Hz, which is greater than twice the maximum.

Table 3.1: System Parameters for Simulation

Parameters	values
Antenna configuration ($N_t \times N_r$)	4×4
FFT size (N_f)	512
CP length (N_{cp})	32
Channel model	ITU Vehicular A model
Channel correlation	Jake's model
Carrier frequency (f_c)	2.0 GHz
Sampling frequency (f_s)	5 MHz
Sampling time (T_s)	0.2 μ s
OFDM symbol time (T)	108.8 μ s
Terminal speed (v)	108 km/h
Maximum Doppler spread (f_d)	200 Hz
Block size (D)	$10 \sim 30$
Predictor order (M)	1, 2, 5, 10
Modulation	64 QAM
Codebook length	2^2, 2^3, 2^4, 2^6

Doppler frequency. For easy comparison, the 6-bit codebook listed in [55] and an AR(10) model are employed. Figure 3.3 shows the BER performance of a precoded SM-MIMO-OFDM system with different feedback delay without channel prediction (Q = 0). It is observed that the BER performance is exacerbated with the increase of the feedback delay. When the delay is over 10 OFDM symbols, the degradation is almost 6 dB at 10^{-4} BER. The BER performance of the system with randomly selected precoding matrices is also shown in Figure 3.3, labeled as "Random PM". If the delay is long enough so that the autocorrelation is close to zero the precoding matrices can be considered to be randomly selected.

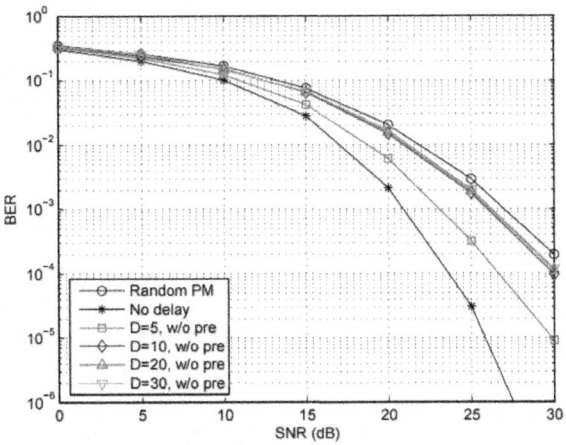

Figure 3.3: BER performance of a precoded SM-MIMO-OFDM system with different feedback delay but without channel prediction.

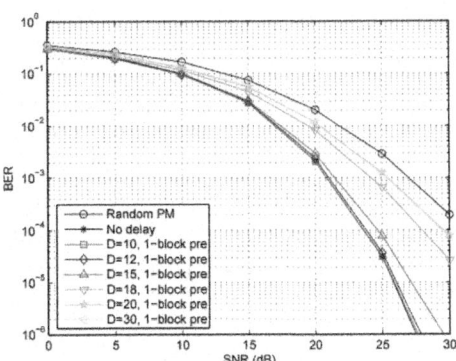

Figure 3.4: BER performance of a precoded SM-MIMO-OFDM system with different feedback delay and 1-block prediction, AR(10).

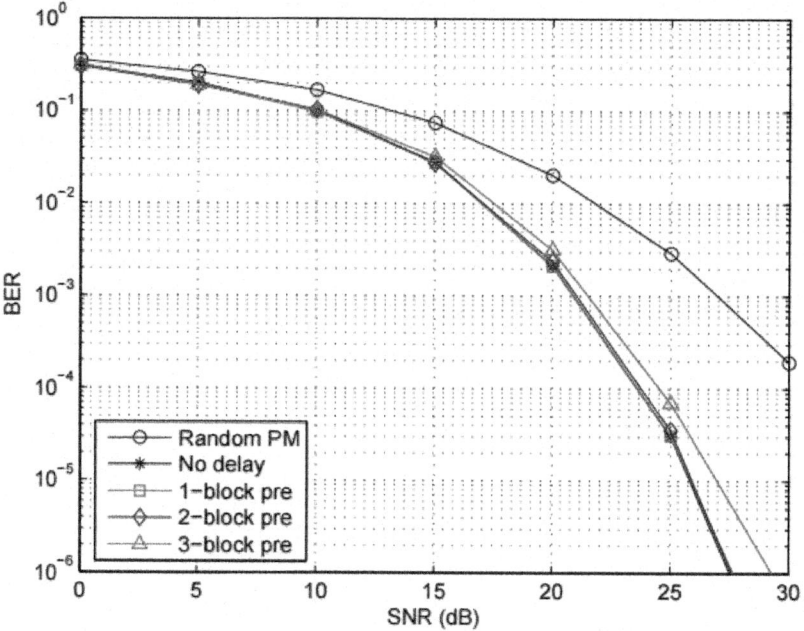

Figure 3.5: BER performance of a limited feedback precoded SM-MIMO-OFDM system with multi-block prediction, D=10, AR(10).

Therefore, the curves labeled as "Random PM" and "No Delay" can be treated as the performance upper bound and lower bound, and are shown in the following figures for comparison. The system performance with different feedback delay and 1-block prediction is shown in Figure 3.4. It is observed that with 1-block prediction, the performance degradation can be successfully mitigated if the feedback delay is less than the period of 12 OFDM symbols. If the feedback delay is longer, performance degradation occurs due to the low sampling rate of the channel matrix. D = 10 is used in the following simulations to ensure accuracy of one-block prediction. Figure 3.5 compares the BER performance of the system with fixed channel sampling rate, D = 10, but with different feedback delays. It is shown that using proposed multi-block iterative prediction, long feedback delay can be well compensated and the enhanced BER performance is close to the lower bound.

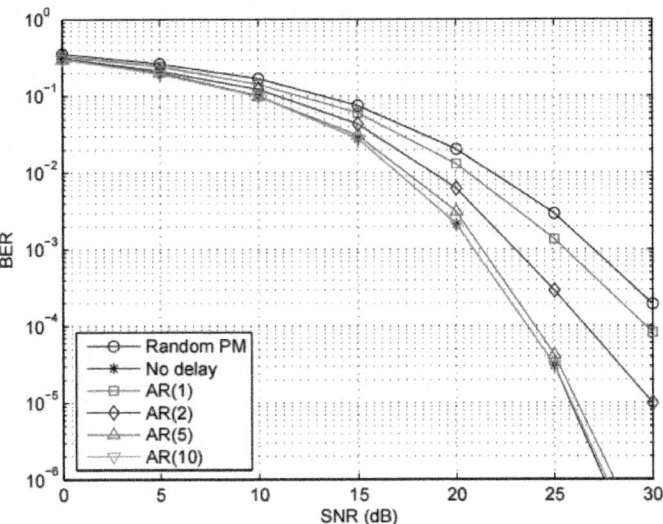

Figure 3.6: BER performance comparison under different orders of AR model, D=10.

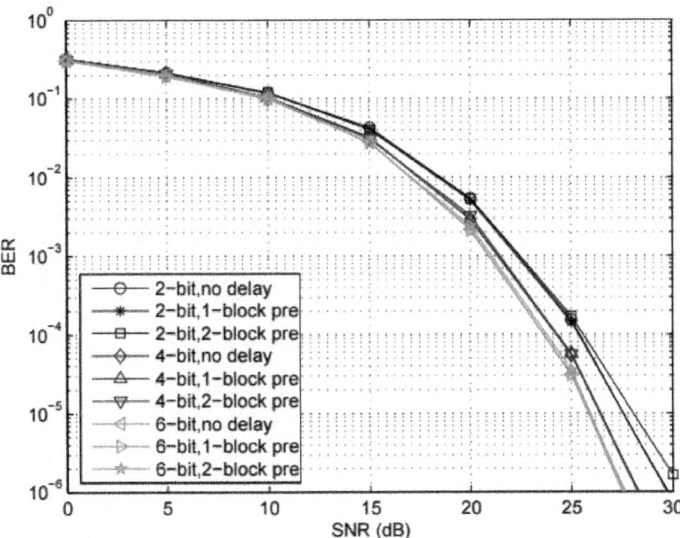

Figure 3.7: BER performance comparison for different size of codebook, D=10, AR(10)

With 3-block prediction, the BER degradation at 10^{-4} is less than 1 dB, resulting of 5 dB more improvement. The effect of the AR-model order is also evaluated by using 1-block channel predictor, shown in Figure 3.6. The performance is improved with the increase of the order number of the AR-model. When the order number is greater than 5, the BER performance approximates to the lower bound. This demonstrates that the AR model is very effective in modeling channel variation. Finally, the BER performance with different codebook sizes is compared in Figure 3.7. It shows that the proposed channel predictor can effectively compensate for the performance degradation of systems using any sizes of codebooks. The improved BER performance is very close to the performance of system without delay.

VIRTUAL CHANNEL-BASED I/Q IMBALANCE COMPENSATION IN MIMO-OFDM SYSTEMS

Although OFDM presents numerous advantages, it suffers from performance degradation due to the hardware component flaws in the analog front-ends of the transceivers. The inphase/quadrature-phase (I/Q) imbalance is a major factor resulting in performance degradation [58, 59]. When the received radio-frequency (RF) signal is downconverted to baseband, the analog front-end imperfections cause imbalances between the I and Q branches, which introduces intercarrier interference (ICI) and frequency-dependent distortion to the received data. This leads to a decrease in the operating signal-to-noise ratios (SNRs) and low data rates. Because the effect of the I/Q imbalances on the system is mixed with the signal distortion from the fading channel, the estimation and compensation of the I/Q imbalances from the received data are critical and challenging in OFDM-based systems. Frequency-independent and frequency-dependent I/Q imbalance models are reported in [58] and [59], respectively. Based on these models, the effects of I/Q imbalances are studied in OFDM systems [60–65] and in MIMO-OFDM systems [66–70]. In [66], the input-output relation of a MIMO-OFDM system with frequency-independent I/Q imbalances at the receiver is derived. Based on this result, an adaptive method is introduced to compensate for the received data. In [68], the effect of frequency-dependent I/Q imbalances on MIMO-OFDM system is studied, using a pilot-based compensation scheme. In both studies, however, the received signals are diversity combined using inaccurate channel state information (CSI) estimated under I/Q imbalances, leading to system performance degradation. We have proposed a method to deal with this problem in [70].

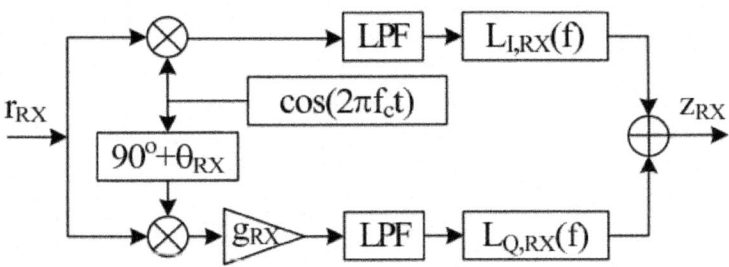

Figure 4.1: Block diagram of frequency-dependent I/Q imbalance at a receiver

By jointly estimating and compensating for multipath fading channels and I/Q imbalances, this method can effectively mitigate the I/Q effects. In this chapter, a novel approach is proposed to analyze MIMO-OFDM wireless communication systems with I/Q imbalances over multipath fading channels [65, 70, 71]. A virtual channel is proposed to bypass the channel estimation under I/Q influence. Based on this approach, the TX and RX I/Q imbalances are treated as parts of the fading channels. The effects of both the fading channels and I/Q imbalances on the system can be jointly estimated before diversity combining and are then employed for diversity combining and signal compensation. A minimum mean square error (MMSE) estimator and a least square (LS) estimator are used to estimate the joint coefficients of the virtual channel. A signal compensation approach based on a zero-forcing algorithm is also provided. The system performance is theoretically analyzed, and the bit error rate (BER) is expressed in closedform. Extensive simulation results verified that the I/Q imbalances at both transmitter and receiver sides can be effectively mitigated by using the virtual channel approach.

System Models

I/Q Imbalance Model

A frequency-dependent I/Q model at the receiver side is shown in Figure 4.1. As described in [59], the I/Q imbalances arise from two effects. One is the effect of the quadrature demodulator, which is frequency-independent and determined by I/Q amplitude imbalance g_{RX} and I/Q phase imbalance θ_{RX}; another is the effect of branch components, which is frequency-dependent and modeled as two filters with frequency response $L_{I,RX(f)}$ and

$L_{Q,RX}(f)$. If the I/Q branches are perfect, then $g_{RX} = 1$, $\theta_{RX} = 0$, and $L_{I,RX}(f) = L_{Q,RX}(f) \equiv 1$.

Assuming $r_{BB}(t)$ is the baseband equivalent signal of the received signal $r_{RX}(t)$, the downconverted signal $z_{RX}(t)$ can be written as

$$z_{RX}(t) = g_{1,RX}(t) \otimes r_{BB}(t) + g_{2,RX}(t) \otimes r_{BB}^*(t)$$

$$(4.1)$$

Where

$$g_{1,RX}(t) = \left[l_{I,RX}(t) + l_{Q,RX}(t)g_{RX}e^{-j\theta_{RX}} \right]/2$$
$$g_{2,RX}(t) = \left[l_{I,RX}(t) - l_{Q,RX}(t)g_{RX}e^{j\theta_{RX}} \right]/2$$

$$(4.2)$$

$l_{I,RX}(t)$ and $l_{Q,RX}(t)$ are the time-domain representations of $L_{I,RX}(f)$ and $L_{Q,RX}(f)$, re-spectively. From the point of view of the frequency domain, the expression in (4.1) can be written as

$$Z_{RX}(f) = G_{1,RX}(f)R_{BB}(f) + G_{2,RX}(f)R_{BB}^*(-f)$$

$$(4.3)$$

where $R_{BB(f)}$ is the Fourier transform of $r_{BB(t)}$, and

$$G_{1,RX}(f) = \left[L_{I,RX}(f) + L_{Q,RX}(f)g_{RX}e^{-j\theta_{RX}} \right]/2$$
$$G_{2,RX}(f) = \left[L_{I,RX}(f) - L_{Q,RX}(f)g_{RX}e^{j\theta_{RX}} \right]/2$$

$$(4.4)$$

In (4.3), the term $R*_{BB}(-f)$ represents the intercarrier interference (ICI) projected from the mirror frequency to the signal frequency. This is called image projection, a major problem caused by I/Q imbalances in signal demodulations. Similarly, the relation between the signal R_{TX} and the transmitted baseband signal Z_{TX} with I/Q mismatch can be written as

$$Z_{TX}(f) = G_{1,TX}(f)R_{TX}(f) + G_{2,TX}(f)R_{TX}^*(-f)$$

$$(4.5)$$

Where

$$G_{1,TX}(f) = \left[L_{I,TX}(f) + L_{Q,TX}(f)g_{TX}e^{j\theta_{TX}} \right]/2$$
$$G_{2,TX}(f) = \left[L_{I,TX}(f) - L_{Q,TX}(f)g_{TX}e^{-j\theta_{TX}} \right]/2$$

$$(4.6)$$

where g_{TX} denotes the I/Q amplitude imbalance at T_X, θ_{TX} represents the I/Q phase imbalance at T_X, and $LI,_T X(f)$ and $L_Q,T_X(f)$ indicate the non-linear frequency characteristics of the I and Q branches at T_X.

MIMO-OFDM Model with I/Q Imbalances

A block diagram of a MIMO-OFDM wireless communication system with Alamouti diversity scheme and frequency-dependent TX and RX I/Q imbalances is shown in Figure 4.2. In this system, there are two transmit antennas and Nr (≥ 1) receive antennas. All signals are represented in the form of space-time coded (STC) blocks in the frequency domain. For example, $S_{i,1}|S_{i,2}$ denotes an STC block data of N $\times$ 2 matrix, where i = 1, 2, $\cdots$ is the STC block index, and N is the number of the used OFDM subcarriers. Vector $[S_{i,j}(-N/2) \cdots S_{i,j}(-1) \ S_{i,j}(1) \cdots S_{i,j}(N/2)]^T$ is an OFDM symbol to be transmitted over the system at the jth, j $\in \{1, 2\}$ time slot of the ith STC block, where $S_{i,j}$(k) denotes the symbol at the $k \subset \{-N/2, \cdots, -1, 1, \cdots, N/2\}$ subcarrier with average symbol energy $E_s/2$. Vector $S_{i,1}$ is transmitted followed by vector $S_{i,2}$. For simplicity, the block index i is omitted in Figure 4.2. Hn,m denotes the channel frequency response between the nth transmit antenna and the mth receive antenna, where $n \in \{1, 2\}$ and $m \in \{1, 2, \cdots, N_r\}$. imbalance parameters are denoted by $G_{1,TX}^{(n)}$ and $G_{2,TX}^{(n)}$, and the RX I/Q , imbalance parameters are denoted by $G_{1,RX}^{(m)}$ and $G_{2,RX}^{(m)}$, where again, $n \in \{1, 2\}$ and $m \in$ {1, 2, $\cdots$, Nr}. Assume two consecutive data symbols, $S_{i,1}(k)$ and $S_{i,2}(k)$, to be transmitted over the $_k$th subcarrier. Based on the Alamouti scheme, $S_{i,1}(k)$ and $-S_{i,2}^*(k)$ are distributed into the kth subcarrier data stream of the first OFDM modulator, while $S_{i,2}(k)$ and $S_{i,1}^*(k)$ are distributed into the kth subcarrier data stream of the second OFDM modulator.

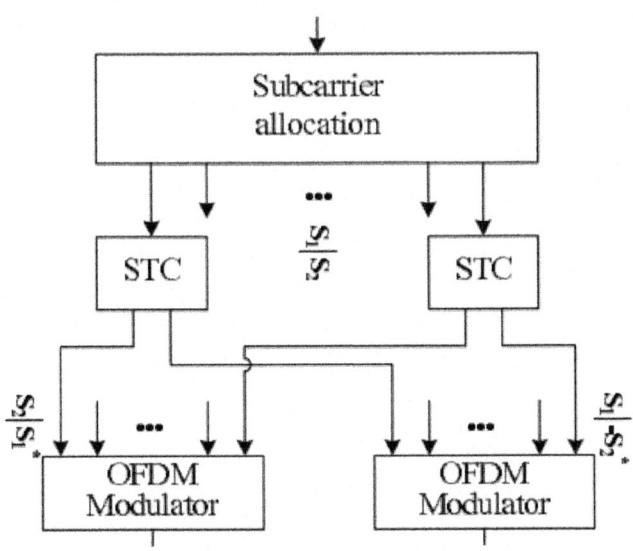

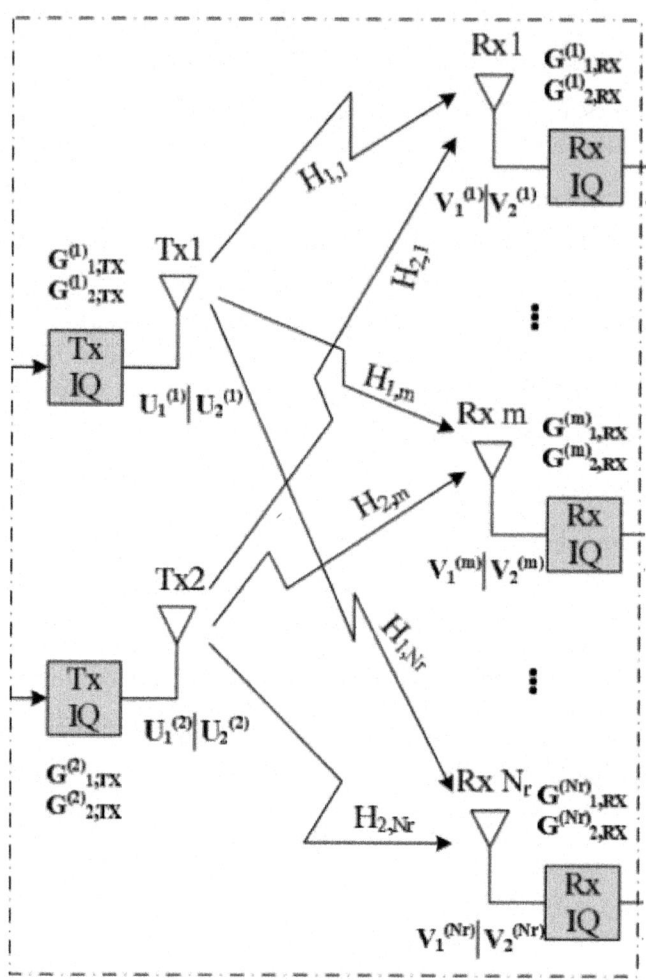

Super Channel

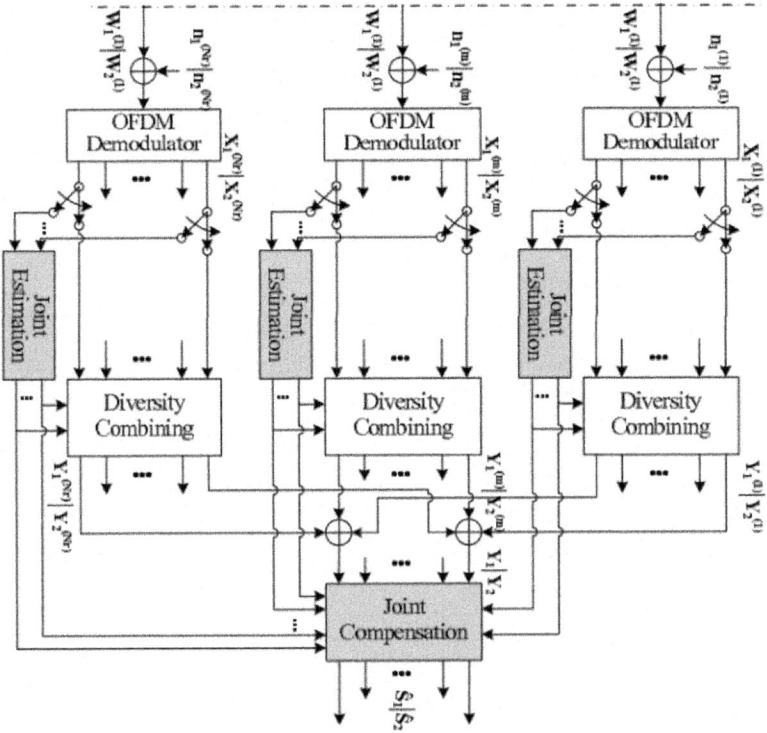

Figure 4.2: Block diagram of a 2×Nr MIMO-OFDM wireless communication system with transmitter and receiver I/Q imbalances. The virtual channel is illustrated in the dashed block

Data streams at all subcarriers are then processed through OFDM modulation, including operations of padding zeros, inverse fast Fourier transform (IFFT), and adding cyclic prefix (CP). Because the IFFT operation only transforms the signal from the frequency domain to the time domain, the signals (viewed in the frequency domain) are not changed after OFDM modulation. Therefore, after OFDM modulation, the two consecutive data symbols at the kth subcarrier of the first transmitter in the frequency domain are still $S_{i,1}(k)$ and $-S_{i,2}^*(k)$. The OFDM-modulated signals are then distorted by TX I/Q imbalances. According to (4.5), the signals to be transmitted via the first transmit antenna are given by

$$
\begin{aligned}
U_{i,1}^{(1)}(k) &= G_{1,TX}^{(1)}(k)S_{i,1}(k) + G_{2,TX}^{(1)}(k)S_{i,1}^*(-k) \\
U_{i,2}^{(1)}(k) &= -G_{1,TX}^{(1)}(k)S_{i,2}^*(k) - G_{2,TX}^{(1)}(k)S_{i,2}(-k)
\end{aligned}
\tag{4.7}
$$

Similarly, the signals to be transmitted via the second transmit antenna are

$$U_{i,1}^{(2)}(k) = G_{1,TX}^{(2)}(k)S_{i,2}(k) + G_{2,TX}^{(2)}(k)S_{i,2}^*(-k)$$

$$U_{i,2}^{(2)}(k) = G_{1,TX}^{(2)}(k)S_{i,1}^*(k) + G_{2,TX}^{(2)}(k)S_{i,1}(-k)$$

$$(4.8)$$

The signals are then transmitted over a multipath fading channel. The received signals at the mth receive antenna are given as

$$V_{i,1}^{(m)}(k) = H_{1,m}(k)U_{i,1}^{(1)}(k) + H_{2,m}(k)U_{i,1}^{(2)}(k)$$

$$V_{i,2}^{(m)}(k) = H_{1,m}(k)U_{i,2}^{(1)}(k) + H_{2,m}(k)U_{i,2}^{(2)}(k)$$

$$(4.9)$$

The received signals at the mth receive antenna are further corrupted by I/Q imbalances in the mth receiver as follows:

$$W_{i,1}^{(m)}(k) = G_{1,RX}^{(m)}(k)V_{i,1}^{(m)}(k) + G_{2,RX}^{(m)}(k)V_{i,1}^{*(m)}(-k)$$

$$W_{i,2}^{(m)}(k) = G_{1,RX}^{(m)}(k)V_{i,2}^{(m)}(k) + G_{2,RX}^{(m)}(k)V_{i,2}^{*(m)}(-k)$$

$$(4.10)$$

Assuming the noise in the system is additive white Gaussian noise (AWGN), then

$$X_{i,1}^{(m)}(k) = W_{i,1}^{(m)}(k) + N_{i,1}^{(m)}(k)$$

$$X_{i,2}^{(m)}(k) = W_{i,2}^{(m)}(k) + N_{i,2}^{(m)}(k)$$

$$(4.11)$$

where $N_{i,j}^{(m)}(k)$ is independently identically distributed (i.i.d.) complex zero-mean Gaussian noise with variance N_0

Combining (4.7), (4.8), (4.9), (4.10), and (4.11), the received data symbols at the mth receive antenna before diversity combining can be written as

$$X_{i,1}^{(m)}(k) = A^{(m)}(k)S_{i,1}(k) + B^{(m)}(k)S_{i,1}^*(-k) + C^{(m)}(k)S_{i,2}(k)$$

$$+ D^{(m)}(k)S_{i,2}^*(-k) + N_{i,1}^{(m)}(k)$$

$$X_{i,2}^{(m)}(k) = C^{(m)}(k)S_{i,1}^*(k) + D^{(m)}(k)S_{i,1}(-k) - A^{(m)}(k)S_{i,2}^*(k)$$

$$- B^{(m)}(k)S_{i,2}(-k) + N_{i,2}^{(m)}(k)$$

$$(4.12)$$

where $A^{(m)}(k)$, $B^{(m)}(k)$, $C^{(m)}(k)$, and $D^{(m)}(k)$ are defined as

$$A^{(m)}(k) \triangleq G_{1,RX}^{(m)}(k)G_{1,TX}^{(1)}(k)H_{1,m}(k) + G_{2,RX}^{(m)}(k)G_{2,TX}^{*(1)}(-k)H_{1,m}^*(-k)$$

$$B^{(m)}(k) \triangleq G_{1,RX}^{(m)}(k)G_{2,TX}^{(1)}(k)H_{1,m}(k) + G_{2,RX}^{(m)}(k)G_{1,TX}^{*(1)}(-k)H_{1,m}^*(-k)$$

$$C^{(m)}(k) \triangleq G_{1,RX}^{(m)}(k)G_{1,TX}^{(2)}(k)H_{2,m}(k) + G_{2,RX}^{(m)}(k)G_{2,TX}^{*(2)}(-k)H_{2,m}^*(-k)$$

$$D^{(m)}(k) \triangleq G_{1,RX}^{(m)}(k)G_{2,TX}^{(2)}(k)H_{2,m}(k) + G_{2,RX}^{(m)}(k)G_{1,TX}^{*(2)}(-k)H_{2,m}^*(-k)$$

$$(4.13)$$

From (4.12) and (4.13), it is observed that the channel frequency response ($H_{n,m}$) and the I/Q effects

$(G_{1,TX}^{(n)}, \ G_{2,TX}^{(n)}, \ G_{1,RX}^{(m)} \ \text{and} \ G_{2,RX}^{(m)})$ are mixed together and produce the coefficients $A^{(m)}(k), \ B^{(m)}(k), \ C^{(m)}(k), \ \text{and} \ D^{(m)}(k).$ If we treat the I/Q effects as parts of the fading channels, we can model the dashed block in Figure 4.2 as a virtual channel with joint coefficients $A^{(m)}(k), \ B^{(m)}(k), \ C^{(m)}(k), \ \text{and} \ D^{(m)}(k).$ Moreover, $A^{(m)}(k)$

and $C^{(m)}(k)$ are critical for data decoding, while the presence of $B^{(m)}(k) \ \text{and} \ D^{(m)}(k)$ may introduce ICI. If the I/Q branches are perfect, then

$$A^{(m)}(k) = H_{1,m}(k), \ B^{(m)}(k) = 0, \qquad C^{(m)}(k) = H_{2,m}(k), \ \text{and} \ D^{(m)}(k) = 0.$$

Furthermore, the received data symbols at the kth subcarrier are distorted by the mixture effects of channel and I/Q imbalances, and interference is introduced by other data symbols within the STC block at the kth and the mirrored (−k) th subcarriers.

To compensate for the received signals, the joint coefficients should be estimated. The estimation methods are described in the next section. Now, assuming the accurate estimate of the joint coefficients are obtained, the received data can be combined to achieve the

$$Y_{i,1}^{(m)}(k) = A^{*(m)}(k)X_{i,1}^{(m)}(k) + C^{(m)}(k)X_{i,2}^{*(m)}(k)$$
$$Y_{i,2}^{(m)}(k) = C^{*(m)}(k)X_{i,1}^{(m)}(k) - A^{(m)}(k)X_{i,2}^{*(m)}(k)$$

$$(44)$$

It should be noted that the combining method is slightly different from the Alamouti scheme, where channel state information is employed. In our scheme, the joint coefficients A$^{(m)}$ (k) and C $^{(m)}$ (k) are used. If the I/Q branches are perfect, (4.14) is reduced to the standard Alamouti diversity combining scheme as

$$Y_{i,1}^{(m)}(k) = H_{1,m}^{*}(k)X_{i,1}^{(m)}(k) + H_{2,m}(k)X_{i,2}^{*(m)}(k)$$
$$Y_{i,2}^{(m)}(k) = H_{2,m}^{*}(k)X_{i,1}^{(m)}(k) - H_{1,m}(k)X_{i,2}^{*(m)}(k)$$

$$(4.15)$$

Substituting (4.12) and (4.13) into (4.14), we obtain

$$Y_{i,1}^{(m)}(k) = Q_{1}^{(m)}(k)S_{i,1}(k) + Q_{2}^{(m)}(k)S_{i,1}^{*}(-k) + Q_{3}^{(m)}(k)S_{i,2}^{*}(-k) + \tilde{N}_{i,1}^{(m)}(k)$$
$$Y_{i,2}^{(m)}(k) = Q_{1}^{(m)}(k)S_{i,2}(k) + Q_{2}^{*(m)}(k)S_{i,2}^{*}(-k) - Q_{3}^{*(m)}(k)S_{i,1}^{*}(-k) + \tilde{N}_{i,2}^{(m)}(k)$$

$$(4.16)$$

Where

$$\begin{aligned}
Q_1^{(m)}(k) &\triangleq |A^{(m)}(k)|^2 + |C^{(m)}(k)|^2 \\
Q_2^{(m)}(k) &\triangleq A^{*(m)}(k)B^{(m)}(k) + C^{(m)}(k)D^{*(m)}(k) \\
Q_3^{(m)}(k) &\triangleq A^{*(m)}(k)D^{(m)}(k) - C^{(m)}(k)B^{*(m)}(k)
\end{aligned}$$

$$\text{(4.17)}$$

And

$$\begin{aligned}
\tilde{N}_{i,1}^{(m)}(k) &= A^{*(m)}(k)n_{i,1}^{(m)}(k) + C^{(m)}(k)n_{i,2}^{*(m)}(k) \\
\tilde{N}_{i,1}^{(m)}(k) &= C^{*(m)}(k)n_{i,1}^{(m)}(k) - A^{(m)}(k)n_{i,2}^{*(m)}(k)
\end{aligned}$$

$$\text{(4.18)}$$

Finally, the received data symbols at multiple receive antennas are combined to obtain receiver diversity, and the raw data symbols are given by

$$\begin{aligned}
Y_{i,1}(k) &= \sum_{m=1}^{N_r}\left[Y_{i,1}^{(m)}(k)\right] \\
&= \underbrace{Q_1(k)S_{i,1}(k)}_{\text{signal}} + \underbrace{Q_2(k)S_{i,1}^*(-k) + Q_3^{(m)}(k)S_{i,2}^*(-k)}_{\text{intercarrier interference}} + \underbrace{\tilde{N}_{i,1}(k)}_{\text{noise}} \\
Y_{i,2}(k) &= \sum_{m=1}^{N_r}\left[Y_{i,2}^{(m)}(k)\right] \\
&= \underbrace{Q_1(k)S_{i,2}(k)}_{\text{signal}} + \underbrace{Q_2(k)S_{i,2}^*(-k) - Q_3^{*(m)}(k)S_{i,1}^*(-k)}_{\text{intercarrier interference}} + \underbrace{\tilde{N}_{i,2}(k)}_{\text{noise}}
\end{aligned}$$

$$\text{(4.19)}$$

Where

$$\begin{aligned}
Q_1(k) &= \sum_{m=1}^{N_r}\left[Q_1^{(m)}(k)\right] \\
Q_2(k) &= \sum_{m=1}^{N_r}\left[Q_2^{(m)}(k)\right] \\
Q_3(k) &= \sum_{m=1}^{N_r}\left[Q_3^{(m)}(k)\right]
\end{aligned}$$

$$\text{(4.20)}$$

are defined as the combined coefficients, and

$$\begin{aligned}
\tilde{N}_{i,1}(k) &= \sum_{m=1}^{N_r}\left[\tilde{N}_{i,1}^{(m)}(k)\right] \\
\tilde{N}_{i,2}(k) &= \sum_{m=1}^{N_r}\left[\tilde{N}_{i,2}^{(m)}(k)\right]
\end{aligned}$$

$$\text{(4.21)}$$

are the combined noise terms. From (4.19), it can be seen that there is ICI in the signals after receiver combining, which is caused by the image projections from the mirrored frequency. To improve system performance, it is necessary to compensate for the received signals.

Estimators and signal compensation

In this section, two training sequence-based estimators are described to show how to estimate the joint coefficients of the virtual channel. The first is a minimal mean square error (MMSE) estimator. Although MMSE estimator is an optimal linear detector, it requires a priori knowledge of the estimated variables and is computationally intensive. An alternative is a least square (LS) estimator, which may significantly reduce the computational complexity at the expense of negligible BER degradation. The estimated joint coefficients can be used to perform diversity combining as in (4.14) and to compensate for the received signals as described at the end of this section.

MMSE Estimator

Assume a total N_{tr} blocks of training sequences are used in this system *. Let $P_{i,j}(k)$, $j \in \{1, 2\}$ denote the jth training symbol in the ith block at the kth subcarrier. According
to (4.12), the input-output relation for one block can be written as

$$\mathbf{u}_i^{(m)}(k) = \mathbf{P}_i(k)\mathbf{v}^{(m)}(k) + \mathbf{n}_i^{(m)}(k), \tag{4.22}$$

Where

$$\mathbf{u}_i^{(m)}(k) = \left[X_{i,1}^{(m)}(k) \ X_{i,2}^{(m)}(k) \right]^T, \tag{4.23}$$

$$\mathbf{P}_i(k) = \begin{bmatrix} S_{i,1}(k) & S_{i,1}^*(-k) & S_{i,2}(k) & S_{i,2}^*(-k) \\ -S_{i,2}^*(k) & -S_{i,2}(-k) & S_{i,1}^*(k) & S_{i,1}(-k) \end{bmatrix}, \tag{4.24}$$

$$\mathbf{v}_i^{(m)}(k) = \left[A^{(m)}(k) \ B^{(m)}(k) \ C^{(m)}(k) \ D^{(m)}(k) \right]^T, \text{ and} \tag{4.25}$$

$$\mathbf{n}_i^{(m)}(k) = \left[N_{i,1}^{(m)}(k) \ N_{i,2}^{(m)}(k) \right]^T. \tag{4.26}$$

For the total T blocks, the input-output relation is written as

$$\mathbf{u}^{(m)}(k) = \mathbf{P}(k)\mathbf{v}^{(m)}(k) + \mathbf{n}^{(m)}(k), \tag{4.27}$$

Where

$$\mathbf{u}^{(m)}(k) = \left[\mathbf{u}_1^{(m)}(k)^T \ \mathbf{u}_2^{(m)}(k)^T \ \cdots \ \mathbf{u}_{N_{tr}}^{(m)}(k)^T \right]^T, \tag{4.28}$$

$$\mathbf{P}(k) = \left[\mathbf{P}_1(k)^T \ \mathbf{P}_2(k)^T \ \cdots \ \mathbf{P}_{N_{tr}}(k)^T \right], \text{ and} \tag{4.29}$$

$$\mathbf{n}^{(m)}(k) = \left[\mathbf{n}_1^{(m)}(k)^T \ \mathbf{n}_2^{(m)}(k)^T \ \cdots \ \mathbf{n}_{N_{tr}}^{(m)}(k)^T \right]^T. \tag{4.30}$$

The MMSE estimate of v is given as [72]

$$\hat{\mathbf{v}}^{(m)}(k) = \mathbf{R}_{vu}\mathbf{R}_u^{-1}\mathbf{u}^{(m)}(k). \tag{4.31}$$

Where

$$\mathbf{R}_{vu} = E\{\mathbf{v}^{(m)}(k)\mathbf{u}^{(m)}(k)^H\} = \mathbf{R}_v\mathbf{P}^H(k), \tag{4.32}$$

$$\mathbf{R}_u = E\{\mathbf{u}^{(m)}(k)\mathbf{u}^{(m)}(k)^H\} = \mathbf{P}(k)\mathbf{R}_v\mathbf{P}^H(k) + \mathbf{R}_n, \tag{4.33}$$

$$\mathbf{R}_n = E\{\mathbf{n}^{(m)}(k)\mathbf{n}^{(m)}(k)^H\} = N_0\mathbf{I}_{2T}, \text{ and} \tag{4.34}$$

$$\mathbf{R}_v = E\{\mathbf{v}^{(m)}(k)\mathbf{v}^{(m)}(k)^H\}. \tag{4.35}$$

LS Estimator

Although MMSE estimator is optimal, it suffers from high computational complexity and requires the knowledge of $\mathbf{R}_v$, which must be estimated using a large amount of transmission data. A simple but effective method is the least square estimator. To further reduce the computational complexity, a special training pattern is design to avoid a matrix inversion operation. In order to utilize this special pattern, training sequences must be transmitted in groups of two blocks. Let s, s, s, and s * be the four consecutive training symbols within the ith and the (i + 1)th STC blocks at the kth subcarrier, where s = p(1 + j) is a complex number with identical real and imaginary parts p. The corresponding training symbols at the −kth subcarrier are also s, s, s, and s * . According to (4.12), the received data within the ith and the (i + 1)th STC blocks at the kth subcarrier can be represented in matrix form as

$$\begin{bmatrix} X_{i,1}^{(m)}(k) \\ X_{i,2}^{(m)}(k) \\ X_{i+1,1}^{(m)}(k) \\ X_{i+1,2}^{(m)}(k) \end{bmatrix} = \begin{bmatrix} s & s^* & s & s^* \\ -s^* & -s & s^* & s \\ s & s^* & s^* & s \\ -s & -s^* & s^* & s \end{bmatrix} \mathbf{v}^{(m)}(k) + \begin{bmatrix} N_{i,1}^{(m)}(k) \\ N_{i,2}^{(m)}(k) \\ N_{i+1,1}^{(m)}(k) \\ N_{i+1,2}^{(m)}(k) \end{bmatrix} \tag{4.36}$$

Thus, the LS estimate of the joint coefficients is given as [73]

$$
\hat{v}^{(m)}(k) =
\begin{bmatrix}
s & s^* & s & s^* \\
-s^* & -s & s^* & s \\
s & s^* & s^* & s \\
-s & -s^* & s^* & s
\end{bmatrix}^{-1}
\begin{bmatrix}
X_{i,1}^{(m)}(k) \\
X_{i,2}^{(m)}(k) \\
X_{i+1,1}^{(m)}(k) \\
X_{i+1,2}^{(m)}(k)
\end{bmatrix}
$$

$$
= \frac{1}{4p}
\begin{bmatrix}
0 & -1-j & 1 & j \\
0 & -1+j & 1 & -j \\
1-j & 0 & j & 1 \\
1+j & 0 & -j & 1
\end{bmatrix}
\begin{bmatrix}
X_{i,1}^{(m)}(k) \\
X_{i,2}^{(m)}(k) \\
X_{i+1,1}^{(m)}(k) \\
X_{i+1,2}^{(m)}(k)
\end{bmatrix}
\tag{4.37}
$$

The estimates of the virtual channel coefficients from different training sequence groups can be averaged to obtain a more accurate estimate

Signal Compensation

Approach Based on the MMSE estimate given by (4.31) or LS estimate given by (4.37) of the joint coefficients, the estimate of the combined coefficients, $\hat{Q}_1(k)$, $\hat{Q}_2(k)$, $\hat{Q}_3(k)$, can be calculated according to (4.17) and (4.20), and then can be used to equalize the raw data symbols. According to (4.19), the raw data symbols at the kth and the (−k)th subcarriers within the ith STC block can be written in matrix form as

$$
\begin{bmatrix}
Y_{i,1}(k) \\
Y_{i,1}^*(-k) \\
Y_{i,2}(k) \\
Y_{i,2}^*(-k)
\end{bmatrix}
=
\begin{bmatrix}
\hat{Q}_1(k) & \hat{Q}_2(k) & 0 & \hat{Q}_3(k) \\
\hat{Q}_2^*(-k) & \hat{Q}_1^*(-k) & \hat{Q}_3^*(-k) & 0 \\
0 & -\hat{Q}_3^*(k) & \hat{Q}_1(k) & \hat{Q}_2^*(k) \\
-\hat{Q}_3(-k) & 0 & \hat{Q}_2(-k) & \hat{Q}_1^*(-k)
\end{bmatrix}
\begin{bmatrix}
S_1(k) \\
S_1^*(-k) \\
S_2(k) \\
S_2^*(-k)
\end{bmatrix}
+
\begin{bmatrix}
\tilde{N}_{i,1}(k) \\
\tilde{N}_{i,1}(-k) \\
\tilde{N}_{i,2}(k) \\
\tilde{N}_{i,2}(-k)
\end{bmatrix}
\tag{4.38}
$$

Thus, the zero-forcing estimates of the transmitted data symbols are given by

$$
\begin{bmatrix}
\hat{S}_1(k) \\
\hat{S}_1^*(-k) \\
\hat{S}_2(k) \\
\hat{S}_2^*(-k)
\end{bmatrix}
=
\begin{bmatrix}
\hat{Q}_1(k) & \hat{Q}_2(k) & 0 & \hat{Q}_3(k) \\
\hat{Q}_2^*(-k) & \hat{Q}_1^*(-k) & \hat{Q}_3^*(-k) & 0 \\
0 & -\hat{Q}_3^*(k) & \hat{Q}_1(k) & \hat{Q}_2^*(k) \\
-\hat{Q}_3(-k) & 0 & \hat{Q}_2(-k) & \hat{Q}_1^*(-k)
\end{bmatrix}^{-1}
\begin{bmatrix}
Y_{i,1}(k) \\
Y_{i,1}^*(-k) \\
Y_{i,2}(k) \\
Y_{i,2}^*(-k)
\end{bmatrix}
\tag{4.39}
$$

Performance Analysis

According to (4.19), the received raw data symbols are contaminated by noise and interference from the signals at the mirrored subcarriers. If the ICI can be successfully canceled by the proposed algorithm, the post-processing SNR at the kth subcarrier, $\eta(k)$, can be calculated as

$$
\begin{aligned}
\eta(k) &= \frac{\mathcal{E}\left\{|Q_1(k)S_{i,1}(k)|^2\right\}}{\mathcal{E}\left\{|\tilde{n}_{i,1}(k)|^2\right\}} = \frac{\mathcal{E}\left\{|Q_1(k)S_{i,2}(k)|^2\right\}}{\mathcal{E}\left\{|\tilde{n}_{i,2}(k)|^2\right\}} \\
&= \sum_{m=1}^{N_r} \left[|A^{(m)}(k)|^2 + |C^{(m)}(k)|^2\right] \frac{E_s/2}{N_0} \\
&= \frac{1}{2}\sum_{m=1}^{N_r} \left[|A^{(m)}(k)|^2 + |C^{(m)}(k)|^2\right] \rho
\end{aligned}
$$

(4.40)

where Es/2 is the average transmit energy per symbol period per antenna and $\rho = E_s/N_0$ can be interpreted as the average SNR for the single-input single-output scheme. This post-processing SNR is determined by the joint effects of channels and I/Q imbalances. For classical i.i.d. channels [7] and perfect I/Q characteristics, the post-processing SNR becomes

$$
\eta(k) = N_r \rho
$$

(4.41)

This shows that the system with perfect I/Q over i.i.d. channel can achieve an array gain of N_r. In OFDM systems, each subcarrier can be treated as a frequency-flat channel. Assuming optimum detection at the receiver, the corresponding symbol error rate for rectangular Mary QAM is given by [74]

$$
P_s(k) = 1 - \left(1 - 2\left(1 - \frac{1}{\sqrt{M}}\right)Q\left(\sqrt{\frac{3\eta(k)}{M-1}}\right)\right)^2
$$

(4.42)

Assuming that only one single bit is changed for each erroneous symbol, the equivalent bit error rate for rectangular M-ary QAM is approximated as [75]

$$
P_b(k) \approx \frac{1}{log_2(M)} P_s(k)
$$

(4.43)

4.5 Simulation Results To evaluate the proposed virtual channel idea, the two estimators, and the signal compensation approach, a Monte Carlo simulation of an OFDM-based 2 × N_r MIMO system with frequency-dependent TX and RX I/Q imbalances is performed. The size of fast Fourier transform (FFT) is 128, the number of used sub-carriers is 96, and the length of CP is 32. The multipath channel is modeled by six independent complex taps with a power delay profile of a 3 dB decay per tap. It should be noted that the actual channel length can be estimated [76]. The simulation bandwidth is set to 20 MHz, leading to a 156.25 KHz subcarrier spacing and a maximum 0.3 µs excess delay. Sixty-four quadrature amplitude modulation (64QAM) is used.

The non-linear frequency characteristics of the I and Q branches, LI (f) and LQ(f), are modeled as two first-order finite impulse response (FIR) filters. A form of parameters $\{g, \theta, [a_I, b_I], [a_Q, b_Q]\}$ is used to describe the I/Q imbalance, where [aI, bI] and [aQ, bQ] are the coefficients of the FIR filters for the I and Q branches, respectively. I/Q parameters of {1.03, 3, [0.01, 0.9], [0.8, 0.02]}, {1.04, 4, [0.8, 0.02], [0.01, 0.9]} are used for the two transmitters, respectively. For simplification of simulation, all receivers use I/Q parameters of {1.05, 3, [0.8, 0.02], [0.9, 0.01]}. Because the estimation of the joint coefficients is performed separately in each receiver, the same I/Q parameters for different I/Q models still lead to generalization of the simulation results. It should be noted that the I/Q imbalance parameters are chosen to be worst case in order to evaluate the robustness of the proposed approach. A frame-by-frame transmission scheme is employed in the simulation. The multipath channel is independently generated for each frame. One frame consists of Ntr STC blocks of training symbols followed by 50 STC blocks of data symbols. A total of 5,000 frames (288 Mbits) are simulated for each scenario at a given SNR. Figure 4.3 shows the frequency response of two of the FIR filters used for simulating the frequency-dependent characteristics of I/Q imbalances. For both filters, the variations of the amplitude are within 0.5 dB, but the average gains are differentiated by 1 dB. It should be noted that the signals after TX I/Q distortions must be normalized in the simulations to compensate for the energy lost due to attenuations of the filters. While the variation of the phase response for the first filter is small, the variation is very large for the second filter. The amplitude and phase differences of the filters are suitable to simulate the frequencydependent characteristics of I/Q imbalances. Figure 4.4 shows the typical constellations of one frame of symbols generated in an ideal channel environment (without multipath fading and noise). Figure 4.4 (a) and (b) show the constellations of the raw data symbols under frequency-independent and frequencydependent I/Q imbalances, respectively. It is observed that the constellation in Figure 4.4

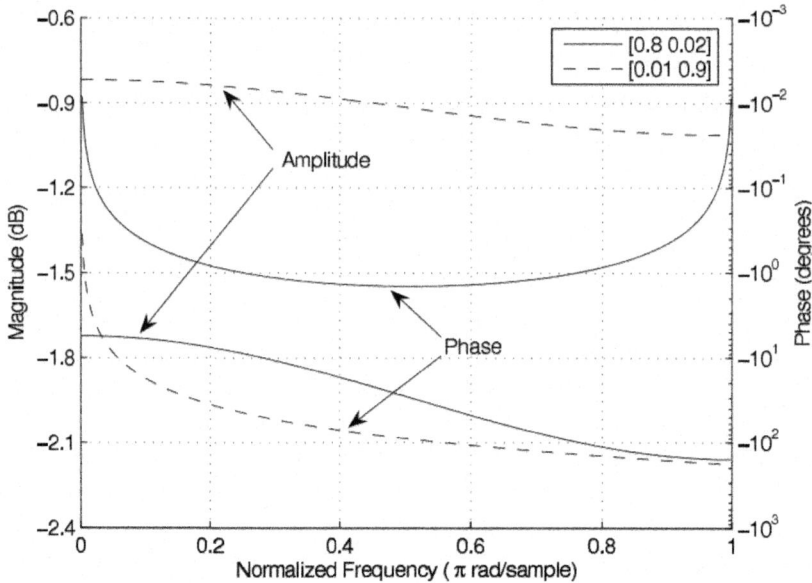

Figure 4.3: Frequency response of two of the FIR filters used for simulating the frequencydependent characteristics of I/Q imbalances.

(b) becomes nearly random compared with Figure 4.4 (a) due to the effect of frequencydependent I/Q imbalances. Figure 4.4 (c) shows the constellation of signals recovered by the proposed LS algorithm, which shows that the proposed algorithm can perfectly compensate for frequency-independent and frequency-dependent I/Q imbalances. The BER performances of the proposed estimators and compensation approach are shown in Figures 4.5-4.10. To make better comparisons, a series of simulations under different scenarios was conducted, including a scenario of perfect I/Q and CSI known at receivers termed as "ideal case", a scenario with frequency-independent I/Q imbalances termed as "indep-IQ", and a scenario with frequency-dependent I/Q imbalances termed as "dep-IQ". The legend term of "NoComp" means that I/Q imbalances are present but no compensation scheme is applied (the CSI is estimated under I/Q imbalances), "MMSE(N)" means that the MMSE estimator with N STC blocks of training sequences (the total $2 \times N$

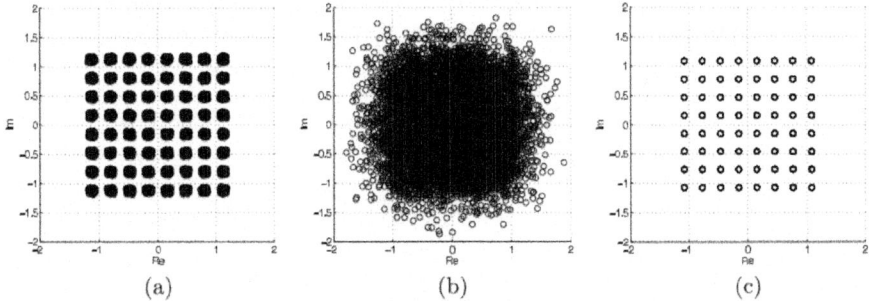

Figure 4.4: Constellations of signals in a frame generated by a 2 ×1 MIMO-OFDM system under ideal channel environment. (a) raw data constellations under frequency-independent I/Q imbalances; (b) raw data constellations under frequency-dependent I/Q imbalances; (c) constellations of the compensated data under frequency-dependent I/Q imbalances.

OFDM symbols) is used, and "LS(N)" means that the LS estimator with N STC blocks of training sequence is used to estimate the joint coefficients. For both MMSE and LS estimators, the proposed compensation approach is applied to compensate for the received raw data symbols. Furthermore, the BER is theoretically calculated according to (6.28) and is shown in the following figures with legend term "analysis". Figure 4.5 and Figure 4.6 show the performance of a 2 × 1 and a 2 × 2 systems with frequency-independent I/Q imbalances, respectively. It is observed from the "NoComp" curves that the I/Q imbalances significantly degrade the system performance, resulting in high error floor. These results agree with the constellation analysis mentioned above. With the proposed estimators and signal compensation approach, the I/Q distortion can be effectively mitigated and the system performance is significantly improved. By using two STC blocks of training sequences, the system performance resulting from both MMSE and LS estimators is close to the ideal case. The performance degradation is less than 1 dB. Although LS estimator performs a little worse than MMSE estimator, the low computational

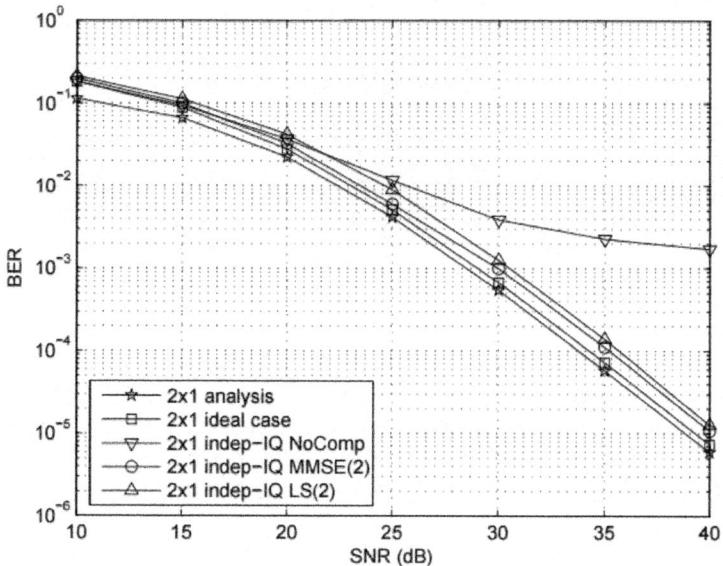

Figure 4.5: Performance of a 2 × 1 MIMO-OFDM system with TX and RX frequencyindependent I/Q imbalances over multipath fading channel.

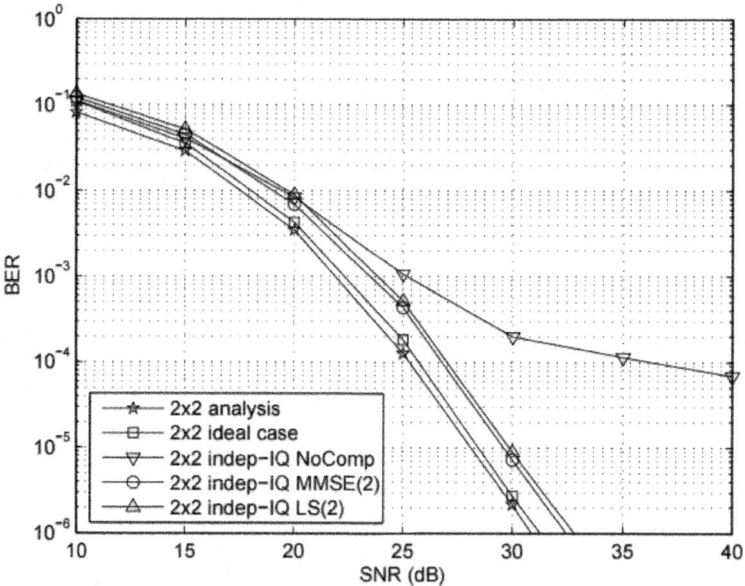

Figure 4.6: Performance of a 2 × 2 MIMO-OFDM system with TX and RX frequencyindependent I/Q imbalances over multipath fading channel.

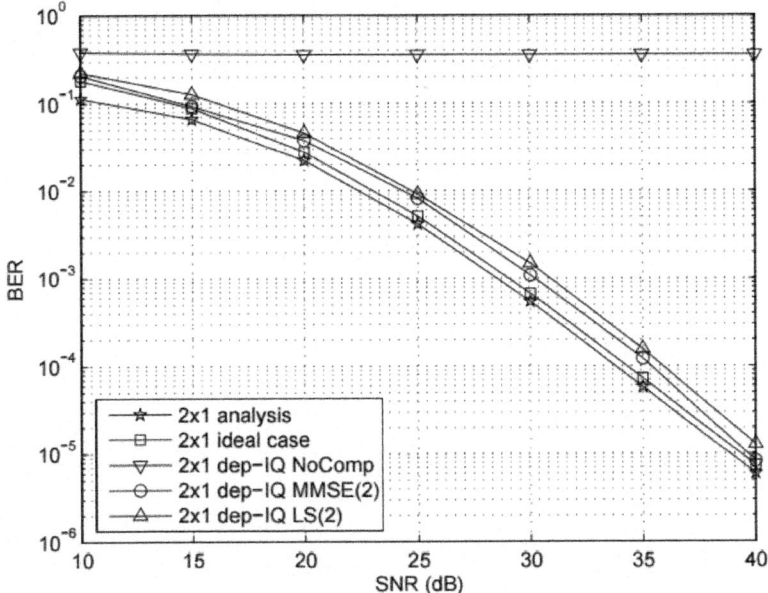

Figure 4.7: Performance of a 2×1 MIMO-OFDM system with TX and RX frequencydependent I/Q imbalances over multipath fading channel.

complexity makes the LS estimator more competitive. By assuming perfect estimation of the joint coefficients and compensation of the received signal, it is reasonable that the analysis results are slightly better than the ideal case. The performance of the systems with frequency-dependent I/Q imbalances is shown in Figure 4.7 and Figure 4.8 for the 2×1 and 2×2 scenarios, respectively. The frequency dependent characteristic of the I/Q imbalances causes fatal influence on the MIMO-OFDM systems. Without compensation, the BER remains one-half regardless of the SNR. Our proposed approach can also successfully combat the deadly effect caused by frequencydependent I/Q imbalances, resulting in good performance that is close to the ideal case. This significant performance improvement is not reported by other literature. An intuitive sense is that a longer training sequence could result in better performance. To demonstrate this, the system performances are compared with different lengths of training symbols in Figure 4.9 and Figure 4.10 for MMSE and LS estimators, respectively. It is observed that BER decreases with the increase of the training symbol numbers.

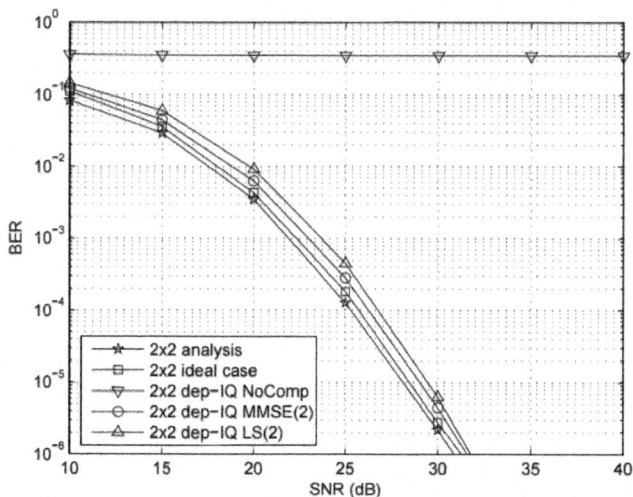

Figure 4.8: Performance of a 2 × 2 MIMO-OFDM system with TX and RX frequencydependent I/Q imbalances over multipath fading channel.

When four blocks of training sequences are utilized, the performance loss compared to the ideal case is negligible.

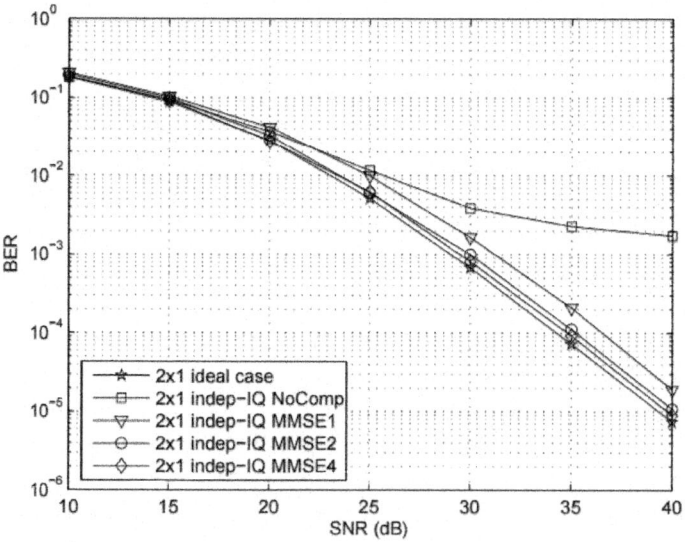

Figure 4.9: Comparison of the performances of the MMSE estimator with different lengths of training symbols for the 2 × 1 MIMO-OFDM system with TX and RX frequencyindependent I/Q imbalances over multipath fading channel.

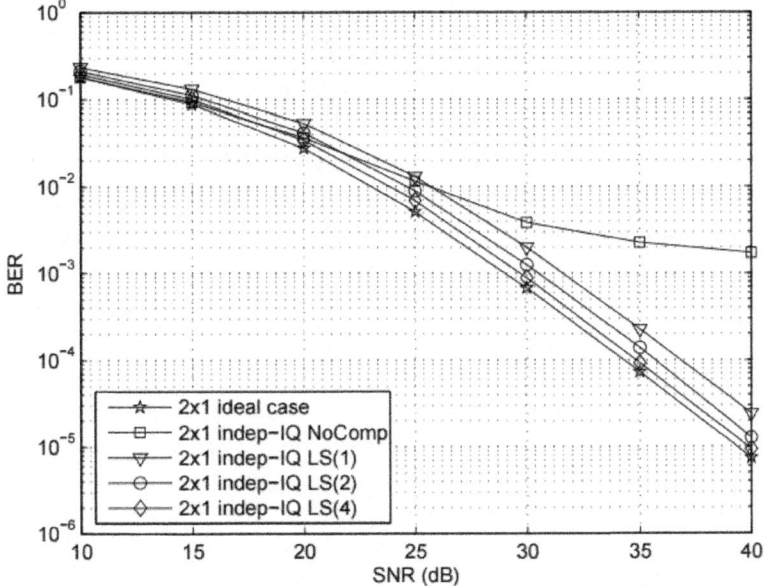

Figure 4.10: Comparison of the performances of the LS estimator with different lengths of training symbols for the 2×1 MIMO-OFDM system with TX and RX frequency-independent I/Q imbalances over multipath fading channel.

DISTRIBUTED MIMO USING COOPERATIVE DIVERSITY

In wireless communications, channel fading, an inherent property of wireless communication links, severely limits the increase of the data rate. The most popular and effective way to combat channel fading is to exploit the diversity from the received signals. By transmitting a signal via multiple independent channels, e.g., at different time slots, different frequency bands, and different spatial directions, the receiver can receive different copies of the signal and thus achieve the time, frequency, and space diversity gains by employing optimal combining schemes. Spatial diversity techniques are particularly attractive since they provide diversity gain without incurring an extra cost of transmission time and bandwidth. Traditionally, spatial diversity is achieved by using multiple antennas at the transmitter and/or receiver, where the antennas are packed together with spacing of the order of wavelength, referred as co-located multiple-input multiple-output (MIMO). Because of the diversity gain, co-located MIMO architectures are effective in increasing system throughput and are capable of combating channel fading. However, the benefits of the co-located MIMO technique are limited in practical systems. The reasons for this limitation are two-fold. First, spatial correlation causes performance degradation. In a co-located MIMO system, antennas at each node have to be placed close to each other. Thus, radio signals at the co-located antennas experience a similar scattering environment, and the channels may be correlated, especially when a line-of-sight (LOS) channel between the transmitter and receiver dominates. The channel matrices could be ill-conditioned, resulting in significant capacity decrease. Second, due to the terminal size limitation, the node cannot be equipped with many antennas. Since the diversity gain is proportional to the number of antennas, the co-located MIMO system with few antennas cannot produce the expected performance. To mitigate the aforementioned drawbacks in the co-located MIMO systems, a new technique named "distributed MIMO" was proposed and has attracted much attention [28]. The major difference between the distributed MIMO and the co-located MIMO is that multiple antennas at the front-end of wireless networks are distributed among widelyseparated radio nodes. In a distributed MIMO system, each node may be only equipped with one antenna. Many nodes at different locations transmit the same information to the receiver. In this manner, multiple nodes form a virtual antenna array that achieves higher spatial diversity gain. This kind of spatial diversity is referred to as user cooperation diversity [28] or, simply, cooperative diversity [29]. This chapter provides a detailed study of the recent advances in distributed MIMO technologies in cooperative wireless networks. The basic concepts of a cooperation system are first introduced, followed by the detailed discussion of relay protocols and cooperative strategies. Simulation results are then presented to demonstrate the effectiveness of the cooperative network with full user cooperation diversity and improved bandwidth efficiency

Cooperation Systems

A cooperative wireless network consists of a source, a number of relays, and a destination, which is illustrated in Figure 5.1. Because we deal with physical layer techniques instead of higher layer protocols in this article, it is reasonably assumed in the network that all nodes use the same multiple-access resources. For example, they may use the same sub-channel in a frequency division multiple access (FDMA) system. Therefore, the network in Figure 5.1 is a subset of a practical network. It could be a cell in a cellular system, or a same-frequency cluster in a mesh network. In Figure 5.1, the source node S is intending to transmit information to the destination node D.

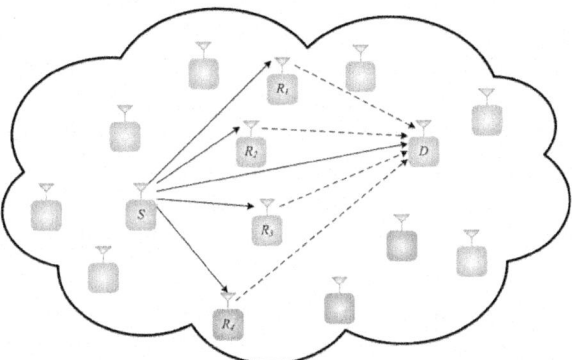

Figure 5.1: A cooperative wireless network.

Other nodes in the networks could be selected as the helpers or relays for the transmission between the source and the destination. Assuming that the nodes R_1, R_2, $\cdots$, and R_n are selected to be the helpers, these helpers are treated as relay nodes, which forward the signals from the source to the destination. The relays together with the source form a virtual transmit array to achieve the spatial diversity. Relays in a cooperation network are different from that in a relay network where the relays only forward signals from other nodes. In cooperation networks, any nodes could function as a source or a relay. When a node collects and transmits information, it works as a source. Otherwise, it can help other nodes as a relay. When a communication link is established for transmitting information, relay nodes are selected based on a set of rules or relay selection strategies. Usually the matrices such as the locations, the loads, and the end-to-end performance of the relays are considered in a network. To complete this task, a cross-layer design is employed, which is beyond the scope of this article. Interested readers can refer to an excellent article [77] for details. The communication between the source and the destination proceeds in two phases: information sharing and cooperative transmission. In the information sharing phase, the source broadcasts its information to the relays and the destination. This step is inevitable because the spatial diversity can only be achieved from the independent transmission of the same information. Through information sharing, all relays get the information from the source and enable an independent data transmission. In the second phase, the information is forwarded by the relays to the destination. In the meantime, the source could either transmit or remain inactive. Because the relays are randomly selected and they are usually separated far away from others, it is highly likely that the channels between each relay and the destination are uncorrelated. In the cooperative transmission, multiple relays and a source node (that are equipped with a single antenna) form a virtual antenna array for a distributed MIMO system. This distributed MIMO provides the benefits of the cooperative system by overcoming the size limitation and ill-conditioned channel in co-located MIMO systems. Co-channel interference is one of the serious problems in the cooperative wireless network. When the information is relayed to the destination, there exists a co-channel interference in that the signals from different relays may interfere with each other at the destination. Although the interference cancellation at the destination is a possible solution, the required algorithm is complex and the performance may not be very satisfying. The common cooperative strategy is to avoid the interference by transmitting the relayed signals in orthogonal subchannels. The orthogonality could be acquired by using repetition-based strategy or by using distributed space-time coding (DSTC). For the repetition-based strategy, the relays forward the signal in different time slots, i.e., in each time slot, only one relay transmits information, while the other relays stay inactive. This strategy is easy to implement, but it results in poor bandwidth efficiency. By using DSTC, the relays can forward information in the same time slot. The orthogonality is constructed in both time and space domains.

Although DSTC-based strategy leads to a more complex network, the bandwidth is utilized more efficiently. Both repetition-based and DSTC-based cooperative strategies can achieve full spatial diversity, i.e., the order of the spatial diversity. However, the achieved diversity is not only decided by the cooperative strategies, but also determined by the methods or relay protocols. The relay protocols are discussed in the next section.

Relay Protocols

Relay protocols significantly affect the system performance in cooperative wireless networks. In this section, an overview of commonly used relay protocols is presented.

Amplify-and-Forward

The simplest relay protocol is amplify-and-forward (AF) [78]. In AF, each relay amplifies the received noisy signals and forwards them to the destination. This simple processing benefits cooperative wireless networks with full spatial diversity at high signal-to-noise ratios (SNRs). Because the noise component is also amplified in AF, the bit-error-rate (BER) performance could be degraded. To optimize the performance in AF systems, a scalar used for amplification is chosen adaptively based on the channel coefficient.

Decode-and-Forward

For decode-and-forward (DF) relay protocol, the relays will first decode the received signals, and then forward the re-encoded signals to the destination node [79]. The decoding can be done fully in bit level or partially in symbol level. When the channel between the source and the relay good quality, DF provides error correlation capability and then is superior to AF. However, when the channel link suffers from deep fading, the decoding could produce errors because no effective method can be applied to combat the fading. These errors will propagate to the destination, leading to worse performance overall. Although DF protocol cannot provide full diversity by itself, it can achieve full diversity when more complex coding

Selection Relaying

To mitigate the effect of the noise amplification in AF and error propagation caused by DF, another relay protocol, named "selection relaying," was proposed by Laneman et al. in [80]. In selection relaying, AF or DF is adopted only when the fading channel has high instantaneous signal-to-noise ratio. Otherwise, the relay suspends forwarding. Selection relaying can offer full spatial diversity. A similar relay protocol, adaptive relaying, was proposed by Li [81]. According to this protocol, either AF or DF is selected based on the decoding result.

Cooperative Strategies

In a cooperative wireless network, when a relay protocol is selected and the received signals are processed, the relay is ready to forward the signals to the destination. As mentioned before, both repetition-based and DSTC-based orthogonal relaying can be adopted as the cooperative strategy. In this section, these two cooperative strategies are discussed in detail.

Repetition-Based Cooperative Strategy

In a repetition-based cooperative strategy, the relays forward signals sequentially, i.e., only one relay is allowed to forward signals at each time slot [80]. Figure 5.2 (a) illustrates the time slot utilization in repetition-based strategy. In phase one, the source broadcasts the information to the destination and the relays at the same time slot. In phase two, each relay forwards the received signal to the destination sequentially. Hence, a total n time slots are required to finish phase two. This repetition scheme takes a long time to complete the forwarding process, leading to inefficient bandwidth utilization. It is easy to verify that the data rate of the repetition-based cooperative strategy is 1/n.

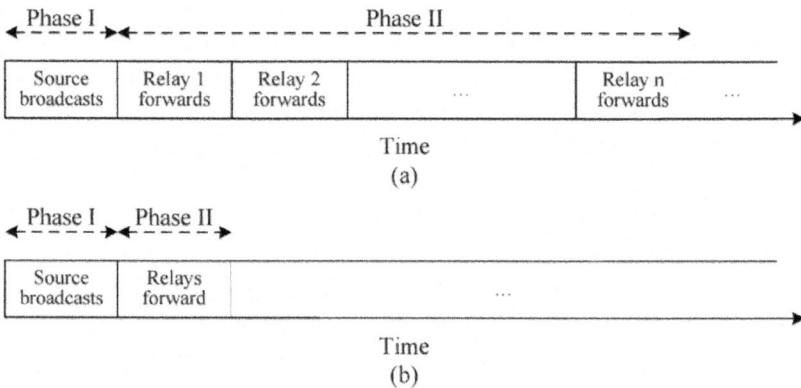

Figure 5.2: Slot assignment for cooperation strategies, (a) repetition-based, (b) DST-Cbased.

To improve the bandwidth efficiency, a novel relay protocol named "incremental relaying" was proposed to reduce the relay repetitions [80]. In incremental relaying, the destination will estimate the signal at the end of phase one and feedback a single bit to the source and the relays to indicate the success or failure of the direct transmission. If the direct transmission succeeds, the relays will not forward the signals and the source will continue to the next time slot of new information transmission. Otherwise, relays will forward the signals to the destination. This protocol improves the bandwidth utilization only at high SNRs

DSTC-Based Cooperative

Strategy Repetition-based cooperative diversity algorithms can achieve full spatial diversity at the expense of decreasing bandwidth efficiency. The utilization of incremental relaying cannot overcome this drawback when the source-destination link is in poor condition. To further improve bandwidth efficiency, distributed space-time coding (DSTC) can be used to enable relays to transmit in the same time slot in cooperative systems [82]. In the DSTC-based cooperative strategy, the source broadcasts information in phase one. Contrary to the above repetition-based cooperative strategy, the source and all relays simultaneously transmit coded signals in phase two. Thus, the full cooperative diversity gain n + 1 is achieved. DSTC-based cooperative strategy is usually applied together with the AF or DF relay protocol [83]. After the signals are amplified or decoded, the relays will re-encode the signal before forwarding. In contrast to the decode-and-forward protocol, the signals are reencoded by using distributed space-time codes in the DSTC-based strategy. The distributed space-time codes can be directly obtained from the conventional space-time codes such that each code is applied to the antenna at each node. The commonly used distributed spacetime codes are distributed space-time block codes [9,10] and distributed space-time trellis codes [84]. These codes guarantee that the signals from different relays are orthogonal, and hence they can be separated at the destination without interference. In this manner, all relays forward the signals at the same time slot, as shown in Figure 5.2 (b). Comparing Figure 5.2 (a) and (b), it is evident that the channel utilization of the DSTC-based strategy is better than that of repetition-based strategy. Improvement of the bandwidth efficiency provided by the DSTC-based strategy comes at the cost of complex signaling and signal processing. To implement distributed spacetime-coding, the relays should have a priori knowledge of the space-time codes assigned to them. This requires extra signaling over the dedicated channels. However, the bandwidth for the extra signaling is negligible compared to the improved bandwidth utilization. While the space-time coding adds some computation complexity to the relays, the one-antenna design simplifies the relays significantly. Consequently, the nodes in a cooperative wireless network are relatively simple compared to the nodes in a traditional MIMO system, which are mounted with multiple antennas and have to process signals from multiple physical channels.

Performance Comparisons

In this section, simulation results are presented that compare the performance of cooperative wireless networks with different relay protocols and cooperative strategies. In [80] and [85], Laneman et al. analyzed the bandwidth efficiency by studying the outage probability $Pr [I < R]$, i.e., given the channel realization, the probability of the mutual information I is less than a given data rate R. The outage probabilities for different relay protocols are compared in Figure 5.3, where x-axis is the signal-to-noise ratio and y-axis is the outage probability Poutage. The data rate is set to R = 1 bit/s/Hz and the number of relay n = 1 in this scenario. Compared with the direct transmission, the AF, selection DF, and incremental AF protocols can achieve full diversity gain (2 in this scenario) because their curves are nearly twice as steep compared to that of the direct transmission. The incremental AF protocol performs even better since the one-bit feedback can effectively reduce the relay repetitions. For example, the incremental AF protocol achieves about 17 dB gain over the direct transmission at the outage probability of 10^{-3}. The DF protocol, however, performs similarly to the direct transmission because it cannot achieve full spatial diversity. In this single relay scenario, the DF protocol cannot achieve any spatial diversity.

The outage capacities of repetition-based cooperative strategy is shown and compared in Figure 5.4, where x-axis is the data rate in bit/s/Hz and y-axis is the outage probability P_{outage}. It is observed that the outage probabilities for the direct transmission and repetition-based cooperative strategy increase with the data rate. This is reasonable based on the definition of outage probability. However, the relationship between the outage probability and the number of relays becomes complicated. At low data rates, the outage probability performance becomes better with more relays, while at high data rates, the outage probability becomes worse when more relays are used. For example, at 1 bit/s/Hz data rate, the outage probability is 3×10^{-4} for one-relay scheme, and it improves to 2×10^{-5} for a three-relay scheme.

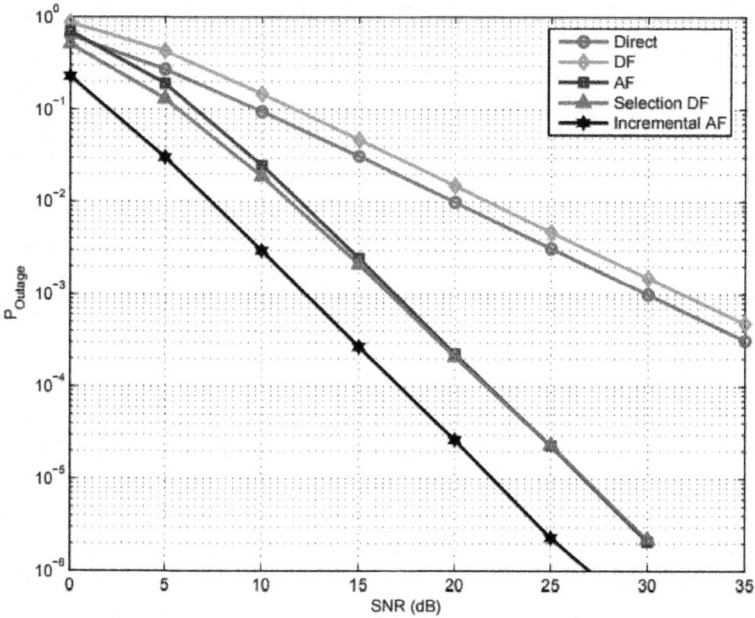

Figure 5.3: Performance comparisons of different relay protocols. One relay is used and data rate is set to 1 bit/s/Hz.

When the data rate is above 1.4 bit/s/Hz, however, the outage probability of a three-relay scheme is higher than that of a one-relay scheme. This phenomenon results from the poor bandwidth efficiency of the repetition-based cooperative strategy. From the above analysis, it is realized that the repetition-based cooperative strategy can guarantee a specific low data rate, but it cannot increase the data rate by adding more relays in the cooperative wireless network. In contrast to repetition-based cooperative strategy, distributed space-time coding based cooperative strategy offers better bandwidth efficiency, as shown in Figure 5.5, where x-axis is the data rate in bit/s/Hz and y-axis is the outage probability P_{outage}. When three relays are used, the outage probability is about 10^{-7} at 1 bit/s/Hz data rate, which is much lower than the outage probability 2×10^{-5} of the repetition-based cooperative strategy. This bandwidth utilization improvement results from the simultaneous transmission of a group of relays. Due to the same reason, adding more relays will not decrease the data rate.

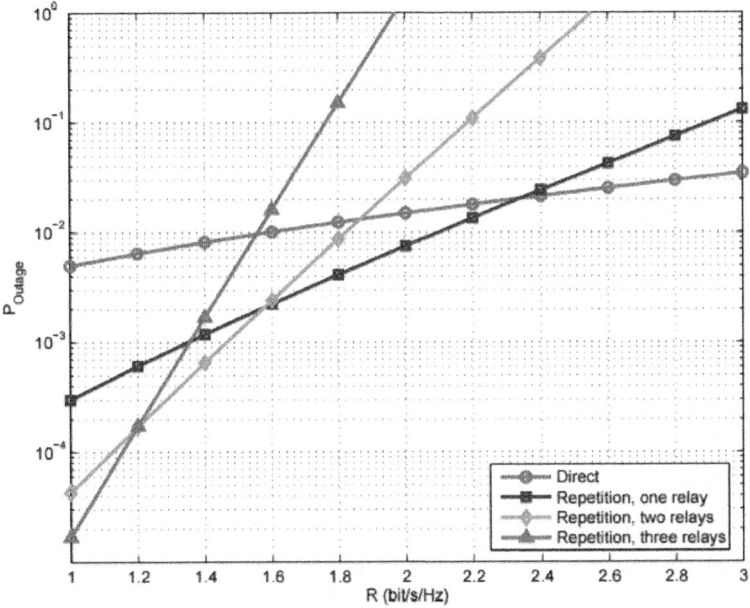

Figure 5.4: Outage probabilities for repetition-based cooperative diversity. Results of = 1, 2, and 3 relays are shown. The signal-to-noise ratio is set to 20 dB.

On the contrary, more relays will increase the data rate due to the higher order of spatial diversity. For example, in Figure 5.5, comparing the three-relay scheme to the one-relay scheme, the outage probability becomes lower by about order of two. It should be noted that the achievable data rate cannot exceed the channel capacity. Therefore, the curves in Figure 5.5 will converge at $P_{outage} = 1$ for sure when the data rate is greater than the channel capacity.

Finally, the BER performance of repetition-based and distributed space-time block coding based cooperative strategies is shown in Figure 5.6, where x-axis is the signal-to-noise ratio in dB and y-axis is the average bit-error rate. The code used in the simulation is single-symbol maximum likelihood decodable distributed STBC proposed in [86]. Three relays and amplify-and-forward relay protocol are used in the simulation. Given the same data rate, DSTC-based strategy outperforms the repetition-based strategy in terms of average BER. For example, at 2 bit/s/Hz data rate in Figure 5.6, the DSTC-based cooperative. strategy acquires about 7 dB gain over the repetition-based cooperative strategy at 10^{-4} average BER.

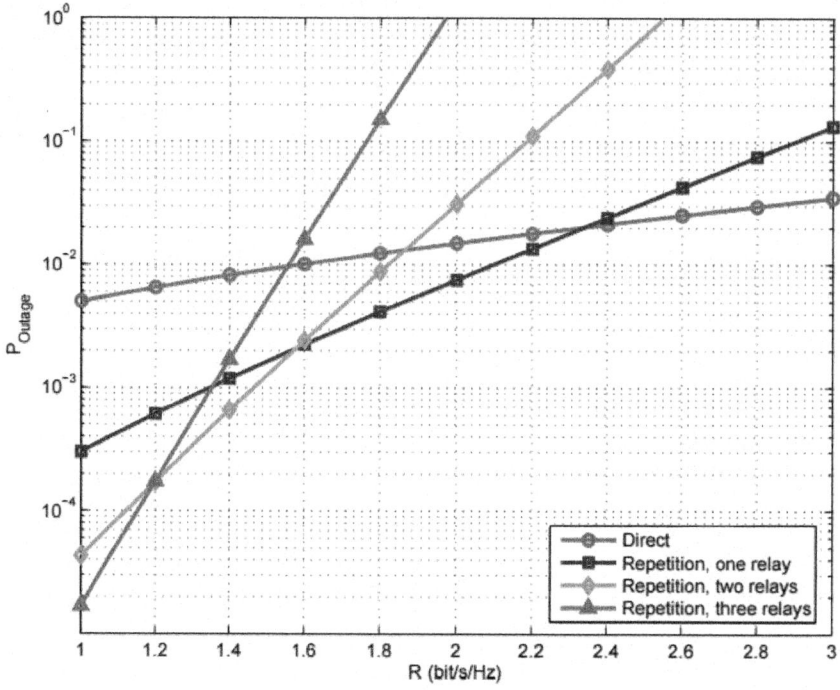

Figure 5.5: Outage probabilities for DSTC-based cooperative diversity. Results for = 1, 2, and 3 relays are shown. The signal-to-noise ratio is set to 20 dB.

When the data rate increases, the BER performance of both strategies degrades. In particular, it is observed that the degradation for repletion-based strategy is more severe. For instance, at average BER 10^{-4}, the BER degradation from 2 bit/s/Hz to 1 bit/s/Hz is about 5 dB for the DSTC-based cooperative strategy but about 10 dB for the repetition-based cooperative strategy. This simulation result shows that repetition-based strategy is bandwidth inefficient.

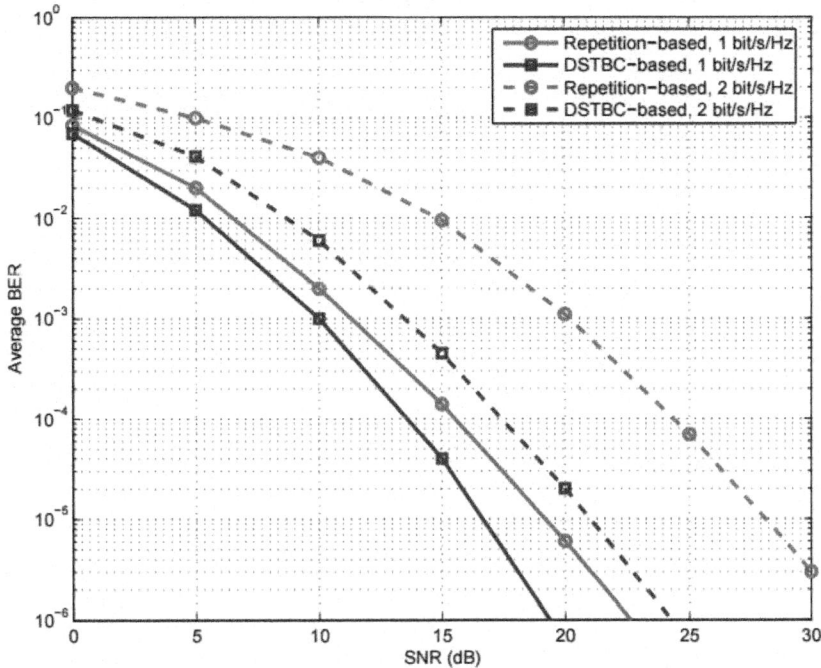

Figure 5.6: Bit error rate performance comparison of repetition-based and DSTC-based cooperative strategies. Amplify-and-forward relay protocol and relays are used in simulation.

ENHANCING SELF-ENCODED SPREAD SPECTRUM WITH MULTI-ANTENNA TRANSMISSION

Conventional direct sequence spread spectrum system employs pseudo-noise (PN) code generators. In contrast, by deriving its spreading sequences from the user data stream, self-encoded spread spectrum (SESS) provides a feasible implementation of random-coded spread spectrum and has a number of unique features, including enhanced transmission security, anti-jamming capability, multi-rate applications, modulation gain and inherent time diversity [87–93]. It has been shown that the modulation memory associated with self-encoding can be exploited with iterative detection to achieve a 3 dB gain and an N-fold time diversity (where N is the spreading length) [92]. These advantages improve the system performance over fading channels In this chapter, MIMO techniques are incorporated into SESS systems to improve the performance [94–96]. MIMO-SESS provides a novel means to combating fading in wireless channels by exploiting diversities in both space and time domains [97]. The

BER performance of the system under Rayleigh fading is analyzed. The closed-from BER expression of the N-fold, time-diversity SESS detector is derived. By approximating the probability density function (pdf) associated with the time-diversity detector by a Dirac delta function, a lower-bound expression is obtained for the BER of the iterative detector.

System Model

In this section, the transceiver of the MIMO-SESS system with iterative detection is described.

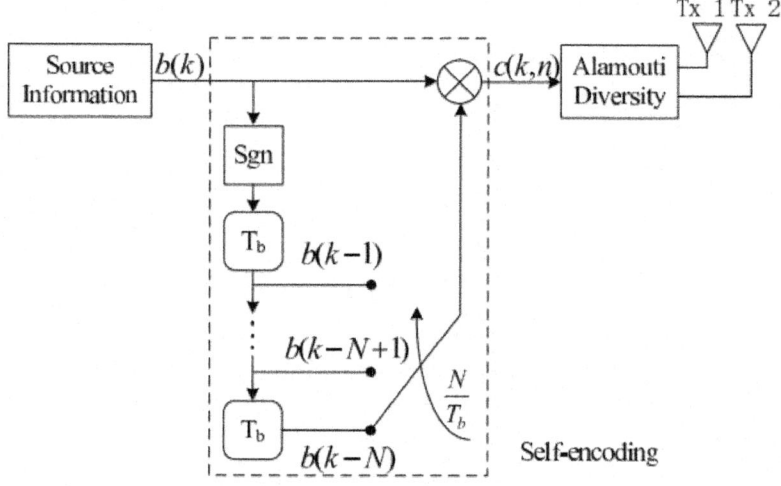

Figure 6.1: Block diagram of the transmitter of the proposed MIMO-SESS system.

Transmitter

The block diagram of the proposed transmitter is shown in Figure 6.1, where the rounded corner blocks represent N delay registers and T_b is the bit interval. Alamouti scheme-based space-time block coding (STBC) is utilized to achieve transmit diversity. The source information b is assumed to be bipolar values of $\pm\sqrt{E_b/2}$, where E_b is the average transmit energy per bit. It should be mentioned that the energy per transmit antenna is one-half of the total transmit energy in order to make the multi-antenna system comparable to a single antenna system. The bits are first spread by the self-encoded spreading sequence of length N at a chip rate of N/T_b. This sequence is constructed from the user's information stored in the delay registers that are updated every T_b.

Thus, with a random input data stream, the sequence is also random and time-varying from one bit to another. For example, the spreading sequence for the kth bit, b(k), is given as

$$\mathbf{s}(k) = [b(k-1)\, b(k-2)\, \cdots\, b(k-N)]^T,$$ (6.1)

and the spreading chips are given as

$$\mathbf{c}(k) = b(k)\mathbf{s}(k).$$ (6.2)

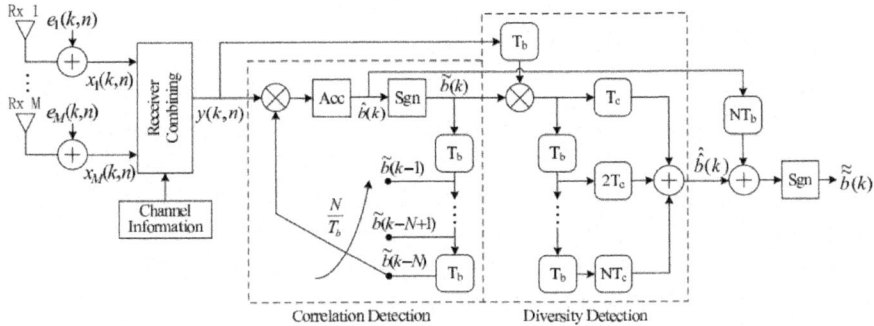

Correlation Detection   Diversity Detection

Figure 6.2: Block diagram of the receiver of the proposed MIMO-SESS system with iterative detection.

To facilitate the description in the sequel, let c(k, n) denote the nth chip of the kth bit, which can be expressed as

$$c(k, n) = b(k)b(k-n).$$ **(6.3)**

The spreading chips are then divided into two streams by applying the Alamouti scheme, where a block of two consecutive chips $, c(k, 2i)$ and $c(k, 2i+1),$ is transmitted by sending $c(k, 2i), -c^*(k, 2i+1)$ to the first antenna, and $c(k, 2i+1), c^*(k, 2i)$

Here $i \in \{0, 1, \cdots, N/2 - 1\}$ is the block index for the chips of the kth bit. The signals are then transmitted over a MIMO fading channel

The channel between each transmit and receive antenna pair is assumed to undergo Rayleigh fading, which remains constant over T_b but is independent from bit to bit [92, 98]. The channels for different transmit/receive antenna pair are independent.

Correlation Detection

Figure 6.2 shows the block diagram of the receiver. For the mth antenna, the received signals within the ith block are given as

$$x_m(k, 2i) = h_{1,m}(k)c(k, 2i) + h_{2,m}(k)c(k, 2i + 1) + e_m(k, 2i),$$

(6.4)

$$x_m(k, 2i + 1) = -h_{1,m}(k)c^*(k, 2i + 1) + h_{2,m}(k)c^*(k, 2i) + e_m(k, 2i + 1),$$

(6.5)

where $h_{1,m}(k)$ and $h_{2,m}(k)$ respectively denote the normalized complex channel impulse response coefficients for the kth bit between the 1st and 2nd transmit antennas and the mth receive antenna; e_m is a Gaussian noise with zero mean and variance $NN_0/2$. It should be noted that the noise here is broadband because it is sampled at chip level. Its variance is thus the narrow-band noise variance $N_0/2$ multiplied by the spreading factor N. It is assumed that the delay registers in the receiver have been synchronized with the transmitter [93], and that perfect channel knowledge is available at the receiver. Under these assumptions, diversity combining from the M receiver antennas is carried out over two consecutive chip intervals according to [7, 23] as

$$
\begin{aligned}
y(k, 2i) &= \sum_{m=1}^{M} \left(h_{1,m}^*(k)x_m(k, 2i) + h_{2,m}(k)x_m^*(k, 2i + 1) \right) + w(k, 2i) \\
&= \alpha(k)c(k, 2i) + w(k, 2i),
\end{aligned}
$$

(6.6)

$$
\begin{aligned}
y(k, 2i + 1) &= \sum_{m=1}^{M} \left(h_{2,m}^*(k)x_m(k, 2i) - h_{1,m}(k)x_m^*(k, 2i + 1) \right) + w(k, 2i + 1) \\
&= \alpha(k)c(k, 2i + 1) + w(k, 2i + 1),
\end{aligned}
$$

(6.7)

Where

$$\alpha(k) = \sum_{m=1}^{M} \left(|h_{1,m}(k)|^2 + |h_{2,m}(k)|^2 \right)$$

(6.8)

And

$$
w(k, n) = \begin{cases}
\displaystyle\sum_{m=1}^{M} \left(h_{1,m}^*(k)e_m(k, n) + h_{2,m}(k)e_m^*(k, n + 1) \right), & \text{if } n \text{ is even} \\
\displaystyle\sum_{m=1}^{M} \left(h_{2,m}^*(k)e_m(k, n - 1) - h_{1,m}(k)e_m^*(k, n) \right), & \text{if } n \text{ is odd}
\end{cases}
$$

(6.9)

Because the channel coefficients and the white noise are independent, the noise term w is a random variable with zero means and variance MNN_0. The combined signals given in (6.6) and (6.7) can be written in one equation as

$$y(k, n) = \alpha(k)c(k, n) + w(k, n),$$

(6.10)

and further in vector form as

$$\mathbf{y}(k) = \alpha(k)\mathbf{c}(k) + \mathbf{w}(k),$$

(6.11)

Where

$$\mathbf{y}(k) = [y(k, 1)\ y(k, 2)\ \cdots\ y(k, N)]^T,$$

(6.12)

And

$$\mathbf{w}(k) = [w(k, 1)\ w(k, 2)\ \cdots\ w(k, N)]^T.$$

(6.13)

After diversity combining, the chips are despread and detected. The correlation estimate of the kth bit is given as

$$
\begin{aligned}
\hat{b}(k) &= \frac{1}{N}\mathbf{y}^T(k)\mathbf{s}(k) \\
&= \alpha(k)b(k) + v_1(k)
\end{aligned}
$$

(6.14)

where

$$v_1(k) = \frac{1}{N}\sum_{n=1}^{N} w(k, n)b(k - n)$$

(6.15)

is a random variable with zero mean and variance MN_0. The conditional SNR of the correlation detection given the channel coefficients is

$$\gamma_1 = \frac{|\alpha(k)b(k)|^2}{\mathcal{E}\{|v_1(k)|^2\}}.$$

(6.16)

Because the channel fading coefficients, the information bits, and the thermal noise are independent of each other, it is easy to see that Equation (6.16) is reduced to

$$\gamma_1 = \frac{E_b}{N_0}\alpha(k).$$

(6.17)

A hard decision is then performed on the correlation output, which in turn is fed back to the receiver delay registers in order to update the despreading sequence for the next bit.

$$\tilde{b}(k) \;=\; sgn[\hat{b}(k)] = \begin{cases} 1 & , \quad \hat{b}(k) > 0 \\ -1 & , \quad \hat{b}(k) < 0 \end{cases}$$

(6.18)

This also provides an estimate of the corresponding bit if no iterative detection is employed

Time-Diversity Detection

SESS provides a unique encoding scheme that can be utilized to achieve temporal diversity. To better illustrate this mechanism, let us write the following N^2 chips in a square matrix as

$$\mathbf{P}(k) = \begin{bmatrix} \alpha(k+1)b(k+1)b(k) & \alpha(k+1)b(k+1)b(k-1) & \cdots & \alpha(k+1)b(k+1)b(k-N+1) \\ \alpha(k+2)b(k+2)b(k+1) & \alpha(k+2)b(k+2)b(k) & \cdots & \alpha(k+2)b(k+2)b(k-N) \\ \vdots & \vdots & \ddots & \vdots \\ \alpha(k+N)b(k+N)b(k+N-1) & \alpha(k+N)b(k+N)b(k+N-2) & \cdots & \alpha(k+N)b(k+N)b(k) \end{bmatrix}$$

(6.19)

where each element represents a chip as given in (6.10) and each row includes the chips of one bit. For simplicity, the noise term is omitted. It is observed that the kth bit, b(k), is present in the diagonal elements of the matrix P(k). This means that the information of the kth bit is effectively transmitted in the next N bits, which is a unique characteristic of the SESS modulation scheme. By defining d(k) = diag (P(k)), the diversity estimate of b(k) can be obtained from the hard decisions of the next N bits $[\tilde{b}(k+1) \; \tilde{b}(k+2) \; \cdots \; \tilde{b}(k+N)]$, as

$$\begin{aligned} \hat{\hat{b}}(k) &= \frac{1}{N}\mathbf{d}(k)[\tilde{b}(k+1) \; \tilde{b}(k+2) \; \cdots \; \tilde{b}(k+N)]^T + v_2(k) \\ &= \frac{1}{N}\sum_{n=1}^{N}\alpha(k+n)b(k) + v_2(k), \end{aligned}$$

(6.20)

Where

$$v_2(k) = \frac{1}{N}\sum_{n=1}^{N} w(k+n, n)\tilde{b}(k+n)$$

(6.21)

is a random variable with zero mean and variance MN_0. Notice that the effect of the possible bit errors from the correlation detection has been ignored - this is justified with sufficient SNR as shown in Section 6.4. The conditional SNR given the channel coefficients for the diversity detection is

$$\gamma_2 = \frac{\left| \frac{1}{N} \sum_{n=1}^{N} \alpha(k+n) b(k) \right|^2}{\mathcal{E}\left\{ |v_2(k)|^2 \right\}}.$$

(6.22)

Because $\alpha(k+n)$, $n \in \{1, 2, \cdots, N\}$, $b(k)$, and $v_2(k)$ are independent of each other, Equation (6.22) reduces to

$$\gamma_2 = \frac{E_b}{N N_0} \sum_{n=1}^{N} \alpha(k+n).$$

(6.23)

It should be noted that the diversity detection introduces a time delay of NT_b and that the structure is similar to a Rake receiver employing N fingers to exploit the temporal diversity of SESS signals.

Iterative Detection

The estimate of b(k) can be improved iteratively with the summation of the correlation estimate, $\hat{b}(k)$ and the diversity estimate $\hat{\hat{b}}(k)$ Thus, let's define

$$\begin{aligned} \breve{b}(k) &= \hat{b}(k) + \hat{\hat{b}}(k) \\ &= \left(\alpha(k) + \frac{1}{N} \sum_{n=1}^{N} \alpha(k+n) \right) b(k) + v(k), \end{aligned}$$

(6.24)

where

$$v(k) = v_1(k) + v_2(k).$$

It is easy to see that the conditional SNR of the iterative detection, conditioned on the channel coefficients, is simply the summation of the SNR of the two independent signals, as follows:

$$\begin{aligned} \gamma &= \frac{\left| \left(\alpha(k) + \frac{1}{N} \sum_{n=1}^{N} \alpha(k+n) \right) b(k) \right|^2}{\mathcal{E}\left\{ |v(k)|^2 \right\}} \\ &= \frac{E_b}{N_0} \left[\alpha(k) + \frac{1}{N} \sum_{n=1}^{N} \alpha(k+n) \right] \\ &= \gamma_1 + \gamma_2. \end{aligned}$$

(6.25)

The final hard decision is then obtained by

$$\check{\tilde{b}}(k) \quad = \quad sgn[\check{b}(k)] = \begin{cases} 1 & , \quad \check{b}(k) > 0 \\ -1 & , \quad \check{b}(k) < 0 \end{cases}$$

(6.26)

Performance

Analysis In this section, we derive the distributions of the SNRs given in (6.17), (6.23) and (6.25), and the corresponding BER expressions. We proceed with the distribution of the normalized channel gain $|h(k)|^2$. It is well known that the coefficients of the Rayleigh fading channel are generated as $h(k) = (h_r + jh_i)/\sqrt{2}$, where h_r and h_i are real values and normally distributed as N (0, 1). The denominator $\sqrt{2}$ is for power normalization and the channel gain $|h(k)|^2 = \left(|h_r|^2 + |h_i|^2\right)/2$ is then gamma-distributed as $\Gamma(1, 1)$ [99].

Because the summation of independent and identically-distributed (i.i.d.) gamma random variables still follows a gamma distribution [99], the spatial diversity gain $\alpha(k)$ given in (6.8) has a $\Gamma(2M, 1)$ distribution. Furthermore, the conditional SNR of the correlation detector given in (6.17) follows $\Gamma(2M, E_b/N_0)$ distribution, i.e.,

$$p_{\gamma_1}(\gamma) = \frac{\gamma^{2M-1} e^{-\frac{\gamma}{E_b/N_0}}}{(2M-1)!(E_b/N_0)^{2M}}, \ \gamma \geq 0.$$

(6.27)

Now the bit-error probability for binary phase shift keying (BPSK) in AWGN with an SNR of E_b/N_0 is given in [98] as

$$P_b = Q\left(\sqrt{\frac{2E_b}{N_0}}\right).$$

(6.28)

Given the conditional SNR of $p_{\gamma_1}(\gamma)$, the BER expression for the correlation detection is given by [98, Eq. 14.4-15]:

$$P_{e,corr} \quad = \quad \int_0^\infty Q\left(\sqrt{2\gamma}\right) p_{\gamma_1}(\gamma) d\gamma$$

$$= \quad \left[\frac{1}{2}(1-\mu_1)\right]^{2M} \sum_{i=0}^{2M-1} \binom{2M-1+i}{i} \left[\frac{1}{2}(1+\mu_1)\right]^i$$

(6.29)

where

$$\mu_1 \triangleq \sqrt{\frac{E_b/N_0}{1 + E_b/N_0}}.$$

(6.30)

A similar analysis would show that the conditional SNR of the time-diversity detector given in (6.23) follows a $\Gamma(2NM, E_b/N_0/N)$ distribution, i.e.,

$$p_{\gamma_2}(\gamma) = \frac{\gamma^{2NM-1} e^{-\frac{\gamma}{E_b/N_0/N}}}{(2NM-1)!(E_b/N_0/N)^{2NM}}, \quad \gamma \geq 0.$$

(6.31)

and the BER expression for the diversity detection can then be obtained as

$$
\begin{aligned}
P_{e,div} &= \int_0^\infty Q\left(\sqrt{2\gamma}\right) p_{\gamma_2}(\gamma) d\gamma \\
&= \left[\frac{1}{2}(1-\mu_2)\right]^{2NM} \sum_{i=0}^{2NM-1} \binom{2NM-1+i}{i} \left[\frac{1}{2}(1+\mu_2)\right]^i
\end{aligned}
$$

(6.32)

where

$$\mu_2 \triangleq \sqrt{\frac{E_b/N_0/N}{1 + E_b/N_0/N}}.$$

(6.33)

Notice that the correlative detector has an M-fold (spatial) diversity whereas the timediversity detector has an MN-fold (spatial and temporal) diversity. For the iterative detection, since the conditional SNR γ is the summation of the independent conditional SNRs γ_1 and γ_2, the pdf of γ is the convolution of the gamma pdfs of γ_1 and γ_2. This is written as follows:

$$
\begin{aligned}
p_\gamma(\gamma) &= p_{\gamma_1}(\gamma) \otimes p_{\gamma_2}(\gamma) \\
&= \frac{\left(\gamma^{M-1} e^{-\frac{\gamma}{E_b/N_0}}\right) \otimes \left(\gamma^{NM-1} e^{-\frac{\gamma}{E_b/N_0/N}}\right)}{(M-1)!(NM-1)!(E_b/N_0)^M(E_b/N_0/N)^{NM}}, \quad \gamma \geq 0.
\end{aligned}
$$

(6.34)

The BER expression can then be obtained by averaging $Q\left(\sqrt{2\gamma}\right)$ over the conditional pdf of γ

The complexity of such a calculation can be obviated with the observation that the pdf $p_{\gamma_2}(\gamma)$ approaches a Dirac delta function as N becomes large. This stems from the fact that the gamma distribution approaches a Gaussian distribution if the degree of freedom, here 2NM, is large [99]. Furthermore, the Gaussian distribution tends toward a Dirac delta function as its variance, here $2NM(E_b/N_0/N)^2 = 2M(E_b/N_0)^2/N$, becomes very small [99]. It follows that if the spreading factor N is sufficiently large, the distribution of $p\gamma^2(\gamma)$ can be approximated by an impulse located at the mean value of $_{\gamma_2}$, which is given as $2NM(E_b/N_0/N) = 2M(E_b/N_0)$ In other words, $p_{\gamma_2}(\gamma)$ is approximatively equal to $\delta(\gamma - 2ME_b/N_0)$.

So as N becomes large, the pdf of γ_2 approaches an impulse and the pdf of γ can be obtained as the shifted pdf of γ_1:

$$
\begin{aligned}
p_\gamma(\gamma) &\approx p_{\gamma_1}(\gamma) \otimes \delta(\gamma - 2ME_b/N_0) \\
&= \begin{cases} \dfrac{(\gamma - 2ME_b/N_0)^{2M-1} e^{-\frac{\gamma - 2ME_b/N_0}{E_b/N_0}}}{(2M-1)!(E_b/N_0)^{2M}}, & \gamma \geq 2M\dfrac{E_b}{N_0} \\ 0 & else \end{cases}
\end{aligned}
$$

(6.35)

The BER of the iterative detector can be calculated as

$$
\begin{aligned}
P_e &= \int_0^\infty Q\left(\sqrt{2\gamma}\right) p_\gamma(\gamma) d\gamma \\
&\approx \int_{2ME_b/N_0}^\infty Q\left(\sqrt{2\gamma}\right) \frac{(\gamma - 2ME_b/N_0)^{2M-1} e^{-\frac{\gamma - 2ME_b/N_0}{E_b/N_0}}}{(2M-1)!(E_b/N_0)^{2M}} d\gamma \\
&= Q\left(\sqrt{4M\frac{E_b}{N_0}}\right) - \sqrt{\frac{\mu_1}{2\pi}} e^{2M} \sum_{k=0}^{2M-1} \frac{N_0^k}{E_b^k k!} \\
&\quad \sum_{l=0}^{k} \binom{k}{l} (\mu_1)^{k-l} \left(-2M\frac{E_b}{N_0}\right)^l \Gamma\left(k - l + \frac{1}{2}, 2M\left(1 + \frac{E_b}{N_0}\right)\right).
\end{aligned}
$$

(6.36)

The complicated expression in 6.36 can be derived as follows: Define

$$
\begin{aligned}
P_e &\approx \int_a^\infty Q\left(\sqrt{2\gamma}\right) \frac{(\gamma - a)^{L-1} e^{-\frac{\gamma - a}{b}}}{(L-1)! b^L} d\gamma \\
&= \int_0^\infty Q\left(\sqrt{2(x+a)}\right) \frac{x^{L-1} e^{-\frac{x}{b}}}{(L-1)! b^L} dx \\
&= \int_0^\infty \frac{1}{\sqrt{2\pi}} \int_{\sqrt{2(x+a)}}^\infty e^{-\frac{u^2}{2}} du \frac{x^{L-1} e^{-\frac{x}{b}}}{(L-1)! b^L} dx
\end{aligned}
$$

(6.37)

By changing the order of the integrals, (6.37) becomes

$$
P_e \approx \frac{1}{\sqrt{2\pi}} \frac{1}{(L-1)! b^L} \int_{\sqrt{2a}}^\infty e^{-\frac{u^2}{2}} P_1 du
$$

(6.38)

where

$$
P_1 = \int_0^{\frac{u^2}{2} - a} x^{L-1} e^{-\frac{x}{b}} dx
$$

(6.39)

Let $z = \frac{x}{b}$, then

$$P_1 = b^L \int_0^{\frac{u^2}{2b} - \frac{a}{b}} z^{L-1} e^{-z} dx$$

Using the indefinite integral equation given in [100, Eq. 7.4.322], P_1 can evaluated as

$$P_1 = -b^L e^{-\frac{u^2}{2b} + \frac{a}{b}} \sum_{k=1}^{L-1} \frac{(L-1)!}{k!} \left(\frac{u^2}{2b} - \frac{a}{b} \right)^k - b^L (L-1)! \left(1 - e^{-\frac{u^2}{2b} + \frac{a}{b}} \right) \tag{6.40}$$

Substituting (6.41) into (6.38) to determine P_e:

$$
\begin{aligned}
P_e &\approx \int_{\sqrt{2a}}^{\infty} \frac{1}{\sqrt{2\pi}} e^{-\frac{u^2}{2}} du - \frac{1}{\sqrt{2\pi}} e^{\frac{a}{b}} \int_{\sqrt{2a}}^{\infty} e^{-\frac{u^2}{2} \left(\frac{1+b}{b} \right)} du \\
&\quad - \frac{1}{\sqrt{2\pi}} e^{\frac{a}{b}} \sum_{k=1}^{L-1} \frac{1}{k!} \left(\frac{1}{2b} \right)^k \int_{\sqrt{2a}}^{\infty} e^{-\frac{u^2}{2} \left(\frac{1+b}{b} \right)} \left(u^2 - 2a \right)^k du \\
&= Q\left(\sqrt{2a} \right) - \frac{1}{\sqrt{2\pi}} e^{\frac{a}{b}} \sum_{k=0}^{L-1} \frac{1}{k!} \left(\frac{1}{2b} \right)^k P_2
\end{aligned}
\tag{6.42}
$$

where

$$P_2 = \int_{\sqrt{2a}}^{\infty} e^{-\frac{u^2}{2} \left(\frac{1+b}{b} \right)} \left(u^2 - 2a \right)^k du \tag{6.43}$$

This can be further transformed with the substitution $t = \frac{u^2}{2} \frac{1+b}{b}$:

$$
\begin{aligned}
P_2 &= 2^{k-1} \sqrt{\frac{2b}{1+b}} \int_{\frac{a(1+b)}{b}}^{\infty} e^{-t} t^{-\frac{1}{2}} \left(\frac{b}{1+b} t - a \right)^k dt \\
&= 2^{k-1} \sqrt{\frac{2b}{1+b}} \int_{\frac{a(1+b)}{b}}^{\infty} e^{-t} t^{-\frac{1}{2}} \sum_{l=0}^{k} \binom{k}{l} \left(\frac{b}{1+b} t \right)^{k-l} (-a)^l dt \\
&= 2^{k-1} \sqrt{\frac{2b}{1+b}} \sum_{l=0}^{k} \binom{k}{l} \left(\frac{b}{1+b} \right)^{k-l} (-a)^l \Gamma\left(k - l + \frac{1}{2}, \frac{a(1+b)}{b} \right)
\end{aligned}
\tag{6.44}
$$

where $\Gamma(a, x)$ denotes the incomplete gamma function with shape a and scale x [101, Eq. 6.5.3]. Substituting (6.44) into (6.42), we have:

$$
\begin{aligned}
P_e &\approx Q\left(\sqrt{2a} \right) - \frac{1}{\sqrt{2\pi}} e^{\frac{a}{b}} \sqrt{\frac{b}{1+b}} \\
&\quad \sum_{k=0}^{L-1} \frac{1}{b^k k!} \sum_{l=0}^{k} \binom{k}{l} \left(\frac{b}{1+b} \right)^{k-l} (-a)^l \Gamma\left(k - l + \frac{1}{2}, \frac{a(1+b)}{b} \right)
\end{aligned}
\tag{6.45}
$$

The final expression for P_e in (6.36) can be obtained with the defined values for a, b, and L:

$$
P_e \approx Q\left(\sqrt{4M\frac{E_b}{N_0}}\right) - \sqrt{\frac{\mu_1}{2\pi}}e^{2M}\sum_{k=0}^{2M-1}\frac{N_0^k}{E_b^k k!}
$$

$$
\sum_{l=0}^{k}\binom{k}{l}(\mu_1)^{k-l}(-2M\frac{E_b}{N_0})^l \Gamma\left(k-l+\frac{1}{2}, 2M\left(1+\frac{E_b}{N_0}\right)\right)
$$

$$(6.46)$$

where μ_1 is given by (6.30). It should be noted that the impulse approximation of the pdf yields a lower bound for the BER, because it can only be approached when the spreading factor N tends to infinity.

Numerical Results

The distributions of the conditional SNRs are first presented and the delta approximation related to the iterative detection is discussed. Figure 6.3 shows the distributions of the conditional SNRs of the correlation detection γ_1, diversity detection γ_2, and iterative detection γ, for different scenarios. Each sub-figure also includes a curve that shows the distribution of $\gamma 1$ that has been right-shifted by $2M\frac{E_b}{N_0}$. The coordinates for different sub-figures have been adjusted in order to illustrate the distributions in detail. As analyzed in Section 6.3, the mean and variance of the conditional SNR of the diversity detection γ_2 are $2M\frac{E_b}{N_0}$ and $\frac{2M}{N}(\frac{E_b}{N_0})^2$ respectively. This means that while the number of receive antennas and the SNR affect both position and shape of the "pulse," the spreading length only influences its shape. Figure 6.3 (a) and (b) show that the position and the shape of the pulse vary with the number of antennas. In both cases, the pulse is not very narrow due to the small value of N = 8, resulting in a discernable difference between the shifted version of the distribution of γ_1 (dotted curve) and the distribution of γ (dashed curve). The "pulse" looks sharper in Figure 6.3 (c) since N = 64 is much larger. This sharper pulse approximates a delta function such that there is negligible difference between the shifted distribution of γ_1 and the distribution of γ. Also, Figure 6.3 (c) and (d) shows that the delta approximation is not very sensitive to SNR, although it is more accurate with lower SNR.

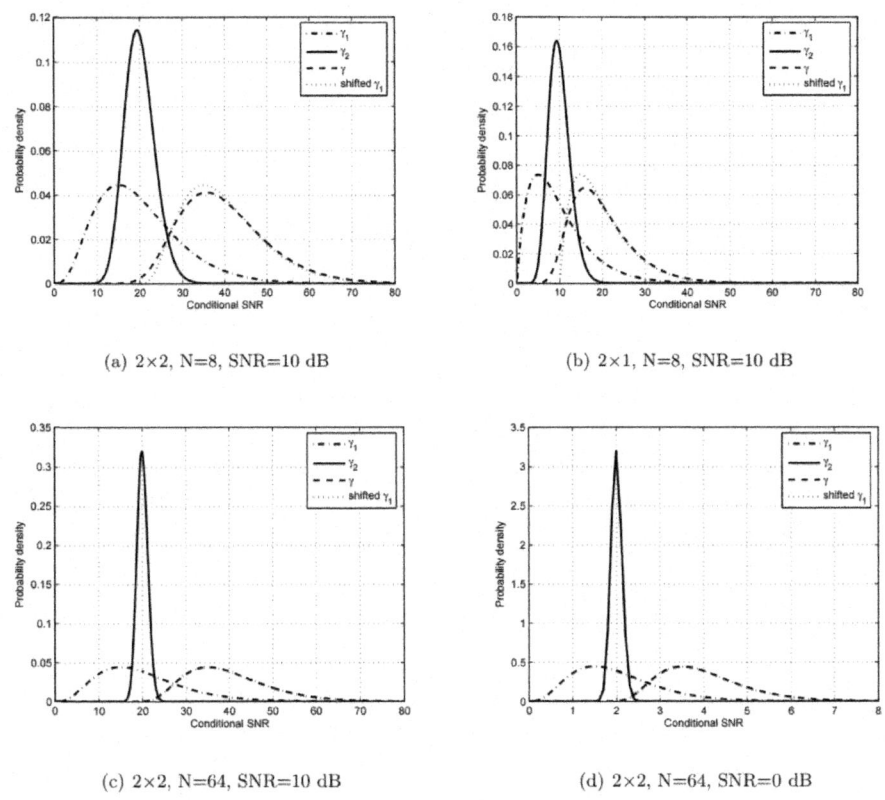

(a) 2×2, N=8, SNR=10 dB

(b) 2×1, N=8, SNR=10 dB

(c) 2×2, N=64, SNR=10 dB

(d) 2×2, N=64, SNR=0 dB

Figure 6.3: Distribution of the conditional SNRs

Next, the BER performance of MIMO-SESS is presented and is compared with a conventional MIMO, PN-coded spread spectrum (MIMO-PNSS) system. The channel between each transmit and receive antenna pair is assumed to be Rayleigh fading that remains constant over T_b but is independent from bit to bit [92, 98]. In addition, the channels for different transmit/receive antenna pair are independent. The length of the spreading sequence is set to N = 64 unless noted otherwise. For each scenario and bit signal-to-noise ratio (SNR), 100 runs of 100,000 bits are simulated to obtain the average bit error rate BER).

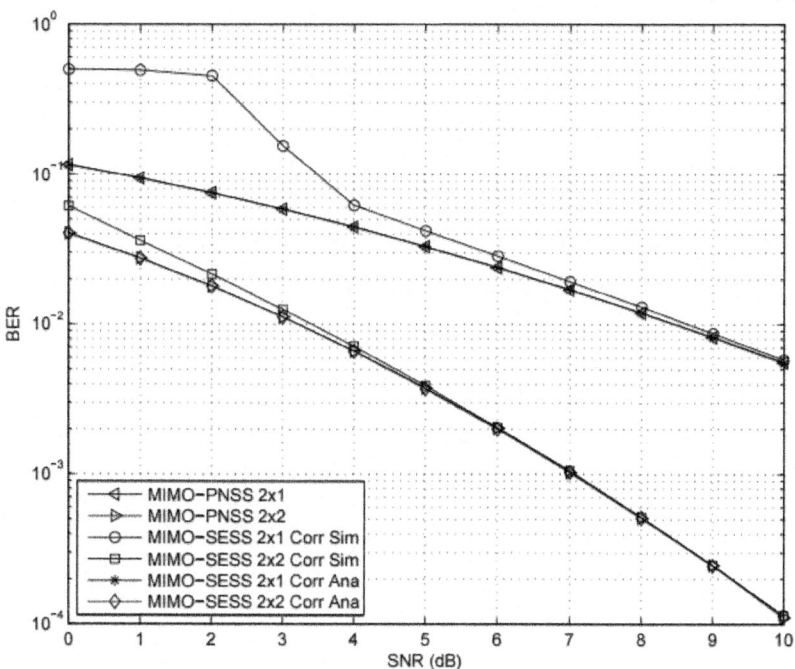

Figure 6.4: Comparison between MIMO-PNSS and MIMO-SESS with correlation detection, N=64

(Figure 6.4 compares the BER of MIMO-SESS with correlation detection to a MIMOPNSS system. It is observed that there is no time diversity gain without iterative detection and the performance with or without self-encoding is the same as expected under high SNR. The performance degradation of MIMO-SESS at low SNR is due to error propagation [87]. In particular, Figure 6.4 shows that the effect of error propagation is quite severe for a BER greater than 10%. The effect of error propagation can be efficiently alleviated with more antennas, as shown by the example 2 × 2 scenario, since larger spatial diversity reduces BER to below 10% even at low SNR. The plots in Figure 6.4 show excellent agreement between the analytical and simulation results at high SNR, where error propagation becomes insignificant. The performance of MIMO-SESS with time-diversity detection is shown in Figure 6.5.

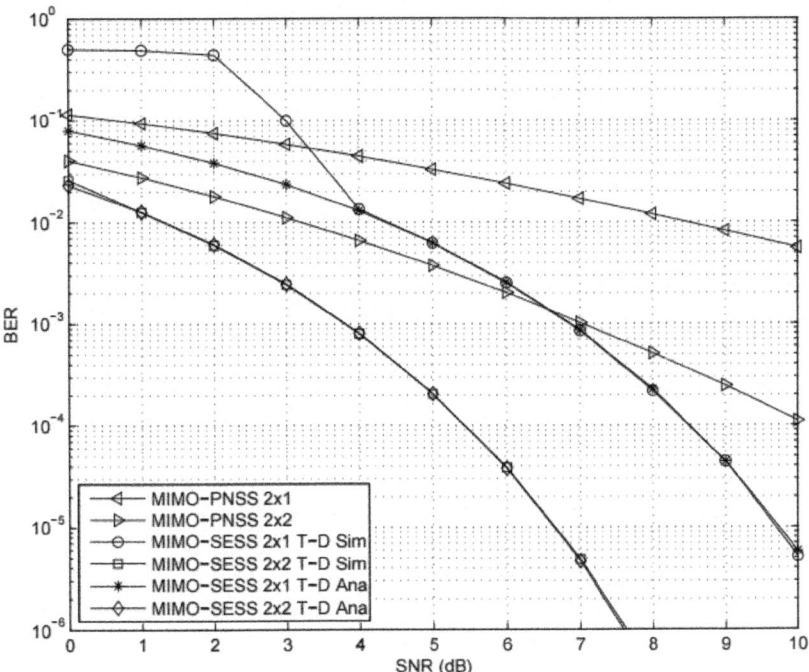

Figure 6.5: Comparison between MIMO-PNSS and MIMO-SESS with diversity detection, N=64.

The plots clearly show that the proposed system significantly outperforms conventional MIMO-PNSS systems. As an example, there is about 4.5 dB gain for the 2×2 configuration at 10^{-4} BER. This performance improvement can be attributed to the N-fold time diversity introduced by self encoding. Due to the significantly improved BER, error propagation is effectively mitigated as shown by the 2 × 2 scenario. Moveover, the agreement with the simulation results verifies that the analysis has been justified in ignoring error propagation. Figure 6.6 compares the BER with the iterative detector. Again, there is excellent agreement between the analytical and simulation results at sufficiently high SNR. The performance gain over conventional MIMO-PNSS systems is almost 7 dB at 10^{-4} BER for the 2 × 2 configuration. The excess gain beyond the expected 3 dB SNR improvement can be attributed to the diversity gain from the time-diversity detection. In fact, the plot shows that a 2 × 2 MIMO-SESS would require an SNR of only 3dB to achieve a 10^{-4}

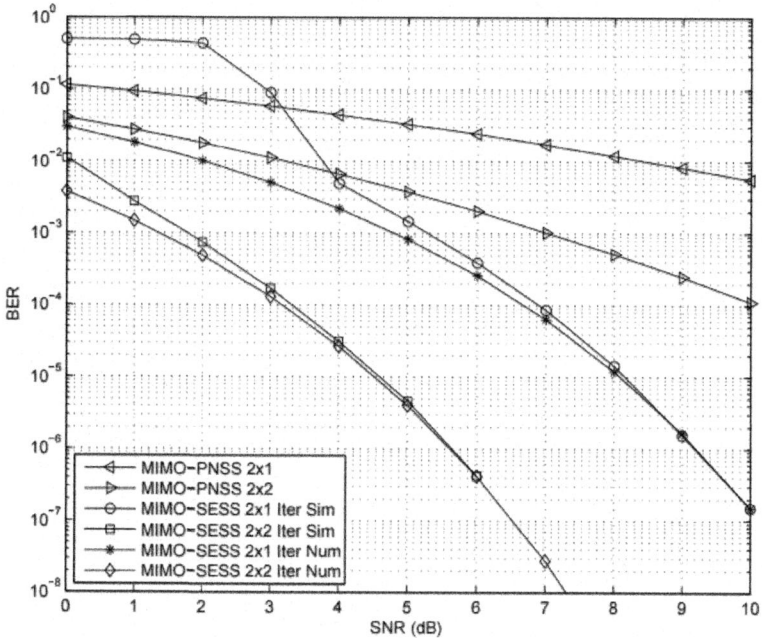

Figure 6.6: Comparison between MIMO-PNSS and MIMO-SESS with iterative detection, N=64.

BER. The results verify the veracity of the performance analysis and demonstrate that MIMO-SESS can completely mitigate the effect of Rayleigh fading. The difference between the numerical and simulation results at low SNR is due to error propagation. This can be easily demonstrated in simulation by feeding the true (transmitted) bits into the delay registers at the receiver. The results of this simulation are compared with the numerical calculations in Figure 6.7. The excellent agreement between the simulation and numerical results demonstrates the validity of the statistical models and approximations. Figure 6.8 shows the numerical results of MIMO-SESS with the iterative detection for different spreading lengths. For both 2 ×1 and 2×2 cases, the performance approaches the lower bound when the spreading length increases. Notice that the lower bound is very tight for N ≥ 64. Also, the lower bound is tighter for the 2 × 1 case and at lower SNR. These results are consistent with the discussion in Section 6.3. It can be seen by comparing the lower bound and the performance of BPSK in AWGN that 2 × 2 MIMO-SESS can provide about 5.2 dB gain over BPSK at 10−4 BER. This significant performance improvement is clearly due to the combined temporal and spatial diversities associated with MIMO-SESS design.

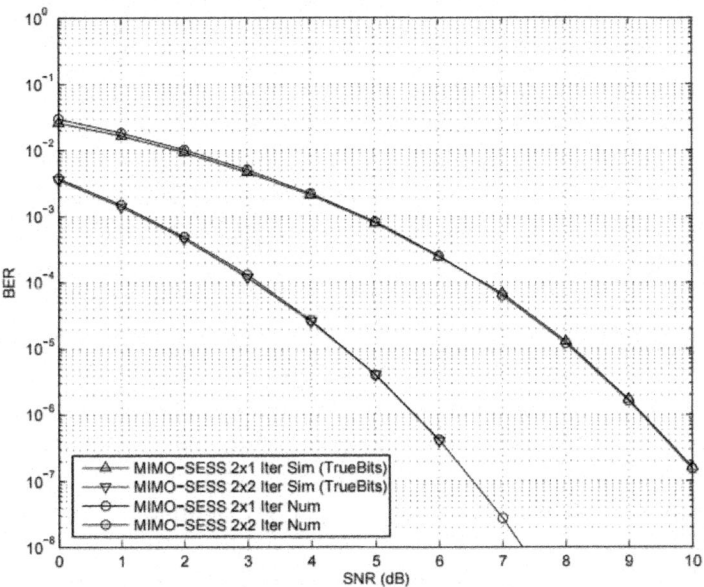

Figure 6.7: Comparison between numerical results and simulation results without error propagation, N=64.

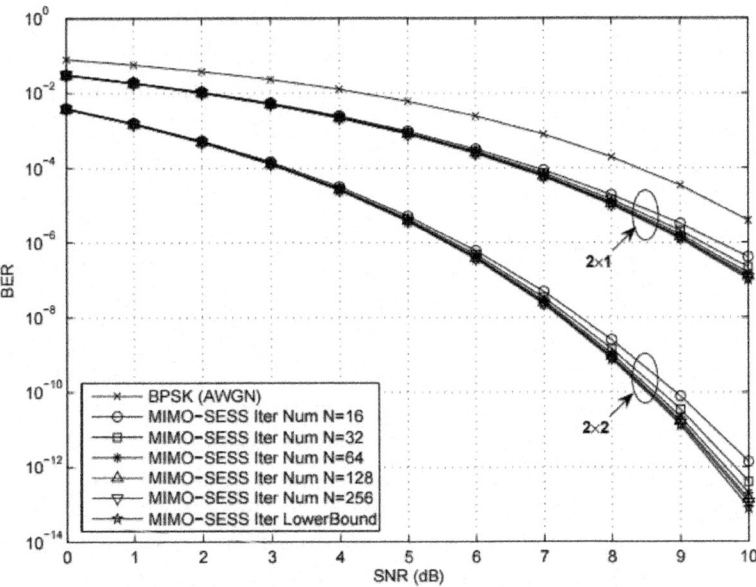

Figure 6.8: Performance of the proposed MIMO-SESS system with various N values.

SECURING WIRELESS COMMUNICATIONS BY EXPLORING PHYSICAL LAYER CHARACTERISTICS

Wireless networks are susceptible to eavesdropping due to their broadcast nature. During the past few decades, security in wireless networks has been mainly considered at higher layers using cryptography [102]. Recently, researchers are seeking security methods at the physical layer. The key idea of physical layer security is to exploit the wireless physical characteristics, e.g., modulation, coding, channel fading, and diversities, to secure wireless communications. The fundamental ability of the physical layer to provide security is characterized by secrecy capacity, which is defined as the maximum achievable rate of information that can be sent in secret from a transmitter to its desired receiver in the presence of an eavesdropper. A feasible solution to this question is to employ cooperative relaying and cooperative jamming. In a cooperative relaying scheme, one or more relays work together with the transmitter to increase the secrecy capacity. For example, relays which utilize decode-andforward (DF) [103] and amplify-and-forward (AF) cooperative strategies [104] have been studied. By optimizing the transmitting power weights among the transmitter and relays, the secrecy capacity was increased subject to a fixed transmit power constraint. In cooperative jamming, a relay transmits jamming signals in order to hide the informationbearing signal or deny the eavesdropper's reception. For instance, a noise-forwarding (NF) strategy was proposed to increase the secrecy capacity of a four-node relay-eavesdropper channel [105]. Based on this NF method, the relay node cooperates with the transmitter by forwarding independent noise to the eavesdropper. However, the problem in the afore-mentioned methods remains that the channel state information (CSI) was assumed globally known, but not at the eavesdropper. Another promising solution to achieving secrecy capacity bound is to utilize multiple antennas. It has been shown that information security and information-hiding capabilities can be enhanced by employing the space-time diversity schemes at the transmitter [106]. A transmission scheme exploiting the redundancy of the transmit antenna array was proposed for multiple-input single-output (MISO) systems [107]. By randomizing the eavesdropper's signals, the secrecy capacity can be increased. By adding artificially generated noise to the information-bearing signal [13, 108], it can successfully degrade the eavesdropper's signal quality without affecting the main channel. The artificial noise is generated based on the null space of the main channel, which provides $M - N$ degrees of freedom for the noise, where M and N denote the numbers of the antennas at the transmitter and the receiver, respectively. This method, therefore, requires more antennas are deployed at the transmitter than that at the receiver [13]. Otherwise, this method becomes invalid because no null spaces exist for such antenna configurations. This chapter proposes to increase secrecy capacity by jamming the eavesdropper's reception with artificially generated noise signal, which is intentionally transmitted together with the information-bearing signal [109–111]. This jamming signal significantly degrades the signal quality at the eavesdropper but not at the intended receiver. Two methods are proposed to generate the jamming noise signal: one is based on the null space of an equivalent channel of the system, and another generates the jamming signal from the eigenvectors of the Hermitian matrix $H^H H$, where H is the channel matrix. The first method is referred to as null space (NS)-based jamming, while the second is called eigenvector (EV)-based jamming. Compared to the artificial noise method presented in [13], our proposed methods demonstrate several significant advantages. First, our approaches overcome the limitation of the number of receive antenna elements. Regardless of the number of the transmit antennas, our methods can be applied in a system with any number of receive antennas.

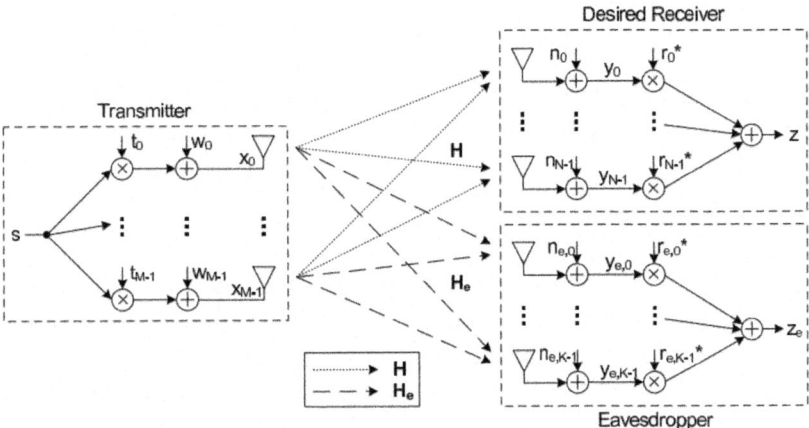

Figure 7.1: Block diagram of a transmit-beamforming diversity system with jamming signal.

Second, the degree of freedom for the jamming noise signal is $M - 1$ and is independent of the number of receive antennas. This facilitates our schemes with the unique ability of randomizing the generated jamming noise signal, significantly increasing the level of information security. Third, our scheme is capable of maintaining the secrecy capacity even if the eavesdropper employs a large size of antenna array or even moves its position closer to the transmitter.

System Model and Problem Statement

The wireless system under our consideration is depicted in Figure 7.1, where a transmitter with M antennas communicates with an intended receiver with N antennas in the presence of an eavesdropper with K antennas. The following describes the functions and assumptions of the transmitter, receiver, and eavedropper.

Transmitter

At the transmitter, the complex data symbol s is modulated by a normalized beamforming weighting vector $\mathbf{t} \in \mathcal{C}^{M \times 1}$ into M branches, where $\mathbf{t}^H \mathbf{t} = 1$.

An intentionally generated jamming signal vector $\mathbf{w} \in \mathcal{C}^{M \times 1}$ is precoded into the information-bearing signal. The transmitted signal $\mathbf{x} \in \mathcal{C}^{M \times 1}$ can be expressed as

$$\mathbf{x} = \mathbf{t}s + \mathbf{w}. \tag{7.1}$$

Both the intended receiver and the eavesdropper are within communication range of the transmitter. The intended receiver receives the signal over a MIMO channel denoted by the channel matrix $\mathbf{H} \in \mathcal{C}^{N \times M}$ while the eavesdropper receives the signal via a MIMO channel represented by the channel matrix $\mathbf{H}_e \in \mathcal{C}^{K \times M}$. Both $\mathbf{H}$ and $\mathbf{H}_e$ re assumed to be quasi-static flat fading channels. In addition, the channel matrix H is assumed to be known to both the transmitter and the desired user, but the channel matrix H_e is only known to the passive eavesdropper. In other words, the eavesdropper is incapable of obtaining the channel matrix between the transmitter and the desired receiver, and the transmitter doesn't need information about the eavesdropper's channel. To satisfy this assumption, both the transmitter and the desired receiver can transmit a training sequence to each other. The CSI is estimated at both ends without feedback. Note that during the channel estimation process, the weighting vector at the transmitter is initially set to be a unit vector

Desired Receiver

In a flat-fading scenario, the received signal vector $y \in \mathcal{C}^{N \times 1}$ at the desired receiver can be written as

$$\mathbf{y} = \mathbf{H}\mathbf{x} + \mathbf{n}, \tag{7.2}$$

Where $\mathbf{n} \in C^{N \times 1}$ is the complex white Gaussian noise vector with zero-mean and covariance matrix $\sigma_n^2 \mathbf{I}_N$. The received signal is then combined using the receive weighting vector

$\mathbf{r} \in C^{N \times 1}$ as

$$z = \mathbf{r}^H \mathbf{y}. \tag{7.3}$$

Combining (7.1), (7.2), and (7.3), the received symbol z at the desired receiver can be written as

$$z = \mathbf{r}^H \mathbf{H} t s + \mathbf{r}^H \mathbf{H} \mathbf{w} + \mathbf{r}^H \mathbf{n}, \tag{7.4}$$

where the second term on the right-hand side, $r^H\mathbf{H}\mathbf{w}$, is the interference caused by the intentionally added jamming signal w. The receive weighting vector is assumed to be normalized, i.e., $\|\mathbf{r}\|^2 = 1$, and the transmit powers for the information signal and the jamming signal are set to be P_s and P_w, respectively (i.e., $E\{|s|^2\} = P_s$ and $\|\mathbf{w}\|^2 = P_w$). The total transmit power is constrained to P, i.e., $E\{\mathbf{x}^H \mathbf{x}\} = P_s + P_w \leq P$. Thus, the achievable rate between the transmitter and the desired receiver is formulated as follows:

$$C_{main} = log\left(1 + \frac{\|\mathbf{r}^H \mathbf{H} t\|^2 P_s}{\|\mathbf{r}^H \mathbf{H} \mathbf{w}\|^2 P_w + \sigma_n^2}\right)$$

$$\text{subject to } E\{\mathbf{x}^H \mathbf{x}\} = P_s + P_w \leq P \tag{7.5}$$

To maximize the achievable rate at the intended receiver, we maximize the signal-tointerference plus noise ratio (SINR) in (7.5). Thus, the transmit weighting vector t is chosen as the eigenvector v corresponding to the largest eigenvalue λ_{max} of $\mathbf{H}^H \mathbf{H}$ (as in MRT) and the receive weighting vector r is chosen to be $\mathbf{r} = \mathbf{H}\mathbf{v}/\|\mathbf{H}\mathbf{v}\|$ (as in MRC) [112] so that we can eliminate the interference component at the desired receiver.

Eavesdropper

Similar to the analysis for the intended receiver, the received signal at the eavesdropper is weighted by r_e^H and given as

$$z_e = \mathbf{r_e}^H (\mathbf{H_e} \mathbf{v} s + \mathbf{H_e} \mathbf{w} + \mathbf{n_e}), \tag{7.6}$$

Where

$\mathbf{r_e} \in \mathcal{C}^{K \times 1}$ and $\mathbf{n_e} \in \mathcal{C}^{K \times 1}$ are the eavesdropper's receive weighting vector and white Gaussian noise vector with zero-mean and covariance matrix $\sigma_{n_e}^2 \mathbf{I}_K.$ Therefore the achievable rate at the eavesdropper is formulated as

$$C_{eave} = log \left(1 + \frac{\|\mathbf{r}_e^H \mathbf{H}_e \mathbf{v}\|^2 P_s}{\|\mathbf{r}_e^H \mathbf{H}_e \mathbf{w}\|^2 P_w + \sigma_{n_e}^2} \right)$$

(7.7)

subject to equal gain combining of r_e and worst case $\sigma^2{}_{ne} \rightarrow 0$

Because the eavesdropper has no information about the beamforming vector w, the MRC method is not applicable. However, to maximize its achievable rate, the eavesdropper may select the suboptimal method of equal gain combining, i.e $, \mathbf{r}_e = \frac{1}{\sqrt{K}}[1\ 1\ \cdots\ 1]^T$ and choose the noiseless receiver

such that $\sigma_{n_e}^2 \simeq 0,$ which may be the worst case for secure wireless communications.

Proposed Approach

It is observed from (7.4) and (7.6) that the combined signal at both the desired receiver and the eavesdropper includes interference caused by the jamming signal w. The secrecy capacity is usually defined as the difference between (7.5) and (7.7), or $C_s = (C_{main} - C_{eave})^+.$ The key idea to increasing the secrecy capacity is to design the transmit weighting factor w such that the interference is removable at the intended receiver but not at the eavesdropper. The method using noise generation in the null space of the main channel provides insight into this concept [13]. This method, however, can only be used in systems where the number of transmit antennas is greater than the number of receive antennas, which is impractical as it imposes configuration constraints for transmitters, receivers, and eavesdroppers. In this section, two methods are presented that can overcome this limitation.

Null Space-Based Jamming

The interference component at the desired receiver resulting from the jamming noise signal in Eq. (7.4) can be written as $\mathbf{a}^H \mathbf{w} / \|\mathbf{H}\mathbf{v}\|$, where $\mathbf{a} \in \mathcal{C}^{M \times 1}$ is defined as $\mathbf{a} = \mathbf{H}^H \mathbf{H}\mathbf{v}$

We can generate a jamming signal vector in the null space of the row vector a ^{H}, which offers M $-$ 1 degrees of freedom regardless of the number of receive antenna elements. Our selection of w will maximize (7.5) while minimizing (7.7). The procedure to generate the jamming noise signal is described as follows: 1. Find the null space matrix

$\Gamma \in \mathcal{C}^{M \times (M-1)}$ of $\mathbf{a}^H$, so that $\mathbf{a}^H \Gamma = (\mathbf{0}_{M-1})^T$ and $\Gamma^H \Gamma = \mathbf{I}_{M-1}$;

Construct a vector $\mathbf{f} \in \mathcal{C}^{(M-1) \times 1}$ with components to be i.i.d. Gaussian with zeromean and variance $\sigma_f^2 = P_w/(M-1)$;

Choose w = $\Gamma \mathbf{f}$, such that the interference term $\mathbf{a}^H \mathbf{w}/\|\mathbf{H}\mathbf{v}\| = 0$ results in a complete removal of the jamming noise signal interference at the desired receiver.

Eigenvector-Based Jamming

The interference component due to the jamming signal at the desired receiver can also be written as $(\mathbf{H}\mathbf{v})^H \mathbf{H}\mathbf{w}/\|\mathbf{H}\mathbf{v}\|$. Instead of generating the jamming signal based on the null space concept, another method is to design the jamming noise signal such that it ensures $\mathbf{H}\mathbf{v} \perp \mathbf{H}\mathbf{w}$.

This design can be realized by taking advantage of the orthogonality between any two different eigenvectors of a normal matrix *. Assume the M eigenvalues of matrix $\mathbf{H}^H\mathbf{H}$ are λ_i, i $\in$ {1, 2, $\cdots$, M}, and their corresponding eigenvectors are $\mathbf{v}_i$, i $\in$ {1, 2, $\cdots$, M}. Because the matrix $\mathbf{H}^H\mathbf{H}$ is normal, its eigenvectors are mutually orthogonal [113]. That is $\mathbf{v}_i^H \mathbf{v}_j = \delta(i - j)$.

Without loss of generality, it is assumed that λ_M and vM are the largest eigenvalue and its corresponding eigenvector, respectively. We may choose w as a linear combination of eigenvectors $\mathbf{v}_i$, $i \in \{1, 2, \cdots, M-1\}$. The procedure to generate the jamming signal is summarized as follows:

Construct an orthogonal set $\mathbf{V} \in \mathcal{C}^{M \times (M-1)}$ with each column as one eigenvector corresponding to one of the least M -1 eigenvalues of HHH, e.g., $\mathbf{H}^H\mathbf{H}$, e.g., $\mathbf{V} = [\mathbf{v}_1 \, \mathbf{v}_2 \, \cdots \, \mathbf{v}_{M-1}]$; Randomly select a vector $\mathbf{f} \in \mathcal{C}^{(M-1) \times 1}$ with components to be i.i.d. Gaussian with zero-mean and variance w = Vf.);

It should be noted that a similar method was applied in a singular value decomposition (SVD)-based MIMO system to find a fixed SINR solution [114]. In this way, the interference component in Eq.(7.4) becomes

$$
\mathbf{r}^H \mathbf{H} \mathbf{w} = \frac{\mathbf{v}_M^H \mathbf{H}^H \mathbf{H} \mathbf{V} \mathbf{f}}{\|\mathbf{H} \mathbf{v}_M\|}
$$

$$
= \frac{[\lambda_1 \mathbf{v}_M^H \mathbf{v}_1 \; \lambda_2 \mathbf{v}_M^H \mathbf{v}_2 \; \cdots \; \lambda_{M-1} \mathbf{v}_M^H \mathbf{v}_{M-1}] \mathbf{f}}{\|\mathbf{H} \mathbf{v}_M\|}
$$

$$
= 0 \quad (for \; \forall \mathbf{f}), \tag{7.8}
$$

resulting in non-influence to the desired receiver. In the aforementioned nullspace (NS) and eigenvector (EV)-based methods, the jamming signal doesn't interfere with the intended receiver. The output SINR at the desired receiver can be written as

$$
\gamma_r = \frac{|\sqrt{\lambda_{max}} s|^2}{E\{|\mathbf{r}^H \mathbf{n}|^2\}} = \frac{\lambda_{max} P_s}{\sigma_n^2}, \tag{7.9}
$$

and the expected main channel capacity is given by

$$
C_{main} = \log\left(1 + \frac{\lambda_{max} P_s}{\sigma_n^2}\right). \tag{7.10}
$$

However, the jamming signal causes significant interference at the eavesdropper, and the output signal-to-interference-plus-noise ratio (SINR) is given by

$$
\gamma_e = \frac{E|\mathbf{r_e}^H \mathbf{H_e} \mathbf{v}|^2 P_s}{E|\mathbf{r_e}^H \mathbf{H_e} \mathbf{w}|^2 P_w + \sigma_{n_e}^2}. \tag{7.11}
$$

The eavesdropper's channel capacity is given by

$$
C_{eave} = \log\left(1 + \frac{E|\mathbf{r_e}^H \mathbf{H_e} \mathbf{v}|^2 P_s}{E|\mathbf{r_e}^H \mathbf{H_e} \mathbf{w}|^2 P_w + \sigma_{n_e}^2}\right). \tag{7.12}
$$

Next, we will evaluate the secrecy capacity based on the aforementioned methods.

Simulation Results

This section investigates the secrecy capacity of the transmit-beamforming diversity system with proposed jamming noise signal under the i.i.d. channel [115] and the 3GPP spatial channel model (SCM) [15]. The performance is first studied by using the i.i.d. channel model. The main channel H and the eavesdropper's channel H_e are generated statistically independently. It is assumed the thermal noise at the desired receiver and the eavesdropper have the same variance, i.e., $\sigma_n^2 = \sigma_{n_e}^2$. The total transmit power P is normalized with respect to the variance. Without the position (and thus the path loss) information of the passive eavesdropper, it is impossible to adaptively optimize the power allocation between the information-bearing signal and the jamming signal. Therefore, we examine two fixed power allocation schemes: one is equal power allocation, i.e., $P_s = P_w = P/2$ and another is without the jamming signal, namely $P_s = P$ and $P_w = 0.$ The antenna configuration is represented by the triple [M, N, K], where M, N, and K denote the number of the transmit antennas, receive antennas, and eavesdropper antennas, respectively. Figure 7.2 shows the main channel capacities associated with the normalized transmit power of the transmit-beamforming system with jamming noise signal. Results under different antenna configurations are compared. As shown in (7.10), the capacity is determined by two factors: the expectation of the largest eigenvalue and the normalized transmit power (the ratio between the transmit power and the noise variance). According to the analysis results in [116], the largest eigenvalue is related to the antenna numbers at the transmitter and the desired receiver. Given a specific antenna configuration, the main channel capacity increases with the normalized transmit power.

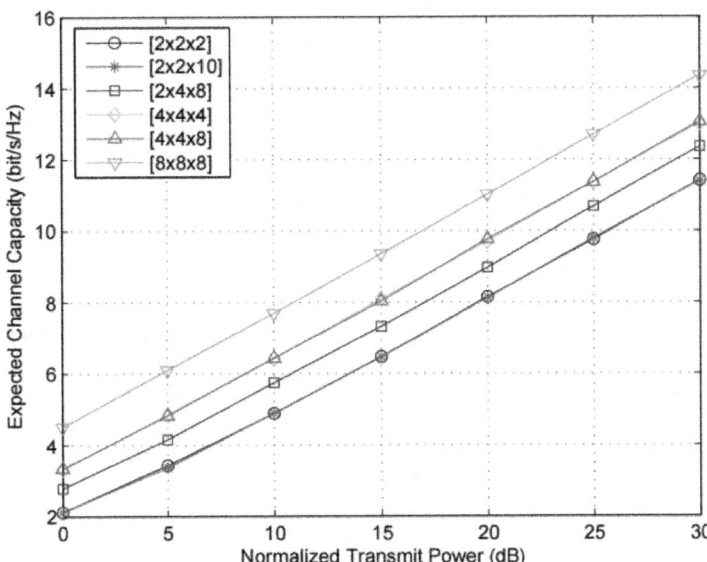

Figure 7.2: Comparison of the main channel capacities for different antenna configurations, with jamming noise signal, i.i.d. channel model.

For fixed normalized transmit power, the channel capacity increases with the transmit and receive antenna numbers. The eavesdropper's channel capacities for transmit-beamforming systems with jamming noise signal under different antenna configurations are shown in Figure 7.3. From this figure, it is observed that the eavesdropper's channel capacity is only determined by the transmit antenna number but is independent of the antenna number at the eavesdropper or the desired receiver. This means that the eavesdropper cannot increase its channel capacity by deploying more antenna elements. This advantage of our proposed scheme is helpful when designing a secure wireless system. When comparing the main channel capacities and the eavesdropper's channel capacities shown in Figure 7.2 and Figure 7.3, it is observed that the secrecy capacity increases significantly, which is shown in Figure 7.4 by the curve set labeled "with jamming." These curves also show that secrecy capacity can be increased with the normalized signal power.

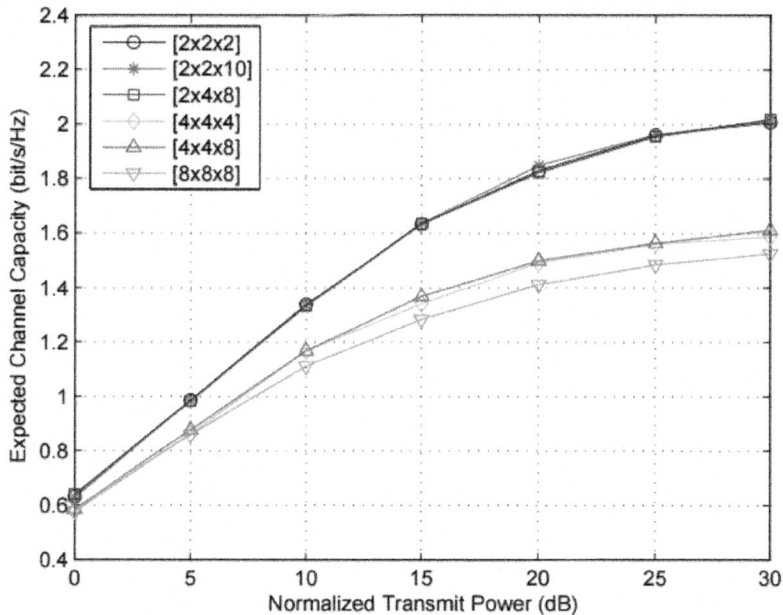

Figure 7.3: Comparison of the eavesdropper's channel capacities for different antenna configurations, with jamming noise signal, i.i.d. channel model.

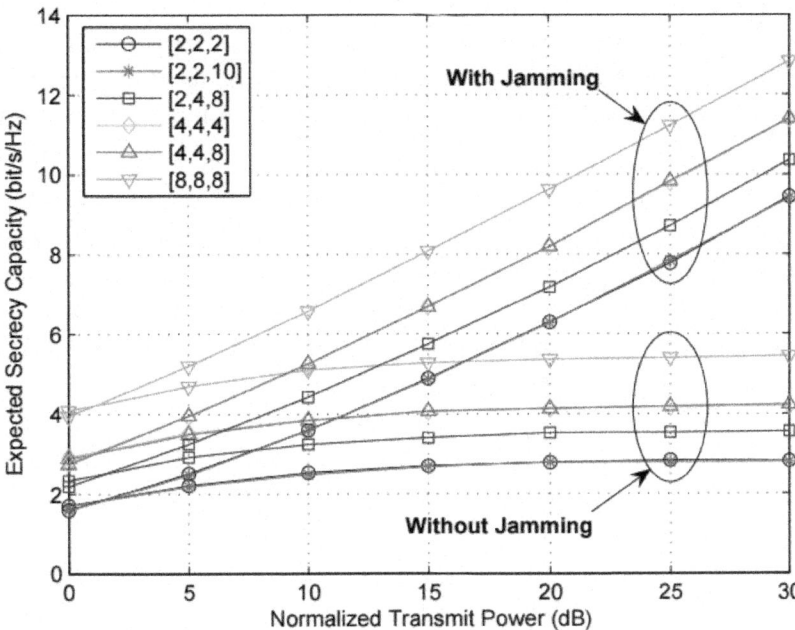

Figure 7.4: Comparison of simulation results with and without jamming noise signal, i.i.d. channel model.

Figure 7.4 also compares the simulation results with and without jamming noise signal. Our results and analysis indicate that the proposed approach significantly increases secrecy capacity. For example, the increase in secrecy capacity is about 7 bits/s/Hz when the normalized transmit power is 30 dB and the antennas are configured as [4, 4, 4]. Next, the performance of the proposed methods is evaluated over the 3GPP spatial channel model. In this scenario, the transmitter is assumed to be a fixed base station, while the desired receiver and the eavesdropper could be at any location within the cell's coverage area. The "urban macro" environment is deliberated, where the transmitter antenna height is 32 m, the receiver and eavesdropper antenna height is 1.5 m, and the carrier frequency is 1900 MHz. The path loss is calculated by $PL = 34.5 + 35 \log_{10}(d)$ (dB), where d is the distance between the transmitter and the receiver. Two scenarios are studied: Scenario 1 The transmitter is fixed in position; the intended receiver and the eavesdropper move around the transmitter, and they are d meters far away from the transmitter.

Scenario 2 The transmitter and intended receiver are fixed in position, and the distance between them is 100 meters; the eavesdropper moves between the transmitter and the intended receiver, and it is 35 to 80 meters away from the transmitter. It is worth noting that the path loss model limits the minimum distance between the transmitter and the receiver to 35 meters. It is assumed that the eavesdropper cannot be very close to the desired receiver, leading to uncorrelated channel matrix H and H_e. The thermal noise at the desired receiver and the eavesdropper is assumed to be -95 dBm, i.e $\sigma_n^2 = \sigma_{n_e}^2 = -95$ The total transmit power P is set as 1 W. As for the i.i.d. model, two power allocation schemes, with (equal power) and without jamming noise signal, are studied. Figure 7.5 shows the results of the ergodic secrecy capacities for scenario 1. The antenna configuration is [4, 4, 4]. For both cases with and without jamming noise signal, the main channel capacity decreases with increasing distance between the transmitter and the receiver.

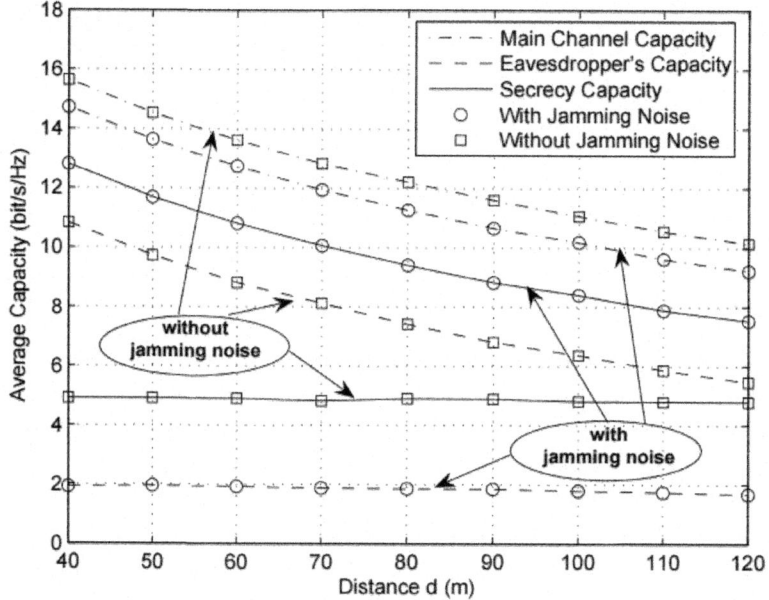

Figure 7.5: Simulation results of scenario 1 of 3GPP SCM channel, [4, 4, 4].

This result is intuitive because the received signal power becomes weaker when the receivers are further away from the transmitter. Since the transmit signal power is reduced to one-half of the total transmit power for the case with jamming noise signal, the main channel capacity is about 1 bit/s/Hz less than that of the case without jamming noise signal. However, the eavesdropper's channel capacities show different trends for the two cases, resulting in different secrecy capacity tendencies: with jamming noise signal, the eavesdropper's channel capacity is low (about 2 bits/s/Hz) and maintains nearly the same level whatever the distance is due to the irremovable jamming noise signal; without jamming noise signal, the eavesdropper's channel presents a relatively high capacity, even though the capacity decreases with increasing distance. Without maximum-ratio combining, the eavesdropper's channel capacity is lower than the main channel capacity. The secrecy capacity is calculated from the main channel and the eavesdropper's channel capacity.

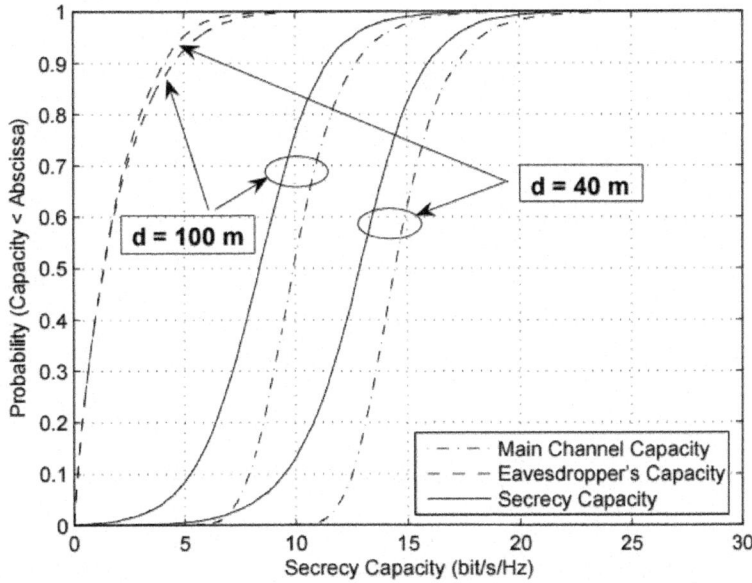

Figure 7.6: Cumulative distribution function of secrecy capacity, scenario 1 of 3GPP SCM channel with jamming noise signal, [4, 4, 4].

Without the jamming noise signal, the secrecy capacity is shown as 5 bits/s/Hz regardless of the distance d, while the secrecy capacity is much higher with jamming noise signal. The increase of the secrecy capacity by adding the jamming noise signal is about 8 bits/s/Hz when the distance d is 40 m and 3 bits/s/Hz at 120 m. The cumulative distribution function (CDF) of the secrecy capacities is shown in Figure 7.6 in order to better evaluate the secrecy capacity distribution. The secrecy capacity shows a wide range of about 18 bits/s/Hz. This is due to the randomly generated jamming noise signal and the random realization of the channels. Nonetheless, outage secrecy capacity remains high. For example, when the distance is 40 m, the 10% outage secrecy capacity is about 10 bits/s/Hz. This guarantees that secret wireless communications are feasible in such a system. In addition, the distributions of the eavesdropper's channel capacities for different eavesdropper locations are very similar. This result further corroborates that the eavesdropper cannot change system secrecy capacity by moving its position.

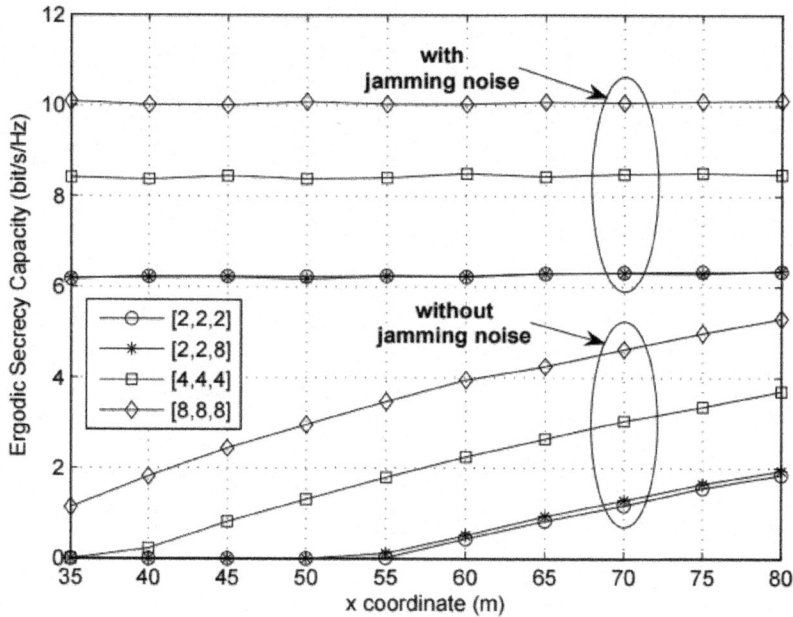

Figure 7.7: Simulation results of scenario 2 of 3GPP SCM channel.

Figure 7.7 shows the simulation results of the second scenario. Without the jamming noise signal, secrecy capacity decreases when the eavesdropper moves closer to the transmitter. If the eavesdropper moves close enough to the transmitter, its channel capacity may increase significantly so that secrecy capacity goes to zero. However, if the jamming noise signal is employed, the eavesdropper hardly acquires more information by moving closer to the transmitter. Moreover, comparing the results of the configurations [2, 2, 2] and [2, 2, 8], it can be seen that the eavesdropper cannot change secrecy capacity by employing more antenna elements.

CONCLUSION AND FUTURE WORK

In this dissertation, the new features of multi-antenna wireless systems are studied. Solutions to long-range channel prediction and destructive I/Q imbalance in MIMO-OFDM systems are proposed. The MIMO technique is incorporated into SESS system to mitigate error propagation and enhance system performance. As a novel idea, multi-antenna technology is also employed to provide physical layer security for wireless communications. In this chapter, we conclude this dissertation by reviewing the novel contributions and discuss possible future work.

Contributions

In this dissertation research, the following contributions have been made:

- A novel multi-block linear channel predictor is proposed to compensate for performance degradation caused by feedback delay in limited feedback precoded spatial multiplexing MIMO-OFDM systems. A publication related to this topic is found in [50].

- A novel virtual channel concept is proposed to compensate for the I/Q imbalances in a MIMO-OFDM wireless communication system over multipath fading channels. The studies have been published in [65, 70, 71].

- MIMO technology is incorporated into the SESS system to mitigate the error propagation and enhance system performance. This new scheme and its performance analysis have been published in [94–97].

- A method based on artificial jamming noise is proposed to secure wireless communications in the physical layer. The related publications are [109–111].

Suggested Future Work

Some of the topics presented in this dissertation can be further studied. First, performance of the physical layer security method proposed in Section 7 highly depends on the CSI at the transmitter. The channel estimation error and the feedback delay may cause the CSI at the transmitter to be inaccurate. The study of the influence of the inaccurate CSI at the transmitter on the system is an interesting topic. Second, the analysis of the MIMO-SESS system is based on BPSK modulation. However, higher order modulation is necessary in a practical system. How to implement higher order modulation, such as M-QAM, in a MIMO-SESS system is a challenging task. Third, the concept of distributed MIMO can be employed in a smart grid to enhance wireless sensor networks. By grouping one-antenna sensors into clusters and making the sensors in each cluster cooperate with each other to form a virtual antenna array, the overall transmission range and the link quality could be dramatically enhanced.

SUMMARY

This chapter introduces the fundamentals of multi-antenna wireless systems. The concepts and the system models will be extended in the following chapters

In this chapter, a novel multi-block linear channel predictor is proposed to compensate for performance degradation caused by feedback delay in the limited feedback precoded spatial multiplexing MIMO-OFDM systems [50]. The time-varying channel is modeled by an autoregressive (AR) model, whose coefficients are obtained by the linear minimum mean square error algorithm. Long-range prediction is enabled by using an iterative prediction of multiple blocks. Simulation results show that 1) performance degradation can be effectively compensated for by using the proposed multi-block predictor; 2) the higher-order AR model is capable of accurately predicting channel variation; and 3) the proposed predictionfeedback mechanism can be utilized with any size of codebooks.

In this chapter, a new virtual channel concept is introduced to analyze the I/Q imbalances in a MIMO-OFDM wireless communication system over multipath fading channels [65,70,71]. The input-output relation is derived in frequency domain, which incorporates the effect of I/Q imbalances with multipath fading channels. The integrated effect can be represented by the joint coefficients of the virtual channel. By using this approach, inaccurate estimation of the channel state information can be avoided at the diversity combining stage. The joint coefficients are estimated by using the proposed MMSE and LS estimators, and are used to compensate for the received signals. Simulation results show that the proposed approach can

effectively mitigate the TX and RX I/Q imbalances in MIMO-OFDM systems. Although only a two transmit antenna scheme is illustrated in this chapter, our proposed approach can be easily extended to any number of transmit antenna configurations by using orthogonal space-time block coding

This chapter presented a detailed discussion of the latest distributed MIMO technologies in cooperative wireless networks. This included the principle of cooperative wireless communications and the popular relay protocols and cooperative strategies. The analysis incorporated a performance comparison of outage probability and bit-error rates. The simulation results indicate that distributed MIMO can provide full user cooperation diversity, and the data rate in the cooperative networks can be significantly increased by using distributed space-time coding.

In this chapter, MIMO techniques are explored to enhance the self-encoded spread spectrum systems [94–97]. The closed-form BER expressions for the correlation detector and time diversity detector are derived, and a lower bound for the iterative detector is obtained. The veracity of the analysis has been confirmed with the numerical calculations based on the analytical expressions, which have demonstrated excellent agreement with the simulation results. The performance analysis has shown that MIMO-SESS offers almost 7 dB of gain at a 10^{-4} BER compared to a 2×2 MIMO PN-coded spread spectrum system. Furthermore, the system requires only about 3 dB SNR to achieve a BER of

10^{-4}. This demonstrates that MIMO-SESS can provide a very effective means to exploit both spatial and temporal diversities in order to achieve robust performance in wireless environments.

In this chapter, jamming noise signals are used to secure wireless communications in transmitbeamforming diversity systems [109–111]. Two jamming noise signal generation methods are proposed based on the null space of an equivalent channel of the system and the eigenvectors of the channel matrix. The proposed approach can be applied in systems with any

antenna configurations. Compared with other classical security methods, our approach can provide more degrees of freedom for jamming noise signals, significantly increasing the level of security. Given the dynamic nature of the wireless channel and the short amount of time available during which to successfully determine the employed jamming noise signal generation parameters, physical layer security can be ensured using our presented approach.

REFERENCES

1. A. J. Paulraj, D. A. Gore, R. U. Nabar, and H. B"olcskei, "An overview of MIMO communications – a key to gigabit wireless," Proceedings of The IEEE, vol. 92, no. 2, pp. 198–218, Feb. 2004.

2. T. S. Rappaport, Wireless Communications: Principles and Practice, 2nd ed. Prentice Hall, 2002.

3. A. J. Viterbi, CDMA: Principles of Spread Spectrum Communication. AddisonWesley, 1995.

4. Z. Tang, R. C. Cannizzaro, and G. L. P. Banelli, "Pilot-assisted time-varying channel estimation for OFDM systems," IEEE Transactions on Signal Processing, vol. 55, no. 5, pp. 2226–2238, May 2007.

5. R. E. Ziemer, R. L. Peterson, and D. E. Borth, Introduction to Spread Spectrum Communications. Prentice Hall, 1995.

6. A. R. S. Bahai, B. R. Saltzberg, and M. Ergen, Multi-carrier Digital Communications: Theory And Applications Of OFDM, 2nd ed. Springer, 2004.

7. A. Paulraj, R. Nabar, and D. Gore, Introduction to Space-Time Wireless Communications. Cambridge Univ. Press: Cambridge, U.K., 2003.

8. 3GPP TR 25.876. Multiple input multiple output in UTRA. V7.0.0. Mar. 2007.

9. 3GPP TR 25.814. Physical layer aspects for evolved universal terrestrial radio access (UTRA). V7.1.0. Sep. 2006.

10. S. Ma, Y. Yang, and H. Sharif, "Distributed MIMO technologies in cooperative wireless networks," to appear in IEEE Communications Magazine, May 2011.

11. S. Cui, A. J. Goldsmith, and A. Bahai, "Energy-efficiency of MIMO and cooperative MIMO techniques in sensor networks," IEEE Journal on Selected Areas in Communications, vol. 22, no. 6, pp. 1089–1098, Aug. 2004.

12. A. Ghosh, R. Ratasuk, B. Mondal, N. Mangalvedhe, and T. Thomas, "LTE-advanced: next-generation wireless broadband technology," IEEE Wireless Communications, vol. 17, no. 3, pp. 10–22, 2010.

13. S. Goel and R. Negi, "Guaranteeing secrecy using artificial noise," IEEE Trans. Wireless Comm., vol. 7, no. 6, pp. 2180–2189, Jun. 2008.

14. R. G. Gallager, Principles of Digital Communication. Cambridge University Press, 2008.

15. 3GPP. Spatial channel model for MIMO simulations. TR 25.996. V7.0.0. Jun. 2007.

16. Online]. Available: http://www.3gpp.org

17. D. S. Baum, J. Salo, G. D. Galdo, M. Milojevic, P. Kyosti, and J. Hansen, "An interim channel model for Beyond-3G systems," in Proc. IEEE VTC 2005 Spring, vol. 1, Stockholm, May 2005.

18. S. Ma, D. Duran, H. Sharif, and Y. Yang, "An extension of the 3GPP spatial channel model in outdoor-to-indoor environments," Berlin, Germany, Mar. 2009, pp. 1064– 1068.

19. T. K. Y. Lo, "Maximum ratio transmission," IEEE Transactions on Communications, vol. 47, no. 10, pp. 1458–1461, Oct. 1999.

20. F. B. Gross, Smart Antennas for Wireless Communications with Matlab. McGrawHill, 2005.

21. 20R. O. Schmidt, "Multiple emitter location and signal parameter estimation," IEEE Transactions on Antennas and Propagation, vol. 34, pp. 276–280, Mar. 1986.

22. M. Harradt and J. A. Nossek, "Unitary ESPRIT: How to obtain increased estimation accuracy with a reduced computational burden," IEEE Transactions on Signal Processing, vol. 43, no. 5, pp. 1232–1242, May 1995.

23. M. Jankiraman, Space-time codes and MIMO systems. Norwood, MA: Artech House, Inc, 2004.

24. S. Alamouti, "A simple transmit diversity technique for wireless communicaitons," IEEE Journal on Selected Areas in Communications,

vol. 16, no. 8, pp. 1451–1458, Oct. 1998.

25. V. Tarokh, H. Jafarkhani, and A. R. Calderbank, "Space-time block codes for wireless communications: Performance results," IEEE Transaction on Information Theory, vol. 45, no. 5, pp. 1456–1467, Jul. 1999.

26. M. Jang and L. Hanzo, "Multiuser MIMO-OFDM for next-generation wireless systems," Proceedings of the IEEE, vol. 95, no. 7, pp. 1430–1469, Jul. 2007.

27. D. J. Love, R. W. Heath, V. K. N. Lau, D. Gesbert, B. D. Rao, and M. Andrews, "An overview of limited feedback in wireless communication systems," IEEE Journal on Selected Areas in Communications, vol. 26, no. 8, pp. 1341–1365, Oct. 2008.

28. IEEE-SA Standards Board. IEEE 802.16 task group m. http://www.wirelessman.org/tgm.

29. A. Sendonaris, E. Erkip, and B. Aazhang, "User cooperation diversity c part I and II," IEEE Transactions on Communications, vol. 51, no. 11, pp. 1927–1948, Nov. 2003.

30. V. Stankovic, A. Host-Madsen, and Z. Xiong, "Cooperative diversity for wireless ad hoc networks," IEEE Signal Processing Magazine, vol. 23, no. 5, pp. 37–49, Sep. 2006.

31. L. Hanzo, M. Muinster, B. Choi, and T. Keller, OFDM and MC-CDMA for Broadband Multi-User Communications, WLANs and Broadcasting. Wiley, 2003.

32. E. T. Standard. Radio broadcast systems: digital audio broadcasting (DAB) to mobile, portable and fixed receivers. ETS 300 401. Feb. 1995.

33. IEEE Standard 802.11a Part 11: Wireless LAN Medium Access Control (MAC) and Physical Layer (PHY) Specifications: High Speed Physical Layer in the 5 GHz Band. 1999.

34. IEEE Standard 802.11g Part 11: Wireless LAN Medium Access Control (MAC) and Physical Layer (PHY) Specifications: Further Higher-Speed Physical Layer Extension in the 2.4 GHz Band. 2003.

35. IEEE Standard for Local and Metropolitan Area Networks, Part 16: Air Interface for Fixed Broadband Wireless Access Systems. IEEE Std 802.16-2004. Oct. 2004.

36. A. Tarighat and A. H. Sayed, "An optimum OFDM receiver exploiting cyclic prefix for improved data estimation," in Proc. IEEE International Conference on Acoustics, Speech, and Signal Processing, vol. 4, Hong Kong, Apr. 2003, pp. 217–220.

37. M. Jiang and L. Hanzo, "Multiuser MIMO-OFDM for next-generation

wireless systems," Proceedings of the IEEE, vol. 95, no. 7, pp. 1430–1469, Jul. 2007.

38. H. Sampath, S. Talwar, J. Tellado, V. Erceg, and A. J. Paulraj, "A fourth-generation MIMO-OFDM broadband wireless system: Design, performance, and field trial results," IEEE Commun. Mag., vol. 40, no. 9, pp. 143–149, Sep. 2002.

39. A. Scaglione, P. Stoica, S. Barbarossa, G. B. Giannakis, and H. Sampath, "Optimal designs for space-time linear precoders and decoders," IEEE Transactions on Signal Processing, vol. 50, no. 5, pp. 1051–1064, May 2002.

40. D. J. Love, R. W. Heath, W. Santipath, and M. L. Honig, "What is the value of limited feedback for MIMO channels?" IEEE Communications Magazine, vol. 42, no. 10, pp. 54–59, Oct. 2004.

41. D. J. Love and R. W. Heath, "Limited feedback precoding for spatial multiplexing systems," in Proc. of IEEE Globecom, San Francisco, CA, Dec. 2003, pp. 1–5.

42. C. Min, N. Chang, J. Cha, and J. Kang, "MIMO-OFDM downlink channel prediction for IEEE802.16e systems using kalman filter," in Proc. of IEEE Wireless Communications and Networking Conference (WCNC'07), Hong Kong, Mar. 2007, pp. 942–946.

43. A. S. Khrwat, B. S. Sharif, C. C. Tsimenidis, and S. Boussakta, "Channel prediction for precoded spatial multiplexing multiple-input multiple-output systems in timevarying fading channels," IET Signal Process, vol. 3, no. 6, pp. 459–466, 2009.

44. P. Zhu, L. Tang, Y. Wang, and X. You, "Quantized beamforming with channel prediction," IEEE Transactions on Wireless Communications, vol. 8, no. 11, pp. 5377–5382, Nov. 2009.

45. A. Duel-Hallen, "Fading channel prediction for mobile radio adaptive transmission systems," Proceedings of the IEEE, vol. 95, no. 12, pp. 2299–2313, Dec. 2007.

46. 3GPP R1-070460, "Further results on the impact of speed, feedback period, and subchannel bandwidth on the performance of downlink closed loop schemes for 4-Tx LTE," Sorrento, Italy, Jan. 2007.

47. A. Duel-Hallen, S. Hu, and H. Hallen, "Long range prediction of fading signals: Enabling adaptive transmission for mobile radio channels," IEEE Signal Processing Magazine, vol. 17, no. 3, pp. 62–75, May 2000.

48. S. Zhou and G. B. Giannakis, "How accurate channel prediciton needs to be for transmit-beamforming with adaptive modulation over Rayleigh

MIMO channels?" IEEE Transactions on Wireless Communications, vol. 3, no. 4, pp. 1285–1294, Jul. 2004.

49.   J. Y. Yun, S. Chung, J. choi, Y. Jang, and Y. H. Lee, "Predictive transmit beamforming for MIMO-OFDM in time-varying channels with limited feedback," in Proc. of International Wireless Communications and Mobile Computing Conference (IWCMC'07), Honolulu, Hawaii, Aug. 2007.

50.   Y. Ma, A. Leith, and R. Schober, "Predictive feedback for transmit beamforming with delayed feedback and channel estimation errors," in Proc. of IEEE International Conference on Communications (ICC'08), Beijing, China, May 2008.

51.   S. Ma, D. Duran, Y. Yang, and H. Sharif, "Performance enhancement in limited feedback precoded spatial multiplexing MIMO-OFDM systems by using multi-block channel prediction," Miami, Florida, USA, Dec. 2010.

52.   J. Choi, B. Mondal, and R. W. Heath, "Interpolation based unitary precoding for spatial multiplexing MIMO-OFDM with limited feedback," IEEE Transactions on Signal Processing, vol. 54, no. 12, pp. 4730–4740, 2006.

53.   S. Coleri, M. Ergen, A. Prui, and A. Bahai, "Channel estimation techniques based on pilot arrangement in OFDM systems," IEEE Transactions on Broadcasting, vol. 48, no. 3, pp. 223–229, 2002.

54.   A. H. Sayed, Adaptive Filters. Hoboken, New Jersey: Wiley-IEEE Press, 2008.

55.   D. J. Love and R. W. Heath, "Limited feedback unitary precoding for spatial multiplexing systems," IEEE Transactions on Information Theory, vol. 51, no. 8, pp. 2967–2976, Aug. 2005.

56.   D. J. Love. http://cobweb.ecn.purdue.edu/ djlove/grass.html.

57.   W. C. Jakes, Microwave Mobile Communications. Hoboken, New Jersey: WileyIEEE, 1974.

58.   Recommendation ITU-R M.1225, "Guidelines for the evaluation of radio transmission technologies for IMT-2000," 1997.

59.   C. L. Liu, "Impacts of I/Q imbalance on QPSK-OFDM-QAM detection," IEEE Trans. Consum. Electron, vol. 44, no. 3, pp. 984–989, Aug. 1998.

60.   M. Valkama, M. Renfors, and V. Koivunen, "Compensation of frequency-selective I/Q imbalances in wideband receivers: models and algorithms," in Proc. of the 3rd IEEE Workshop on Signal Processing Advances in Wireless Communications (SPAWC '01), Taiwan, China, Mar. 2001, pp.

42–45.

61. A. Tarighat, R. Bagheri, and A. H. Sayed, "Compensation schemes and performance analysis of IQ imbalances in OFDM receivers," IEEE Transactions on Signal Processing, vol. 53, no. 8, pp. 3257–3268, Aug. 2005.

62. J.-K. Hwang, W.-M. Chen, Y.-L. Chiu, S.-J. Lee, and J.-L. Lin, "Adaptive baseband compensation for I/Q imbalance and time-varying channel in OFDM system," in Intelligent Signal Processing and Communications (ISPACS '06), Japan, Dec. 2006, pp. 638 – 641.

63. J. Tubbax, B. Come, L. van der Perre, S. Donnay, M. Engles, and C. Desset, "Joint compensation of IQ imbalance and phase noise," IEEE Semiannual on Vehicular Technology Conference, vol. 3, pp. 1605–1609, Apr. 2003.

64. D. Tandur and M. Moonen, "Joint compensation of OFDM frequency-selective transmitter and receiver IQ imbalance," Eurasip Journal on Wireless Communications and Networking, vol. 2007, no. 68563, pp. 1–10, May 2007.

65. L. Anttila, M. Valkama, and M. Renfors, "Frequency-selective I/Q mismatch calibration of wideband direct-conversion transmitters," IEEE Transactions on Circuits and Systems II, vol. 55, no. 4, pp. 359–363, Apr. 2008.

66. S. Ma, D. Duran, and Y. Yang, "Estimation and compensation of frequency-dependent I/Q imbalances in OFDM systems over fading channels," in To appear in proceedings of the 5th International Conference on Wireless Communications, Networking and Mobile Computing (WiCOM), Beijing, China, Sep. 2009.

67. A. Tarighat, W. M. Younis, and A. H. Sayed, "Adaptive MIMO OFDM receivers: Implementation impairments and complexity issues," in Proc. IFAC workshop on Adaptation and Learning in Control and Signal Processing, Yokohama, Japan, Aug. 2004.

68. Y. Zou, M. Valkama, and M. Renfors, "Digital compensation of I/Q imbalance effects in space-time coded transmit diversity systems," IEEE Transactions on Signal Processing, vol. 56, no. 6, pp. 2496–2508, Jun. 2008.

69. ——, "Analysis and compensation of transmitter and receiver I/Q imbalances in space-time coded multiantenna OFDM systems," Eurasip Journal on Wireless Communications and Networking, no. 391025, pp. 1–10, Jan. 2008.

70. D. Tandur and M. Moonen, "Compensation of RF impairments in MIMO

OFDM systems," in Proc. of IEEE International Conference on Acoustics, Speech and Signal Processing (ICASSP), Las Vegas, Nevada, Apr. 2008, pp. 3097–3100.

71. S. Ma, D. Duran, H. Sharif, and Y. Yang, "An adaptive approach to estimation and compensation of frequency-dependent I/Q imbalances in MIMO-OFDM systems," Honolulu, Hawaii, USA, Nov. 2009.

72. S. Ma, D. D. Duran, Y. Yang, and H. Sharif, "A novel approach to analyzing MIMOOFDM systems with I/Q imbalances in fading channels," International Journal of Communications Network and System Sciences, vol. 3, no. 9, pp. 711–721, 2010.

73. A. H. Sayed, Adaptive Filters. Wiley-IEEE Press, 2008.

74. D. Manolakis, V. K. Ingle, and S. M. Kogon, Statistical and Adaptive Signal Processing: Spectral Estimation, Signal Modeling, Adaptive Filtering and Array Processing. Norwood, MA: Artech House Publishers, 2005.

75. J. G. Proakis, Digital Communications, 4th ed. New York, NY: McGraw-Hill, 2000.

76. R. L. Peterson, R. E. Ziemer, and D. E. Borth, Introduction to Spread-Spectrum Communications, 3rd ed. Prentice Hall, 1995.

77. X. Wang, H.-C. Wu, S. Y. Chang, Y. Wu, and J.-Y. Chouinard, "Efficient non-pilotaided channel length estimation for digital broadcasting receivers," IEEE Transactions on Broadcasting, vol. 55, no. 3, pp. 633–641, Sep. 2009.

78. P. Liu, Z. Tao, Z. Lin, E. Erkip, and S. Panwar, "Cooperative wireless communications: a cross-layer approach," IEEE Wireless Communications, vol. 13, no. 4, pp. 84–92, Aug. 2006.

79. T. Issariyakul and V. Krishnamurthy, "Amplify-and-forward cooperative diversity wireless networks: model, analysis, and monotonicity properties," IEEE/ACM Transactions on Networking, vol. 17, no. 1, pp. 225–238, Feb. 2009.

80. K.-S. Hwang, Y.-C. Ko, and M.-S. Alouini, "Outage probability of cooperative diversity systems with opportunistic relaying based on decode-and-forward," IEEE Transactions on Wireless Communications, vol. 7, no. 12, pp. 5100–5107, Dec. 2008.

81. J. N. Laneman, D. N. C. Tse, and G. W. Wornell, "Cooperative diversity in wireless networks: efcficient protocols and outage behavior," IEEE Transactions on Information Theory, vol. 50, no. 12, pp. 3062–3080, Dec. 2004.

82. Y. Li and B. Vucetic, "On the performance of a simple adaptive relaying protocol for wireless relay networks," in VTC-Spring 2008, Singapore, May 2008.

83. Y. Jing and B. Hassibi, "Distributed space-time coding in wireless relay networks," IEEE Transactions on Wireless Communications, vol. 5, no. 12, pp. 3524–3536, Dec. 2006.

84. R. U. Nabar, H. Bolcskei, and F. W. Kneubuhler, "Fading relay channels: performance limits and space-time signal design," IEEE Journal on Selected Areas in Communications, vol. 22, no. 6, pp. 1099–1109, Aug. 2004.

85. J. Yuan, Z. Chen, Y. Li, and L. Chu, "Distributed space-time trellis codes for a cooperative system," IEEE Transactions on Wireless Communications, vol. 8, no. 10, pp. 4897–4905, Oct. 2009.

86. J. N. Laneman and G. W. Wornell, "Distributed space-time-coded protocols for exploiting cooperative diversity in wireless networks," IEEE Transactions on Information Theory, vol. 49, no. 10, pp. 2415–2425, Oct. 2003.

87. Z. Yi and I.-M. Kim, "Single-symbol ML decodable distributed STBCs for cooperative networks," IEEE Transactions on Information Theory, vol. 53, no. 8, pp. 2977–2985, Aug. 2007.

88. L. Nguyen, "Self-encoded spread spectrum communications," in Proc. of Military Communications Conference Proceedings (MILCOM 1999), vol. 1, Oct. 1999, pp. 182–186.

89. ——, "Self-encoded spread spectrum and multiple access communications," in Proc. of IEEE Sixth International Symposium on Spread Spectrum Techniques and Applications, vol. 2, Parsippany, NJ, Sep. 2000, pp. 394–398.

90. W. Jang and L. Nguyen, "Capacity analysis of m-user self-encoded multiple access system in AWGN channels," in Proc. of IEEE Sixth International Symposium on Spread Spectrum Techniques and Applications, vol. 1, Sep. 2000, pp. 216–220.

91. Y. Kong, L. Nguyen, and W. Jang, "Self-encoded spread spectrum modulation with differential encoding," in Proc. of IEEE Seventh International Symposium on Spread Spectrum Techniques and Applications, vol. 2, Sep. 2002, pp. 471–474.

92. W. M. Jang, L. Nguyen, and M. Hempel, "Self-encoded spread spectrum and turbo coding," Journal of communication and networks, vol. 6, no. 1, pp. 9–18, Mar. 2004.

93. Y. S. Kim, W. Jang, Y. Kong, and L. Nguyen, "Chip-interleaved self-encoded multiple access with iterative detection in fading channels," Journal of Communication and Networks, vol. 9, no. 1, pp. 50–55, Mar. 2007.

94. K. Hua, L. Nguyen, and W. Jang, "Synchronisation of self-encoded spread spectrum system," Electronics Letters, vol. 44, no. 12, pp. 749–751, Jun. 2008.

95. S. Ma, L. Nguyen, W. M. Jang, and Y. Yang, "Multiple-input multiple-output selfencoded spread spectrum system with iterative detection," in Proc. IEEE International Conference on Communications (ICC'10), Cape Town, South Africa, May 2010.

96. ——, "Performance enhancement in MIMO self-encoded spread spectrum systems by using multiple codes," in Proc. of the 33rd IEEE Sarnoff Symposium, Princeton, NJ, Apr. 2010.

97. ——, "Bit-error rates of MIMO-SESS over rayleigh fading channels," Taichung, Taiwan, Oct. 2010.

98. ——, "MIMO self-encoded spread spectrum with iterative detection over rayleigh fading channels," Hindawi Journal of Electrical and Computer Engineering, Volume 2010, Article ID 492079, 2010.

99. J. G. Proakis, Digital Communications, 4th ed. New York: McGraw-Hill, 1989.

100. P. G. Hoel, S. C. Port, and C. J. Stone, Introduction to Probability Theory. Boston: Houghton Mifflin, 1971.

101. L. Rade and B. Westergren, Mathematics handbook for science and engineering. Cambridge, MA, U.S.A: Birkhauser Boston, Inc., 1995.

102. M. Abramowitz and I. A. Stegun, Handbook of Mathematical Functions With Formulas, Graphs, and Mathematical Tables. New York: Dover Publications Inc., 1965.

103. N. Sklavos and X. Zhang, Wireless Security and Cryptography: Specifications and Implementations. Boca Raton, FL: CRC Press, 2007.

104. L. Dong, Z. Han, A. Petropulu, and H. V. Poor, "Secure wireless communications via cooperation," in Proc. Allerton Conference on Communication, Control, and Computing, Monticello, IL, USA, Sep. 2008.

105. ——, "Amplify-and-forward based cooperation for secure wireless communications," in Proc. of IEEE Intl Conf. Acoust. Speech Signal Proc., Taipei, Taiwan, Apr. 2009.

106. L. Lai and H. E. Gamal, "The relay-eavesdropper channel: Cooperation

for secrecy," IEEE Trans. Inform. Theory, vol. 54, no. 9, pp. 4005–4019, Sep. 2008.

107. A. O. Hero, "Secure space-time communication," IEEE Transactions on Information Theory, vol. 49, no. 12, pp. 3235–3249, Dec. 2003.

108. X. Li, J. Hwu, and E. P. Ratazzi, "Using antenna array redundancy and channel diversity for secure wireless transmissions," Journal of Communications, vol. 2, no. 3, pp. 224–32, May 2007.

109. R. Negi and S. Goelm, "Secret communication using artificial noise," in Proc. of IEEE Vehicular Tech. Conf, vol. 3, Dallas TX, Sep. 2005, pp. 1906–1910.

110. S. Ma, M. Hempel, Y. Yang, and H. Sharif, "A new approach to null space-based noise signal gneration for secure wireless communications in transmit-receive diversity systems," Beijing, China, Jun. 2010.

111. ——, "A novel approach to secure wireless communications using randomized eigenvector-based jamming signals," Caen,France, Jun. 2010, pp. 1172–1176.

112. ——, "Securing wireless communications in transmit beamforming diversity systems by precoding jamming signals," Security and Communication Networks, 2011, major revision.

113. P. A. Dighe and R. K. M. S. S. Jamuar, "Analysis of transmit-receive diversity in Rayleigh fading," IEEE Transactions on Communications, vol. 51, no. 8, pp. 694– 703, Apr. 2003.

114. D. S. Bernstein, Matrix Mathematics: Theory, Facts, and Formulas with Application to Linear System Theory. Princeton, New Jersey: Princeton University Press, 2005.

115. A. L. Swindlehurst, "Fixed SINR solutions for the MIMO wiretap channel," in Proc. of IEEE International Conference on Acoustics, Speech and Signal Processing, Taipei, Taiwan, Apr. 2009, pp. 2437–2440.

116. P. Almers, E. Bonek, A. Burr, and et al., "Survey of channel and radio propagation models for wireless MIMO systems," EURASIP Journal on Wireless Communications and Networking, vol. 2007, Article ID 19070, 19 pages, 2007.

117. M. Kang and M.-S. Alouini, "Largest eigenvalue of complex wishart matrices and performance analysis of MIMO MRC systems," IEEE Journal on Selected Areas in Communications, vol. 21, no. 3, pp. 418–426, Apr. 2003.

Chapter 6

PERFORMANCE ANALYSIS FOR COOPERATIVE COMMUNICATION SYSTEM WITH QC-LDPC CODES CONSTRUCTED WITH INTEGER SEQUENCES

Yan Zhang[1,2] Feng-fan Yang,[1] and Weijun Song[2]

[1]College of Electronic and Information Engineering, Nanjing University of Aeronautics and Astronautics, Nanjing, Jiangsu 210016, China

[2]Nanjing College of Information Technology, Nanjing, Jiangsu 210046, China

ABSTRACT

This paper presents four different integer sequences to construct quasi-cyclic low-density parity-check (QC-LDPC) codes with mathematical theory. The paper introduces the procedure of the coding principle and coding. Four different integer sequences constructing QC-LDPC code are compared with LDPC codes by using PEG algorithm, array codes, and the Mackey codes, respectively. Then, the integer sequence QC-LDPC codes are used in coded cooperative communication. Simulation results show that the integer sequence constructed QC-LDPC codes are effective, and overall performance is better than that of other types of LDPC codes in the coded cooperative communication. The performance of Dayan integer sequence constructed QC-LDPC is the most excellent performance.

INTRODUCTION

In 1979, Cover and El Gamal reached substantial progress in cooperative communication technology [1, 2]. Amplify-and-forward (AF) [3], decode-and-forward (DF) [4], and coded cooperation (CC) [5] are three commonly user cooperation modes. In coded cooperation, the relay user does not send signals of cooperative users repeatedly. With features of coding, this method sends different parts of codes, respectively. Due to the fact that different parts in codes have innate correlated characteristics, if the base station receives different parts of codes by mutually opposed channels, that is, if the information of each code is sent by two channels, the transmission diversity is realized and the coding gain is acquired. As the application of channel, coded cooperation mode is superior to the previous methods.

According to the characteristics of low-density parity-check (LDPC) codes coded cooperative system, many scholars had achieved fruitful research on designing LDPC codeword construction in the coded cooperative system [6–8]. In single-source single relay scenario, the literature [9] proposed that using computer search structure could optimize LDPC code sets. In the multisource coded cooperative system, the literature [10] introduced that LDPC codes are optimized by distribution function without taking into account the complexity of encoding and decoding, memory, and other resource consumption. This optimized distribution function is long enough to clearly show the advantages of its performance. As the length of LDPC is large, the complexity of coding and decoding is too high and the memory consumption is larger in hardware implementation.

Quasi-cyclic low-density parity-check (QC-LDPC) codes coded cooperative system is an important subclass of LDPC. Compared with randomly constructed LDPC, QC-LDPC can achieve a simple shift register and linear coding complexity, but the parameters of the code are not flexible enough, and they cannot meet the needs of practical application.

In point-to-point (noncooperative) system, the literature [11] proposed various design methods of QC-LDPC; for example, the literature [12] designed the QC-LDPC with the ring length of at least 8 through the greatest common divisor inequality when the subblock length is greater than a lower bound. The literature [13] designed the QC-LDPC in coded cooperative system by finite field (Galois Field, GF) of the multiplicative inverses, but the check matrix of QC-LDPC is limited to the lower triangular structure.

For this problem, this paper introduces the QC-LDPC constructed by four different integer sequences, compared with traditional QC-LDPC; the class QC-LDPC can save more space used with linear encoding complexity and quasi-cyclic structure. Through simulating, the results show that the QC-LDPC constructed by integer sequences is without the girth-4 length and is excellent in performance.

The rest of the paper is organized as follows. Section 2 introduces fundamentals of general coded relay cooperation. Section 3 presents QC-LDPC constructed by four different integer sequences. The simulation results are shown in Section 4. Finally, the conclusions are reached in Section 5.

FUNDAMENTALS OF GENERAL CODED RELAY COOPERATION

In this paper, 3-nodes relay channel model is a common model to describe the cooperative communication system. It is shown in Figure 1. It is comprised of three fundamental communication units such as the source S, the relay R, and the destination D. All these terminals are assumed to have a single antenna and they communicate with each other in half-duplex mode.

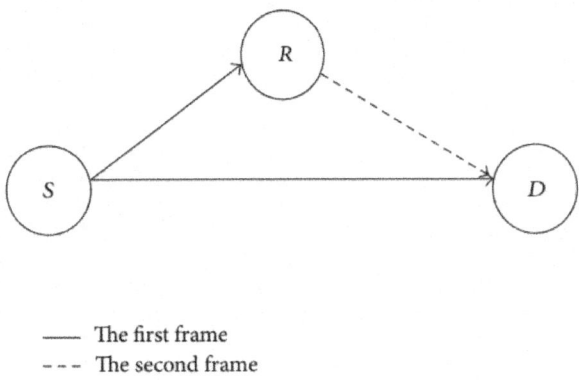

——— The first frame
- - - The second frame

Figure 1: Coded cooperative communication model.

The transmission of information data is divided into two frames. In the first frame, the source node (S) encodes information using the code c_1 and sends its information to the relay node (R) and the destination node (D) simultaneously. During the first frame the received signal y_{SR} over a Rayleigh fading channel at the relay is given as

$$y_{SR} = h_{SR}x_A + n_{SR},$$

(1)

where n_{SR} is complex additive Gaussian noise with each component as a zero-mean complex Gaussian variable with independent real and imaginary parts of equal variance $N_0/2$. h_{SR} is also complex Gaussian variables with zero-mean 1/2- variance for the real and imaginary parts of each component. Similarly, the received signal y_{SD} over a Raleigh fading channel at the destination during the first frame is described as

$$y_{SD} = h_{SD}x_A + n_{SD},$$

(2)

where n_{SD} and h_{SD} are corresponding to Gaussian noise and fading, respectively.

In the second frame, if the relay node can decode the information successfully, then using BPSK modulation the relay sends the coded information c_2 to the destination. Otherwise, the relay node does not work. $h_{i,j}$ and $n_{i,j}$ are were taken representatively as channel gain and additive noise of each channel. Each channel is mutually independent. The received signal y_{RD} over a Rayleigh fading channel at the destination during the second frame is given as

$$y_{RD} = h_{RD}x_R + n_{RD}, \tag{3}$$

where n_{RD} and h_{RD} are associated with Gaussian noise and fading, respectively. x_R represents the information sent by relay node. Finally, the overall received signal y at the destination is given as

$$y = |y_{SD} \mid y_{RD}|, \tag{4}$$

where $|y_{SD} \mid y_{RD}|$ is the series concatenation of ySD and yRD received in the first frame and second frame, respectively. The signal y is jointly decoded to recover the transmitted information at the source. In the information transmitting procedure, two different L_{DPC} codes $C_1(K_1, N_1)$ and $C_2(K_2, N_2)$ were used to represent the source node S and relay node R. Ki and Ni, respectively, represent the dimension and length of codes. The length of coded message sequence of source node S is K_1 and the check matrix is H_1. C_1 and C_2 can be designed for simple systematic codes. The code of C_1 is (I, P_1), where $P1$ is the parity bits with the length of $N_1 - K_1$. If relay node succeeds to code after receiving information from C_1, relay node will generate check matrix H_2 by adding new parity bits P_2 with the length of $N_2 - K_2$. The destination node D would receive information from source node S and relay node R. After the combination, the code (I, P_1, P_2) is obtained. The joint coding is conducted by the check matrix H comprised of $H1$ and $H2$ to acquire source information. The entire code of C_1 and parity bit of C_2 respectively, reach the destination node through S-D channel and R-D channel, which can be obtained by the check relationship of codes. Consider

$$H_1C_1 = 0, \qquad H_2C_2 = 0. \tag{5}$$

As can be seen from (5), the entire code of coded cooperative communication system can meet the following check relationship:

$$HC = 0. \tag{6}$$

QC-LDPC Constructed by Four Different Integer Sequences

The check matrix H of regular (J, L) QC-LDPC codes is expressed by $J \times L$ array of $p \times p$ cyclic permutation matrix. Its length is $n = LP$, which can be defined as

$$H = \begin{bmatrix} I(p_{0,0}) & I(p_{0,1}) & \cdots & I(p_{0,L-1}) \\ I(p_{1,0}) & I(p_{1,1}) & \cdots & I(p_{1,L-1}) \\ \vdots & \vdots & \vdots & \vdots \\ I(p_{J-1,0}) & I(p_{J-1,1}) & \cdots & I(p_{J-1,L-1}) \end{bmatrix},$$

(7)

where $0 \le j \le J - 1$; $0 \le l \le L - 1$, $I(p_{j,l})$) is the $p \times p$ cyclic permutation matrix, and each row and line of the matrix contain only one 1. The last line of the matrix is ring shifted right by one element so as to acquire the line one of the matrix; each column of the matrix is ring shifted downward by one element to obtain the next column of the matrix; the last column of the matrix is ring shifted downward by one element to obtain the first column of the matrix. For example, matrix P is the 5×5 cyclic permutation matrix:

$$P = \begin{bmatrix} 1 & 0 & 0 & 0 & 0 \\ 0 & 1 & 0 & 0 & 0 \\ 0 & 0 & 1 & 0 & 0 \\ 0 & 0 & 0 & 1 & 0 \\ 0 & 0 & 0 & 0 & 1 \end{bmatrix}.$$

(8)

$p_{i,}$ represents the shift value of this cyclic permutation matrix; that is, 1 of rth of this matrix was in the column $(r+p_{j,l})$ mod p, $0 \le r \le p-1$, while the other columns of this row are 0. (0)represents the unit matrix with the size of $p \times p$, and $I(p_{j,l})$ shows the matrix after the ring shift right of $I(0)$ by $p_{i,j}$. Assume the shift matrix B of check matrix H formed by $(p_{i,j})$ with the right shifting number pi,j as follows:

$$B = \begin{bmatrix} p_{0,0} & p_{0,1} & \cdots & p_{0,L-1} \\ p_{1,0} & p_{1,1} & \cdots & p_{1,L-1} \\ \vdots & \vdots & \vdots & \vdots \\ p_{J-1,0} & p_{J-1,1} & \cdots & p_{J-1,L-1} \end{bmatrix}.$$

(9)

The above QC-LDPC codes check matrix contains a cycle, which could be expressed by the sequence of $p \times p$ cyclic permutation matrix; the cycle of QC-LDPC codes with the length of $2i$ can be expressed as $(j_0, l_0), (j_1, l_1), \cdots$, $(j_k, l_k), \ldots, (j_{i-1}, l_{i-1}), (j_0, l_0)$, where (j_i, l_i) is $I(p_{j,k})$ of H matrix at row j_k and column l_k. To accurately express a cycle, $j_k \ne j_{k+1}$ and $l_k \ne l_{k+1}$ should be met. The literature [14] proposed that the necessary and sufficient condition for a code without girth 4 is

$$\sum_{k=0}^{i-1} \left(p_{j_k,l_k} - p_{j_{k+1},l_k} \right) \ne 0 \bmod p,$$

(10)

where $j_i = j_0$; $j_k \ne j_{k+1}$; $l_k \ne l_{k+1}$.

QC-LDPC Codes Constructed by Fibonacci Array

Definition 1 (see [15]). For an array (n), n is a nonnegative integer. If the array meets

$$f(n) = f(n-1) + f(n-2) \quad (n \geq 2),$$

$$f(0) = 1, \qquad f(1) = 1. \tag{11}$$

Then, this array is known as Fibonacci array, and its element is Fibonacci value [16]. For example, a typical Fibonacci array is as follows: 1, 1, 2, 3, 5, 8, 13, 21, 34, 55, According to the check matrix in (7), the shift value of cyclic permutation matrix at row j column l is

$$p_{j,l} = f(j+l+2) + j. \tag{12}$$

Literature [17] verified that girth 4 of H matrix designed by Fibonacci array can be removed. According to the above rules, it can be assumed that in the design of QC-LDPC of Fibonacci array of $(10, 1, 2)$, each item of shift matrix was

$$p_{0,0} = f(0+0+2)+0 = f(2) = 2,$$

$$p_{0,1} = f(0+1+2)+0 = f(3) = 3. \tag{13}$$

Shift matrix B can be written as follows: $B = [2\ 3]$. p can flexibly select any value greater than the maximal Fibonacci value and p is identified as shift. The maximal value in the matrix is +2 and $p=5$, and according to the definition of cyclic permutation, one has

$$H(10,1,2) = [I(p_{0,0})\ I(p_{0,1})]$$

$$= \begin{bmatrix} 0 & 0 & 1 & 0 & 0 & 0 & 0 & 0 & 1 & 0 \\ 0 & 0 & 0 & 1 & 0 & 0 & 0 & 0 & 0 & 1 \\ 0 & 0 & 0 & 0 & 1 & 1 & 0 & 0 & 0 & 0 \\ 1 & 0 & 0 & 0 & 0 & 0 & 1 & 0 & 0 & 0 \\ 0 & 1 & 0 & 0 & 0 & 0 & 0 & 1 & 0 & 0 \end{bmatrix}. \tag{14}$$

The Design of QC-LDPC Codes with Dayan

Integer Sequence Definition 2 (see [18]). For an array (n), when n is a nonnegative integer and an odd number, one has

$$f(n) = \frac{(n \times n - 1)}{2}, \tag{15}$$

and when n is an even number, one has

$$f(n) = \frac{(n \times n)}{2}. \tag{26}$$

The sequence that satisfied the recurrence relation is known as Dayan integer sequence, and its element is the item of Dayan integer sequence. A typical Dayan integer sequence is as follows: 0, 2, 4, 8, 12, 18, 24, 32, 40, 50, ...

In a Dayan integer sequence, the new sequence formed by Dayan integer sequence items of numbers with the same item number difference shows monotone increasing. Besides, QCLDPC constructed by Dayan integer sequence can remove girth 4, and the verification process is shown in Literature [19]. In accordance with check matrix in (7), the shift equation of cyclic permutation matrix at row j and line l is

$$p_{j,l} = f(j + l + l) + j.$$

$$(17)$$

For example, to construct a check matrix (30, 1, 2), each item of corresponding shift matrix $B(1, 2)$ is as follows:

$$p_{0,0} = f(1 + 1 + 1) + 1 = f(3) + 1$$

$$= \frac{(3 \times 3 - 1)}{2} + 1 = 4 + 1 = 5,$$

$$p_{0,1} = f(1 + 2 + 2) + 1 = f(5) + 1$$

$$= \frac{(5 \times 5 - 1)}{2} + 1 = 12 + 1 = 13.$$

$$(18)$$

The form to the shift matrix is $(1, 2) = [5 \ 13]$.

QC-LDPC Constructed by Differential Sequence

Definition 3 (see [20]). Let $h_0, h_1, h_2,...,h_n,....$ As a digital sequence, thus, we define a new sequence $\Delta h_0, \Delta h_1, \Delta h_2, . . . , \Delta h_n,...$, known as first-order difference sequence, where

$$\Delta h_n = h_{n+1} - h_n, \quad (n \geq 0).$$

$$(19)$$

In (20), each item of the difference sequence is the difference value of adjacent items in sequence (19). After the construction of difference sequence (20), the second-order difference sequence can be easily obtained:

$$\Delta^2 h_0, \Delta^2 h_1, \Delta^2 h_2,..., \Delta^2 h_n,....$$

$$(20)$$

Hence, one has

$$\Delta^2 h_n = \Delta\left(\Delta h_n\right) = \Delta h_{n+1} - \Delta h_n$$

$$= \left(h_{n+2} - h_{n+1}\right) - \left(h_{n+1} - h_n\right)$$

$$= h_{n+2} - 2h_{n+1} + h_n \quad (n \geq 0),$$

(21)

Where

$$\Delta^m h_n = \Delta(\Delta^{m-1} h_n).$$

Hence, m-order differential sequence is the first-order differential sequence of $m-1$-order differential sequence. We define the zero-order differential sequence as the digital sequence, as $\Delta^0 h_n = h_n$, $(n \geq 0)$. Assume $(3, k)$ QC-LDPC codes, known as DS-LDPC codes. The check matrix of this kind of codes consisted of zero-, first-, and second-order differential sequences generated by the quadratic polynomial. The design process of check matrix is shown in the literature [21].

QC-LDPC Constructed by Hoey Sequence

Definition 4 (see [22]) In a Hoey sequence, h_i shows monotone increasing and meets the minimal nonnegative integer sequence with the sum of the two being different.

For $i = 0, 1, \ldots, 49$, the top 50 elements of h_i are

0, 1, 3, 7, 12, 20, 30, 44, 65, 80, 96, 122, 147, 181, 203, 251, 289, 360, 400, 474, 564, 661, 774, 821, 915, 969, 1015, 1158, 1311, 1394, 1522, 1571, 1820, 2028, 2253, 2509, 2779, 2924, 3154, 3353, 3590, 3796, 3997, 4296, 4432, 4778, 4850.   (22)

Let the form of shift matrix B of QC-LDPC codes be as follows:

$$B = \begin{bmatrix} h_0 & h_1 & \cdots & h_{L-1} \\ 0 & 0 & \cdots & 0 \end{bmatrix}.$$

(23)

Elements in the first row of shift matrix B are derived from top L elements in Hoey sequence and each element in the second row is 0. Shift matrix B is a $2 \times L$ matrix, and the corresponding check matrix H is a $2 \times L$ array formed by $p \times p$ cyclic permutation matrix, where $p > \max_{0 \leq i \leq L-1}\{h_i\}$ and the null space of H presents a $(2, L)$ regular QC-LDPC code with the length of pL and the code rate of at least $(pL - 2p + 1)/pL$. The check matrix constructed by Hoey sequence does not include girth 4, which was verified by the literature [23]. With the example of the minimum matrix, due to the fixed pattern of base matrix of Hoey sequence, we obtain (23). Thus, the minimal matrix is $2*4$ matrix. Consider

$$B = \begin{bmatrix} 0 & 1 & 3 & 7 \\ 0 & 0 & 0 & 0 \end{bmatrix}.$$

(24)

SIMULATION RESULTS

In this section, we select the (3, 6) Fibonacci array with code rate $R = 1/2$ and a length of 354 to construct QC-LDPC codes ($p = 59$), as well as the (5, 10) Fibonacci with the length of 9970 to construct QC-LDPC codes ($p = 997$), which were compared with array LDPC codes [24] with the same length in the Gaussian white noise channel. The maximum number of iterations is 50. As can be seen from Figure 2, performance of QC-LDPC codes constructed by Fibonacci with the length of 354 is better than that of array LDPC codes. When BER = 4×10^{-6}, the performance of Fibonacci-constructed QC-LDPC codes is improved by about 1.6 dB compared to the performance of array LDPC codes. The QC-LDPC codes constructed by Fibonacci with the length of 9970 are superior to array LDPC codes with the same parameters.

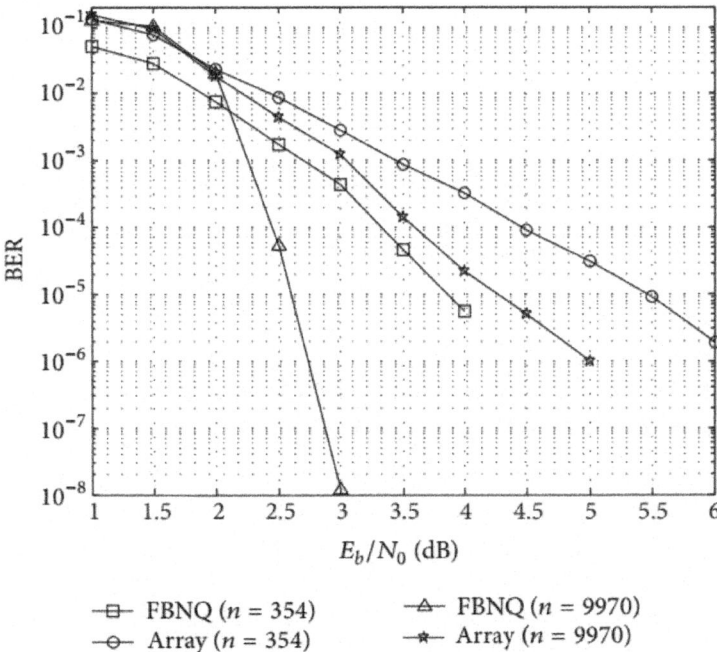

-□- FBNQ ($n = 354$)   -△- FBNQ ($n = 9970$)
-○- Array ($n = 354$)   -*- Array ($n = 9970$)

Figure 2: The curves of Fibonacci-constructed QC-LDPC and QC-LDPC codes BER (bit error rate) constructed by array sequence.

For the simulation of QC-LDPC code constructed by differential sequence, a quadratic polynomial $hn = 2n^2 +3n+1$ is found by the search algorithm. As $\max_{0\leq i\leq 5, j\in\{0,1,2\}}\{\Delta^j h_i\} = 66$, $p = 67$, and a 67×67 cyclic permutation matrix formed 3×6 array H. H matrix is a 201×402 matrix. The column weight is 3 and the row weight is 6. In order to facilitate the comparison, the PEG algorithm [25] is used to construct the pseudorandom (402, 201) PEG codes. Besides, the random (402, 201) MacKay codes (see [26]) and a (402, 203) array codes are constructed (see [27]), and the performance is compared as in Figure 3.

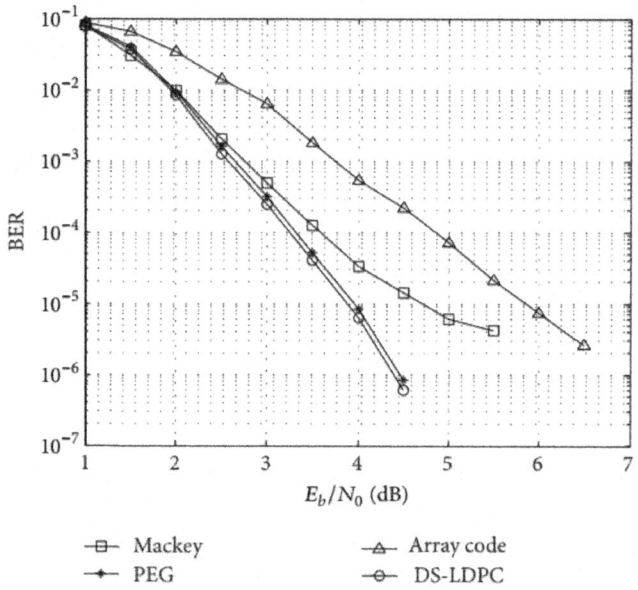

Figure 3: The BER curve comparison between QC-LDPC constructed by differential sequences, Mackay, PEG algorithm, and array codes.

As can be seen from Figure 3, at BER = 10^{-5}, Mackay codes began to show platform, whereas DS-LDPC codes did not show platform at BER = 10^{-7}. At BER = 10^{-5}, DS-LDPC codes comparing with Mackay codes and array codes, there are about 0.7 dB and 2.0 dB coding gain. The performance of differential sequence is better than that of MacKay, PEG algorithm, and array codes in the condition of the same parameters.

With the performance simulation of QC-LDPC codes constructed by Hoey sequence, let $L = 12$ and $p = 123$, and a 2×12 shift matrix can be constructed to make a 2×12 array H constructed by a 123×123 cyclic permutation matrix.

H is a 246 × 1476 matrix, with the column weight and row weight of 2 and 12, respectively. Its null space provides (1476, 1231) binary QC-LDPC codes with the code rate of 0.83. The performance comparison and analysis of (1476, 1231) pseudorandom PEG codes and the (1476, 1231) array codes are as in Figure 4.

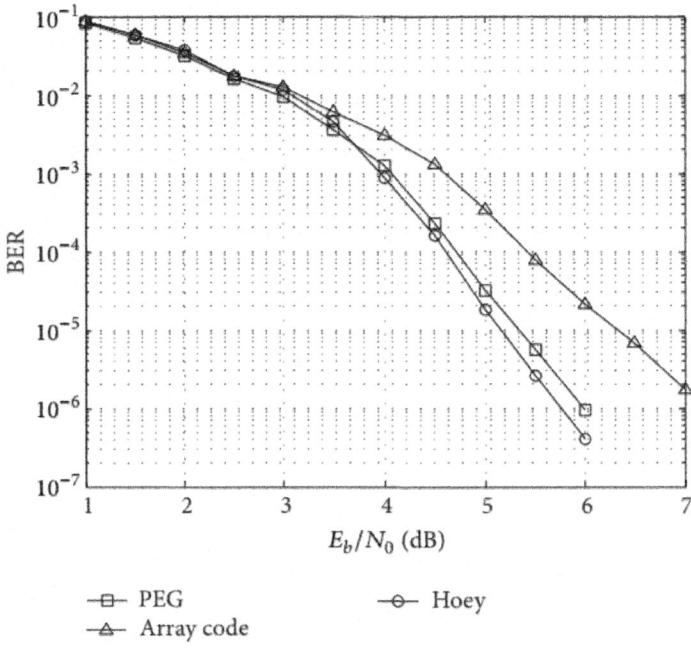

Figure 4: The BER curve comparison of QC-LDPC constructed by Hoey sequence, PEG algorithm codes, and array codes.

As can be seen from Figure 4, at BER = 10^{-5}, Hoey code compared with array code improves about 1.1 dB coding gain. The performance of QC-LDPC codes constructed by Hoey sequence is better than that of the QC-LDPC codes constructed by PEG algorithm and that of QC-LDPC codes constructed by array code in the condition of the same parameters.

QC-LDPC constructed by four different integer sequences, respectively, is used in the code cooperative communication system. The system model is shown in Figure 1. Assume the channel is a Rayleigh channel, and the relevant parameters are shown in Table 1.

Table 1: The parameters of Figure 5

Coding type	Source node length	Code rate	Relay length
Mackey	5098	0.5	5098
Dayan	5568	0.5	5568
Hoey	4930	0.1174	4570
Difference	4068	0.5	4068
Fibonacci	5664	0.5	5310

Figure 5 shows that the BER performance of the Mackey code which is random configuration method is inferior to the structured LDPC code. When the length of Mackey code is longer, its generating matrix and parity check matrix are more complex. Integer sequence constructed QC-LDPC codes are structural characteristics and effectively avoid the parity check matrix of the short loop. The results show that the performance of Dayan constructed QC-LDPC code is the most outstanding one.

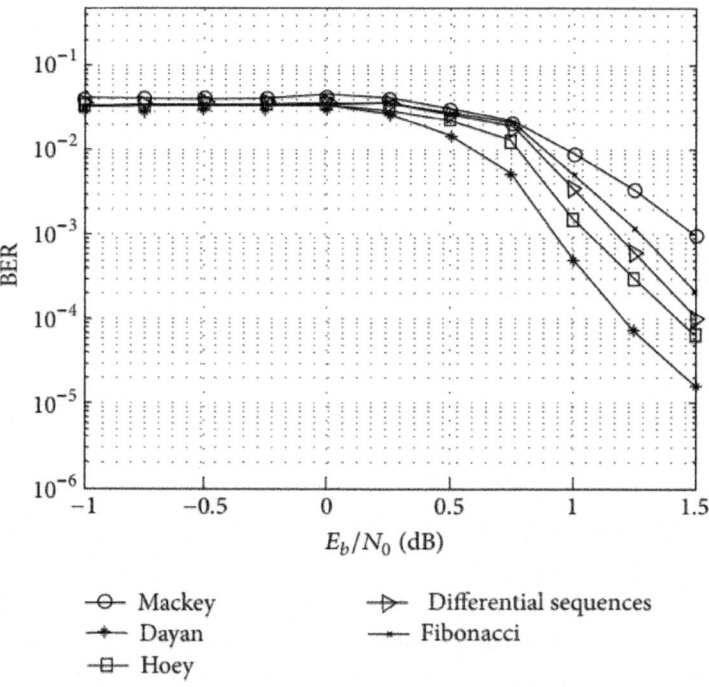

Figure 5: The BER performance comparison curve of QC-LDPC constructed by four different integer sequences and Mackey code in the code cooperative communication system.

CONCLUSIONS

In this paper, four different integer sequences were introduced and compared. The design of QC-LDPC constructed by four different sequences was used in the coded cooperative communication. Through the theoretical analysis and simulation results, it can be seen that the application of QC-LDPC codes constructed by integer sequence in the cooperative communication system is effective. Besides, the coding structure is simple and the overall performance is better than that of the other types of LDPC codes. Moreover, QC-LDPC constructed by Dayan integer sequence presents the best performance.

CONFLICT OF INTERESTS

The authors declare that there is no conflict of interests regarding the publication of this paper.

ACKNOWLEDGMENTS

This work is supported by the Science and Technology on Avionics Integration Laboratory and National Aeronautical Science Foundation of China under Contract no. 20105552.

REFERENCES

1. E. C. van der Meulen, "Three-terminal communication channels," Advances in Applied Probability, vol. 3, pp. 120–154, 1971.

2. T. M. Cover and A. A. El Gamal, "Capacity theorems for the relay channel," IEEE Transactions on Information Theory, vol. 25, no. 5, pp. 572–584, 1979.

3. Sendonaris, E. Erkip, and B. Aazhang, "User cooperation diversity. Part I: system description," IEEE Transactions on Communications, vol. 51, no. 11, pp. 1927–1938, 2003.

4. J. N. Laneman, D. N. C. Tse, and G. W. Wornell, "Cooperative diversity in wireless networks: efficient protocols and outage behavior," IEEE Transactions on Information Theory, vol. 50, no. 12, pp. 3062–3080, 2004.

5. T. E. Hunter and A. Nosratinia, "Cooperation diversity through coding," in Proceedings of the IEEE International Symposium on Information Theory, pp. 220–221, July 2002.

6. P. Razaghi and W. Yu, "Bilayer low-density parity-check codes for decode-and-forward in relay channels," IEEE Transactions on Information Theory, vol. 53, no. 10, pp. 3723–3739, 2007.

7.   J. P. Canees and V. Meghdadi, "Optimized low density parity check codes designs for half duplex relay channels," IEEE Transactions on Wireless Communications, vol. 8, no. 7, pp. 3390–3395, 2009.

8.   C. X. Li, G. S. Yue, M. A. Khojastepour, X. Wang, and M. M. Madihian, "LDPC-coded cooperative relay systems: performance analysis and code design," IEEE Transactions on Communications, vol. 56, no. 3, pp. 485–496, 2008.

9.   D. Bing and Z. Jun, "Design and optimization of joint network-channel LDPC code for wireless cooperative communications," in Proceedings of the 11th IEEE Singapore International Conference on Communication Systems (ICCS '08), pp. 1625–1629, November 2008.

10. L. Tang, F. F. Yang, S. W. Zhang, and H. Xu, "Joint iterative decoding for LDPC-coded multi-relay cooperation with receive multi-antenna in the destination," IET Communications, vol. 7, no. 1, pp. 1–12, 2013.

11. G. H. Zhang, R. Sun, and X. M. Wang, "Construction of girth-eight QC-LDPC codes from greatest common divisor," IEEE Communications Letters, vol. 17, no. 2, pp. 369–372, 2013.·

12. X. Jiang and M. H. Lee, "Large girth quasi-cyclic LDPC codes based on the chinese remainder theorem,"IEEE Communications Letters, vol. 13, no. 5, pp. 342–344, 2009.

13. N. J. Sloane, "The on-line encyclopedia of integer sequences," Notices of the American Mathematical Society, vol. 50, no. 8, pp. 912–915, 2003.

14. S. Zhang, F. Yang, and P. Zong, "LDPC-coded cooperative system based on joint iterative decoding,"Journal of Southwest Jiaotong University, vol. 46, no. 3, pp. 469–487, 2011.

15. M. P. C. Fossorier, "Quasi-cyclic low-density parity-check codes from circulant permutation matrices,"IEEE Transactions on Information Theory, vol. 50, no. 8, pp. 1788–1793, 2004.

16. G. Zhang, R. Sun, and X. Wang, "Construction of girth-eight QC-LDPC codes from greatest common divisor," IEEE Communications Letters, vol. 17, no. 2, pp. 369–372, 2013.

17. X. Q. Jiang and M. H. Lee, "Large girth quasi-cyclic LDPC codes based on the chinese remainder theorem," IEEE Communications Letters, vol. 13, no. 5, pp. 342–344, 2009.

18. Y. H. Liu, X. M. Wang, R. W. Chen, and Y. He, "Generalized combining method for design of quasi-cyclic LDPC codes," IEEE Communications Letters, vol. 12, no. 5, pp. 392–394, 2008.·

19. J. Y. Kang, Q. Huang, L. Zhang, B. Zhou, and S. Lin, "Quasi-cyclic

LDPC codes: an algebraic construction," IEEE Transactions on Communications, vol. 58, no. 5, pp. 1383–1396, 2010.

20. W. M. Tam, F. C. M. Lau, and C. K. Tse, "A class of QC-LDPC codes with low encoding complexity and good error performance," IEEE Communications Letters, vol. 14, no. 2, pp. 169–171, 2010.

21. M. P. Fossorier, "Quasi-cyclic low-density parity-check codes from circulant permutation matrices,"IEEE Transactions on Information Theory, vol. 50, no. 8, pp. 1788–1793, 2004.

22. W. Zhan, G. X. Zhu, and L. Peng, "Design of QC-LDPC codes using Fibonacci sequence," Journal of Huazhong University of Science and Technology (Natural Science), vol. 36, no. 10, pp. 63–65, 2008.

23. L. Zhu, H. Wang, Y. Shi, and T. Xing, "Research on base matrix construction of LDPC code," Journal of Computational Information Systems, vol. 8, no. 4, pp. 1515–1521, 2012.

24. B. Li, L. Zhang, and L. L. Cheng, "A class of (3, k) quasi-cyclic LDPC codes from difference sequences with girth 8," in Proceedings of the 4th IET International Conference on Wireless, Mobile and Multimedia Networks (ICWMMN '11), pp. 108–113, November 2011. ·

25. N. J. A. Sloane, "The on-line encyclopedia of integer sequences," in Towards Mechanized Mathematical Assistants, vol. 4573 of Lecture Notes in Computer Science, pp. 130–134, Springer, Berlin, Germany, 2007.

26. L. Bin, Design of QC-LDPC codes based on integer sequence [M.S. thesis], Beijing Jiaotong University, Beijing, China, 2011.

27. L. Xingcheng and C. Haohui, "Study on the construction method of quasi-cyclic LDPC codes based on progressive edge-growth (PEG) algorithm," Journal of Circuits and Systems, vol. 14, pp. 115–119, 2009.

Chapter 7

RELIABILITY EVALUATION OF DATA COMMUNICATION SYSTEM BASED ON DYNAMIC FAULT TREE UNDER EPISTEMIC UNCERTAINTY

Rongxing Duan and Jinghui Fan

School of Information Engineering, Nanchang University, Nanchang 330031, China

ABSTRACT

Fault tree analysis is a well-structured, precise, and powerful tool for system evaluation. However, the conventional approach has been found to be inadequate to deal with the absence of fault data, failure dependency, and uncertainty problems. This paper presents a comprehensive study on the evaluation of data communication system (DCS) using dynamic fault tree approach based on fuzzy set. It makes use of the advantages of the dynamic fault tree for modelling, fuzzy set theory for handling uncertainty, and Bayesian network (BN) for inference ability. Specifically, it adopts expert elicitation and fuzzy set theory to evaluate the failure rates of the basic events for DCS and uses a dynamic fault tree model to capture the dynamic failure mechanisms. Furthermore, some reliability parameters can be calculated by mapping a dynamic fault tree into an equivalent BN. The results show that the proposed method is more flexible and adaptive than conventional fault tree analysis for fault diagnosis and reliability estimation of DCS.

INTRODUCTION

Data communication system (DCS) is a key subsystem of urban rail transit and its reliability has a direct impact on the stability and safety of the train operation system. For fast technology innovation, the performance of key equipment in the DCS of urban mass transit has been greatly improved with the wide application of high technology on one hand, but, on the other hand, its complexity of technology and structure increasing significantly raise challenges in system reliability evaluation and maintenance. These challenges are displayed as follows. (1) Lack of sufficient fault data: fault data integrity has significant influence on the system reliability analysis. However, it is very difficult to obtain mass fault samples which need lots of case studies in practice due to some reasons. One reason is the imprecise knowledge in an early stage of new product design. The other factor is the changes of the environmental conditions which may cause that the historical fault data cannot represent the future failure behaviours. (2) Failure dependency of components: DCS adopts many redundancy units and fault tolerance techniques to improve its reliability. So, the behaviours of components in the system and their interactions, such as failure priority, sequentially dependent failures, functional dependent failures, and dynamic redundancy management, should be taken into consideration. (3) High levels of uncertainty: DCS is usually operated in a dynamic environment and is greatly affected by the technical, human, and operational malfunctions that may lead to hazardous incidents.

Fault tree analysis (FTA) has been widely used to calculate reliability of complex systems. It is a logical and diagrammatic method for evaluating the possibility of an accident resulting from combinations of failure events. However, the conventional FTA, which is commonly assuming that components of a complex system are described by precise probability distributions describing their reliability characteristics, has been found to be inadequate to deal with these challenges mentioned above. Therefore, fuzzy set theory has been introduced as a useful tool to handle challenges (1) and (3). The fuzzy fault tree analysis model employs fuzzy set and possibility theory and deals with ambiguous, qualitatively incomplete, and inaccurate information. Several researchers successfully used the fuzzy fault tree technique in various areas, including the nuclear safety assessment [1], risk analysis [2, 3], and reliability of gas power plant [4]. They treated basic events probabilities as fuzzy numbers and applied the fuzzy extension principle to compute the top event probability. However, these approaches use the static fault tree to model the system fault behaviours and cannot cope with challenge (2). Dynamic fault tree analysis has been introduced [5], which takes into account not only the combination of failure events but also the order in which they occur. Meshkat et al. analysed the dependability of systems with on-demand and active failure modes using dynamic fault tree and solved it to get some reliability results by Markov chains (MC) model [6]. However, this method has two well-known problems: one is the ineffectiveness in solving large dynamic fault tree; that is, MC-based approach has the infamous state space explosion problem. The other is the ineffectiveness in handling uncertainty of failure data; that is, the failure rates of the system components are considered as crisp values. Hence, Li et al. proposed a fuzzy dynamic fault tree to analyse the fuzzy reliability of the CNC machining centre [7]. Nevertheless, the solution for the fuzzy dynamic fault tree is still based on the MC model. In order to solve a larger dynamic fault tree, a discrete-time Bayesian network (DTBN) was proposed for the reliability analysis of dynamic fault tree in [8, 9]. They converted dynamic logic gates to DTBN and calculated the reliability results by a standard Bayesian network (BN) inference algorithm. However, this is an approximate solution and requires huge memory resources to obtain the joint probability distribution accurately. An innovative algorithm has been introduced to reduce the dimension of conditional probability tables by an order of magnitude. However, this method cannot perform probability updating [10]. Montani et al. proposed a translation of the dynamic fault tree into a dynamic Bayesian network (DBN) [11]. The DBN model is essentially applicable to Markov processes and the result of the calculation gives the approximated probabilities.

Motivated by the problems mentioned above, this paper presents a reliability evaluation for DCS based on fuzzy set and dynamic fault tree. It pays special attention to meet the above three challenges. We adopt expert elicitation and fuzzy set theory to deal with insufficient fault data and uncertainty problem by treating the failure rates as fuzzy numbers. In addition, we use a dynamic fault tree model to capture the dynamic behaviours of DCS failure mechanisms and calculate some reliability results using BN and algebraic technique in order to avoid the aforementioned problems.

The objective of this paper is to evaluate the reliability of DCS using fuzzy set and dynamic fault tree. This paper is organized as follows. Section 2 provides a brief introduction on DCS and its dynamic fault tree model. Section 3 describes estimation of failure rates for the basic events. Section 4 presents a novel dynamic fault tree solution which uses BN and algebraic technique. The outcomes of the research and future research recommendations are presented in the final section.

DYNAMIC FAULT TREE OF DCS

DCS

DCS is one of the key components of the train control system and is a medium for transmitting data among the modules in the automatic train control system. It mainly includes ground wire backbone communication networks and train-ground communication networks shown in Figure 1. The ground wire backbone communication networks are mainly used to connect zone controller, computer based interlocking system, automatic train supervision system data storage unit, and so on. As for the ground wire backbone communication networks, we usually adopt bidirectional self-healing loop industrial Ethernet. In particular, when one device fails, the communication networks will not interrupt. The train-ground communication networks have experienced a point-type electromagnetic induction communication, point-type wireless communication, and continuous wireless communication. The wireless communication based train control can not only decrease the ground units but also satisfy the requirements of mass train-ground information transmission and secure communication and thus improve the operational capability of the urban rail transport system.

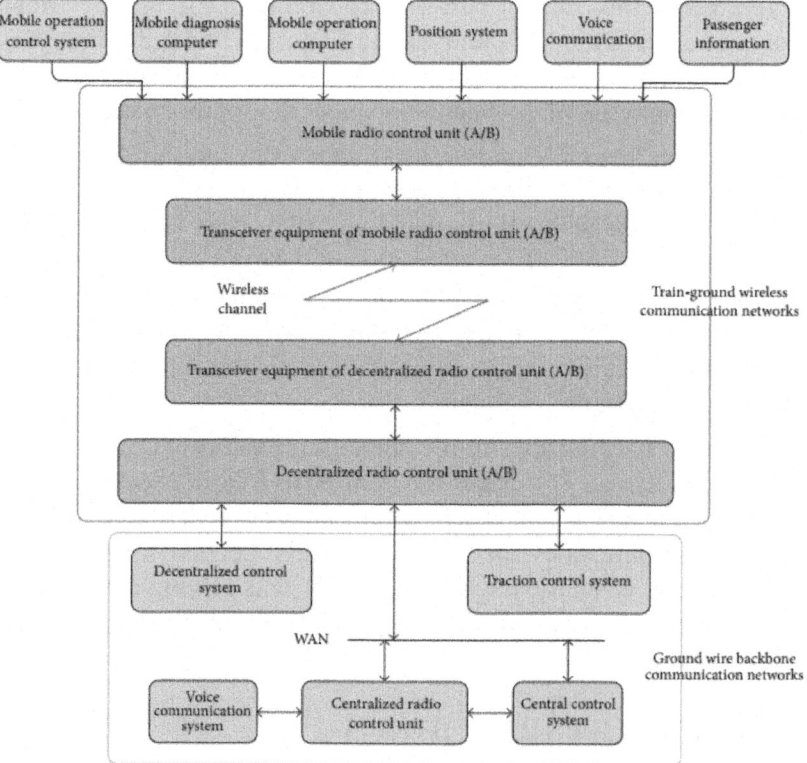

Figure 1: A system block diagram of DCS.

The train-ground communication networks consist of the train-ground access devices and the train-ground communication transmission system. The train-ground access devices are responsible for information acquisition, information composition, information decomposition, information encoding, information decoding, and information transmission security mechanism. This can guarantee a safe, reliable, and real-time information transmission. Specifically, the train-ground access devices include the following.

(i)   Centralized Radio Control Unit (CRCU). CRCU, located in the control center, is primarily responsible for transmitting diagnostic information, passenger travel information, and speech information.

(ii)   Decentralized Radio Control Unit (DRCU). DRCU, located in the decentralized control center, offers the interface between the decentralized control system and the traction power supply system. In addition, it also performs the most important task such as information acquisition, composition, decomposition, encoding, and decoding among the decentralized control system, the vehicle control system, localization system, and the traction power supply system.

(iii)   Mobile Radio Control Unit (MRCU). MRCU, located on opposite ends of the train, not only offers the interface between the vehicle control system and the localization system, but also implements information processing among the vehicle control system, the localization system, the decentralized control system, and the traction power supply system.

Dynamic Fault Tree for DCS

DCS of urban mass transit is a complex system and adopts redundancy technique to ensure higher reliability. For example, the hardware redundancy technique is adopted in designing CRCU, DRCU, and MRCU. High coupling degree together with complicated logic relationships exists between these modules. So, the behaviours of components in these modules and their interactions, such as failure priority, sequentially dependent failures, functional-dependent failures, and dynamic redundancy management, should be taken into consideration. Obviously, traditional static fault tree is unsuitable to model these dynamic fault behaviours. So, we use the dynamic fault tree model to capture the dynamic behavior of system failure mechanisms such as sequence-dependent events, spares and dynamic redundancy management, and priorities of failure events. Taking the decentralized traction control failure as the top event, the dynamic fault tree of DCS is established in Figure 2. The failure events and different components of DCS are represented by different symbols which are presented in Table 1.

Table 1: The basic events of DCS

Node symbol	Description
X1	Software failure
X2	Regional traction power supply 1
X3	Regional traction power supply 2
X4	Regional control system 1
X5	Regional control system 2
X6	Vehicle location system 1
X7	Vehicle location system 2
X8	Switch failure
X9	Fiber network failure
X10	Vehicle base station 1
X11	Vehicle base station 2
X12	Ground base station 1
X13	Ground base station 2
X14	Power board 1 for MRCU
X15	Power board 2 for MRCU
X16	MRCU multiplexer board 1
X17	MRCU multiplexer board 2
X18	MRCU demultiplexer board 1
X19	MRCU demultiplexer board 2
X20	Power board 1 for DRCU
X21	Power board 2 for DRCU
X22	DRCU multiplexer board 1
X23	DRCU multiplexer board 2
X24	DRCU demultiplexer board 1
X25	DRCU demultiplexer board 2
X26	Traction control system

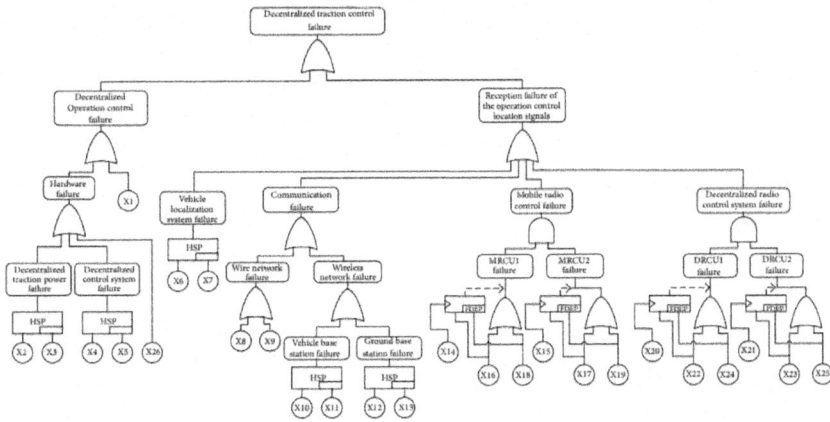

Figure 2: A dynamic fault tree for decentralized traction control failure of DCS.

ESTIMATION OF FAILURE RATES FOR BRAKING SYSTEM

In order to evaluate the reliability of DCS, failure rates of the basic events must be known. However, it is very difficult to estimate a precise failure rate due to lack of insufficient data or vague characteristic of the events, especially for the new equipment. In this study, the expert elicitation through several interviews and questionnaires and fuzzy set theory are used to determine the fault rates of the basic events.

Selecting Experts to Form Evaluation Committee

Experts are selected from different fields, such as design, installation, maintenance, operation, and management of the braking system, to judge failure rates of the basic events. They are more comfortable justifying event failure likelihood using qualitative natural languages based on their experiences and knowledge about the braking system, which capture uncertainties rather than by expressing judgments in a quantitative manner. The granularity of the set of linguistic values commonly used in engineering system safety is from four to seven terms. In this paper, the component failure rate is defined by seven linguistic values, that is, very high, high, reasonably high, moderate, reasonably low, low, and very low.

Converting Linguistic Terms to Fuzzy Numbers

After experts' evaluation, a numerical approximation system was proposed to systematically map linguistic terms into trapezoidal fuzzy numbers. Each predefined linguistic value has a corresponding mathematical representation.

The shapes of the membership functions to mathematically represent linguistic variables in engineering systems are shown in Figure 3. To eliminate bias coming from an expert, eleven experts are asked to justify how likely a basic event will fail in the system under investigation. So, it is necessary to combine or aggregate these opinions into a single one. There are many methods to aggregate fuzzy numbers. An appealing approach is the linear opinion pool [12]:

$$M_i = \sum_{j=1}^{n} \omega_j A_{ij}, \quad i = 1, 2, 3, \ldots, m,$$

(1)

where m is the number of basic events; A_{ij} is the linguistic expression of a basic event i given by expert j; n is the number of the experts; ω_{ij} is a weighting factor of the expert j; and M_i represents combined fuzzy number of the basic event i. Usually, an α-cut addition followed by the arithmetic averaging operation is used for aggregating more membership functions of fuzzy numbers. The membership function of the total fuzzy numbers from n experts' opinion can be computed as follows:

$$f(z) = \max_{z = x_1 + x_2 + \cdots + x_n} [\omega_1 f_1(x) \wedge \omega_2 f_2(x)$$

$$\wedge \cdots \wedge \omega_n f_n(x)],$$

(2)

where $f_n(x)$ is the membership function of a trapezoidal fuzzy number from expert n and $f(z)$ is the membership function of the total fuzzy numbers.

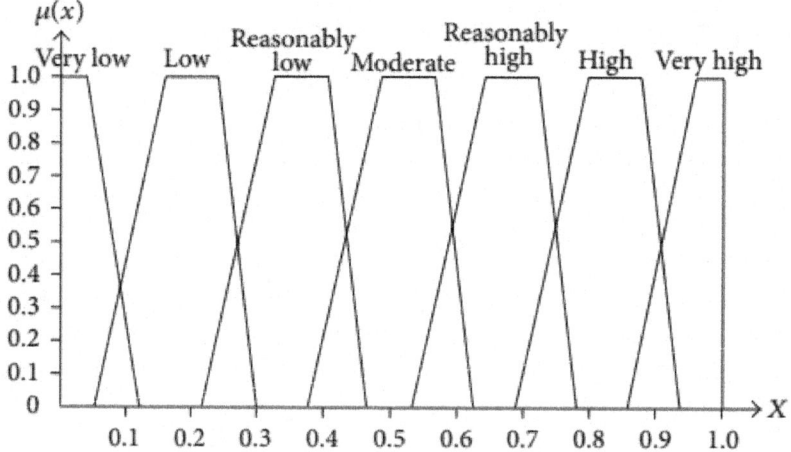

Figure 3: Fuzzy numbers used for representing linguistic value.

Calculating Fuzzy Fault Rate of the Basic Events

Obviously, the final ratings of the basic events are also fuzzy numbers and cannot be used for fault tree analysis because they are not crisp values. So, fuzzy number must be converted to a crisp score, named as fuzzy possibility score (FPS), which represents the most possibility that an expert believe occurring of a basic event. This step is usually called defuzzification. There are several defuzzification techniques [13]: area defuzzification technique, the left and right fuzzy ranking defuzzification technique, the centroid defuzzification technique, the area between the centroid point and the original point defuzzification technique, and the centroid-based Euclidean distance defuzzification technique. In this paper, an area defuzzification technique is used to map the fuzzy numbers into FPS because it has the lowest relative errors and has the closest match with the real data. If (a,b,c,d 1) is a trapezoidal fuzzy number, then its area defuzzification technique is as follows:

$$
\begin{aligned}
\text{FPS} &= \big((a + 2b - 2c - d)\big((2a + 2b)^2 + (c + d)(-3a + 2c - d) \\
&\quad -2c(3b + d) - 4ab\big)\big) \qquad\qquad \times \big(18(a + b - c - d)^2\big)^{-1}.
\end{aligned}
\tag{3}
$$

The event fuzzy possibility score is then converted into the corresponding fuzzy failure rate, which is similar to the failure rate. Based on the logarithmic function proposed by Onisawa [14], which utilizes the concept of error possibility and likely fault rate, the fuzzy failure rate can be obtained by (4). Table 2 shows the fuzzy failure rates of the basic events for the braking system:

$$
\text{FFR} = \begin{cases} \dfrac{1}{10^{[(1-\text{FPS})/\text{FPS}]^{1/3} \times 2.301}}, & \text{FPS} \neq 0 \\ 0, & \text{FPS} = 0. \end{cases}
\tag{4}
$$

Table 2: Basic events' FPS and FFR

Basic events	Fuzzy numbers	FPS	FFR
X1	[0.1602, 0.2093, 0.3104, 0.6988]	0.0741	4.6e − 6
X2, X3	[0.2501, 0.3003, 0.5499, 0.9331]	0.1235	3.8e − 5
X4, X5	[0.2201, 0.2598, 0.4001, 0.9503]	0.1005	2e − 5
X6, X7	[0.2002, 0.2397, 0.3596, 0.8395]	0.0887	1e − 5
X8	[0.1396, 0.1801, 0.2802, 0.6812]	0.0714	3.9e − 6
X9	[0.1203, 0.1501, 0.2497, 0.6030]	0.0653	2.6e − 6
X10, X11	[0.2604, 0.2899, 0.5765, 0.9369]	0.1326	5e − 5
X14, X15, X20, X21	[0.2396, 0.2801, 0.5006, 0.9983]	0.1209	3.5e − 5
X16, X17, X22, X23	[0.2801, 0.3099, 0.6601, 0.9024]	0.1439	6.8e − 5
X18, X19, X24, X25	[0.2704, 0.3006, 0.6199, 0.9001]	0.1367	5.6e − 5
X26	[0.1795, 0.2203, 0.3504, 0.7792]	0.0856	8.6e − 6

DYNAMIC FAULT TREE ANALYSIS USING BN AND ALGEBRAIC TECHNIQUE

Mapping Static Fault Tree into BN

There is a clear correspondence between static fault tree and BN. The fault tree can be seen as a deterministic particular case of the BN. Conceptually, it is straightforward to map a fault tree into a BN: one only needs to "redraw" the nodes and connect them while correctly enumerating reliabilities. Figure 4 shows the conversion of OR and AND gates into equivalent nodes in a BN. Parent nodes A and B are assigned prior probabilities, which coincident with the failure probability of the corresponding basic nodes in the fault tree, and child node C is assigned its conditional probability table (CPT). Since the OR and AND gates represent deterministic causal relationships, all the entries of the corresponding CPT are either 0 or 1. The detailed algorithm of converting a fault tree into a BN was proposed in [15, 16].

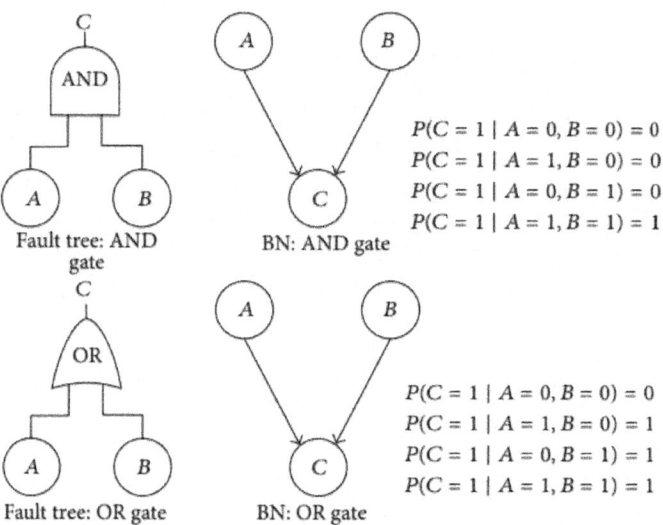

Figure 4: The equivalent BN of OR and AND gates.

Fault Probability of a Module with Sequence Dependence

Let us consider an event sequence composed of n events, e_1, e_2,...,$_n$, including several spare events. An event in the sequence is denoted by $e\ i\ j$, which means that the event that failed in the jth order of the sequence is designated a spare of an event that failed in the ith order. e^i_j denotes an event that was originally in active mode. e^i_j ($i > 0$, $i < j$) has a dormancy factor $0 \leq \alpha_i \leq 1$. The sequence probability of $\langle e^{i1}_1, e^{i2}_2, \ldots, e^{in}_n \rangle$ can be calculated using the n- tuple integration as

$$\Pr\left(\left\langle e^{i1}_1, e^{i2}_2, \ldots, e^{in}_n \right\rangle\right)(t)$$

$$= \int_0^t \int_0^{x_n} \cdots \int_0^{x_2} \prod_{e^0_j \in S_a} f_j(x_j) \prod_{e^i_j \in S_{ss}} f_{j\alpha}(x_j)$$

$$\times \prod_{e^i_j \in S_{sa}} \overline{F}_{j\alpha}(x_i) f_j(x_j - x_i) dx_1 dx_2 \cdots dx_n,$$

(5)

where xj indicates the occurrence time of e^i_j, $f_j(x)$ is the probability distribution function of e^i_j, and $\overline{F}_{j\alpha}(x)$ is the

survival function of e^i_j in standby mode. S_a is a set of events that were originally in active mode and S_{sa} (S_{ss}) is a set of spare events that fail in active (standby) mode [17]. When the failure time of e^i_i in active mode follows an exponential distribution with λ^i_j, the sequence probability is

$$\Pr\left(\left\langle e^{i1}_1, e^{i2}_2, \ldots, e^{in}_n \right\rangle\right)(t)$$

$$= \prod_{e^i_j \in S_{ss}} \alpha_j L^{-1} \left\{ \frac{1}{s} \prod_{i=1}^n \left(\frac{\lambda_i}{s + a_i} \right) \right\},$$

(6)

where

$$a_i = \sum_{k=i}^n \lambda_k - \sum_{\substack{k=i \\ e^i_k \in S_{ss}}}^n (1 - \alpha_k) \lambda_k - \sum_{\substack{k=i \\ e^i_k \in S_{sa}}}^n (1 - \alpha_j) \lambda_j,$$

(7)

for $a_i > 0$, and L^{-1} is the inverse Laplace transform operator. If every a_i in the above equation is distinct from the other, the sequence probability is

$$\Pr\left(\left\langle e^{i1}_1, e^{i2}_2, \ldots, e^{in}_n \right\rangle\right)(t)$$

$$= \prod_{e^i_j \in S_{ss}} \alpha_j \prod_{i=1}^n \lambda_i \sum_{k=0}^n \frac{e^{-a_k t}}{\prod_{j=0, j \neq k}^n (a_j - a_k)},$$

(8)

where $a_0 = 0$.

Mapping Dynamic Fault Tree into BN

Dynamic fault tree extends traditional fault tree by defining special gates to capture the components' sequential and functional dependencies. Currently there are six types of dynamic gates defined: the functional dependency gate (FDEP), the cold, hot, and warm spare gates (CSP, HSP, WSP), the priority AND gate (PAND), and the sequence enforcing gate (SEQ). Here, we briefly discuss the FDEP and the WSP gates as they will be later used in our examples.

WSP Gate

The WSP gate has one primary input and one or more alternate inputs. The primary input is initially powered on and the alternate inputs are in standby mode. When the primary fails, it is replaced by an alternate input, and, in turn, when this alternate input fails, it is replaced by the next available alternate input, and so on. In standby mode, the component failure rate is reduced by a factor α called the dormancy factor. α is a number between 0 and 1. A cold spare has a dormancy factor $\alpha=0$ and a hot spare has a dormancy factor $\alpha=1$. The WSP gate output is true when the primary and all the alternate inputs fail. Figure 5 shows the WSP gate and its equivalent DTBN. Table 3 shows the CPT of node A. Suppose that A and S follow the same exponential distribution with λ. Here, $p_1(t)$ and $p_2(t)$ in this table can be derived as

$$p_1(t) = P(A = 1 \mid S = 0)$$

$$= \frac{P(S = 0, A = 1)}{P(S = 0)} = 1 - e^{-\lambda_A \alpha t},$$

$$p_2(t) = P(A = 1 \mid S = 1) = \frac{P(S = 1, A = 1)}{P(S = 1)}$$

$$= \frac{P\left(\langle P, A^S \rangle\right)(t) + P(\langle A, S \rangle)(t)}{F_S(t)}.$$

$$\text{(9)}$$

$P(\langle P, A^S \rangle)(t)$ and $P(\langle A, S \rangle)(t)$ are sequence probabilities calculated by (8). Consider

Table 3: The CPT of node A.

	$S = 0$	$S = 1$
$A = 0$	$1 - p_1(t)$	$1 - p_2(t)$
$A = 1$	$p_1(t)$	$p_2(t)$

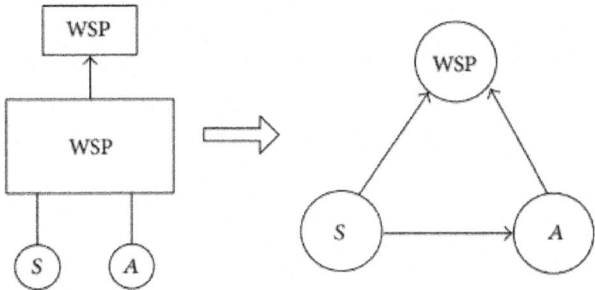

Figure 5: WSP and its equivalent BN.

$$P\left(\langle P, A^S \rangle\right)(t) + P\left(\langle A, S \rangle\right)(t)$$

$$= 1 - e^{-\lambda t} + \frac{e^{-(\lambda + \lambda \alpha)t} - e^{-\lambda t}}{\alpha}.$$

$$(10)$$

The output of node WSP is an AND gate whose CPT is shown in Figure 4.

FDEP Gate

FDEP is used for modelling situations where one component's correct operation is dependent upon the correct operation of some other component. It has a single trigger input, which could be another basic event or the output of another gate, a nondependent output reflecting the status of the trigger, and one or more dependent basic events. Figure 6 shows functional dependency gate and its equivalent BN. Table 4 shows the CPT of node. A. Here, $p_3(t)$ in this table can be derived as

$$p_3(t) = P(A = 1 \mid T = 0) = 1 - e^{-\lambda_A t}.$$

$$(11)$$

Table 4: The CPT of node

	$T = 0$	$T = 1$
$A = 0$	$1 - p_3(t)$	0
$A = 1$	$p_3(t)$	1

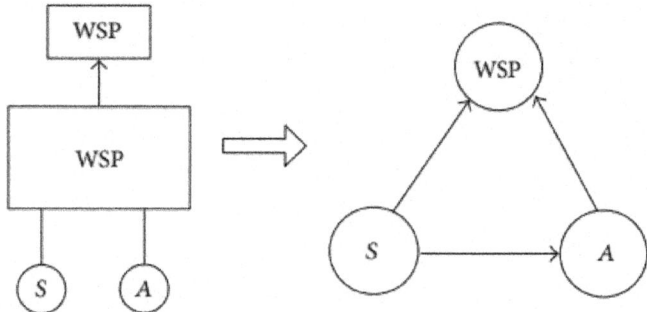

FIGURE 5: WSP and its equivalent BN.

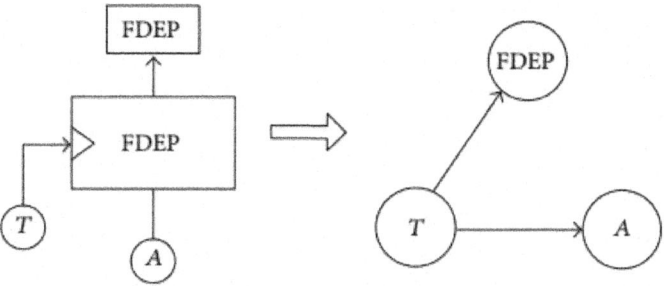

Figure 6: FDEP and its equivalent BN.

The CPT of output node FDEP is shown in Table 5.

Table 5: The CPT of node FDEP.

	$T = 0$	$T = 1$
FDEP = 0	1	0
FDEP = 1	0	1

RELIABILITY ANALYSIS OF DCS

Calculating Reliability

According to the dynamic fault tree shown in Figure 2 and the basic failure data shown in Table 1, we can map the dynamic fault tree into an equivalent BN using the proposed method. Once the structure of a BN is known and all the probability tables are filled, it is straight forward to compute the fault probability of DCS using the inference algorithm. BN has already had some relatively mature accurate and approximate inference algorithms such as the variable elimination algorithm, the search-based algorithm, the conditioning algorithm, the jointree algorithm, and the differential algorithm. Here, we use the jointree algorithm to calculate the reliability indices of DCS. Table 6 shows the unreliability of DCS at the different mission time using some different methods for the dynamic fault tree solution. As we can see in Table 6, the accuracy of DTBN method increases when n increases. Although the DTBN method (n=5) is almost in agreement with the proposed method in this paper, the difference becomes larger with the memory of CPT and execution time.

Table 6: The unreliability of DCS

Mission time (hours)	Fault probability of DCS		
	DTBN ($n = 2$)	DTBN ($n = 5$)	Proposed method
300	0.0121	0.0109	0.0107
600	0.0315	0.0305	0.0303
1000	0.0661	0.0645	0.0649
1500	0.1243	0.1231	0.1229

Sensitivity Analysis

Sensitivity analysis allows the designer to quantify the importance of each of the system's components and the impact the improvement of component reliability will have on the overall system reliability. Here, we show how one can perform sensitivity through the usage of sensitivity index [18]. The sensitivity index of the i th basic event is defined as

$$\alpha_{SI,i} = \frac{\gamma_i}{\lambda_{max}} \quad i = 1, 2, \ldots, m,$$

$$\gamma_i = 1 - \frac{P(S \mid \bar{i})}{P(S)},$$

$$\gamma_{max} = \max\{\gamma_1, \gamma_2, \ldots, \gamma_m\}, \tag{12}$$

where p(s) is the probability of the top event failure; $P(S \mid \bar{i})$ is the probability that the top event has occurred given that the basic event has not occurred.

Table 7 shows the sensitivity index of all basic events for DCS. According to Table 7, we know that the MRCU multiplexer board and DRCU multiplexer board have the maximum sensitivity index, which means that they are the key components. So, we should improve their reliability at the stage of product design in order to decrease the failure probability of DCS by several approaches.

Table 7: The sensitivity index of DCS at the given time of 600 hours

Ranking	Components	Sensitivity index
1	X16, X17, X22, and X23	1
2	X26	0.919
3	X18, X19, X24, and X25	0.866
4	X1	0.558
5	X14, X15, X20, and X21	0.326
6	X8	0.308
7	X9	0.291
8	X10, X11, X12, and X13	0.151
9	X2, X3	0.099
10	X4, X5	0.017
11	X6, X7	0.002

Performing Diagnosis

Diagnosis is an obvious capability of the framework due to the use of BN. We can conveniently calculate some importance parameters by BN and perform diagnosis to locate the system failure. The diagnostic importance factor (DIF) is the corner stone of reliability based diagnosis methodology. DIF is defined conceptually as the probability that an event has occurred given that the top event has also occurred. This quantitative measure allows us to discriminate between components by their importance from a diagnostic point of view. Components with larger DIF are checked first. This assures a reduced number of system checks while fixing the system. Consider

$$DIF_C = P(C \mid S),$$ (13)

where is a component in system .

Suppose the system has failed, we would like to know what is the most probable cause that took the system down. So, we enter the evidence that the braking system has failed; that is, $P(S = \text{state} 1) = 1$, and we solve the BN using the jointree algorithm. Table 8 gives the components' DIF. We should check the component with larger DIF firstly one by one to locate the DCS failure. According to Table 8, the MRCU multiplexer board and DRCU multiplexer board have the maximum DIF, which means that they are the most unreliable components. So, when DCS fails, we should diagnose them firstly to locate the failure of DCS. Furthermore, proper measures should be allocated for these components to improve their reliability at the stage of product design in order to decrease the failure probability of DCS.

Table 8: DIF of components for DCS at the given time of 600 hours.

Ranking	Components	Components' DIF
1	X16, X17, X22, and X23	0.221
2	X18, X19, X24, and X25	0.196
3	X26	0.165
4	X1	0.099
5	X14, X15, X20, and X21	0.077
6	X8	0.066
7	X10, X11, X12, and X13	0.058
8	X9	0.053
9	X2, X3	0.038
10	X4, X5	0.017
11	X6, X7	0.007

CONCLUSION

In this work, we have discussed the use of fuzzy set theory, dynamic fault tree, and BN to evaluate the reliability of DCS. Specifically, it has emphasized three important issues that arise in engineering diagnostic applications, namely, the challenges of insufficient fault data, uncertainty, and failure dependency of components. In terms of the challenge of insufficient fault data and uncertainty, we adopt expert elicitation and fuzzy sets theory to evaluate the failure rate of the basic events for DCS. In terms of the challenge of failure dependency, we use a dynamic fault tree to model the dynamic behaviours of system failure mechanisms. Furthermore, we calculate some reliability parameters of DCS using BN and algebraic technique in order to avoid the state space explosion problem and huge memory resources. As it can be seen from Tables 7 and 8, the MRCU multiplexer board and DRCU multiplexer board have the most contribution to the top event probability. So, we should improve their reliability at the stage of product design in order to decrease the failure probability of DCS by several approaches. The proposed method makes use of the advantages of the dynamic fault tree for modelling, fuzzy set theory for handling uncertainty, and BN for inference ability, which is especially suitable for reliability evaluation and fault diagnosis of the complex system.

In the future work, we will focus on the common cause failures to optimize the dynamic fault tree model and establish a method of dynamic fault tree solution without the exponential distribution assumption.

CONFLICT OF INTERESTS

The authors declare there is no conflict of interests.

ACKNOWLEDGMENTS

This work was supported by the National Natural Science Foundation of China under Grant 61074139 and Science and Technology Foundation of Department of Education in Jiangxi Province under Grant GJJ14166.

REFERENCES

1.   J. H. Purba, J. Lu, G. Zhang, and D. Ruan, "Failure possibilities for nuclear safety assessment by fault tree analysis," International Journal of Nuclear Knowledge Management, vol. 5, no. 2, pp. 162–177, 2011.

2.   R. Ferdous, F. Khan, R. Sadiq, P. Amyotte, and B. Veitch, "Fault and event tree analyses for process systems risk analysis: uncertainty handling formulations," Risk Analysis, vol. 31, no. 1, pp. 86–107, 2011. ·

3.   E. Jafarian and M. A. Rezvani, "Application of fuzzy fault tree analysis for evaluation of railway safety risks: an evaluation of root causes for passenger train derailment," Proceedings of the Institution of Mechanical Engineers F: Journal of Rail and Rapid Transit, vol. 226, no. 1, pp. 14–25, 2012.

4.   M. Verma, A. Kumar, and Y. Singh, "Fuzzy fault tree approach for analysing the fuzzy reliability of a gas power plant," International Journal of Reliability and Safety, vol. 6, no. 4, pp. 354–371, 2012.

5.   J. B. Dugan, S. J. Bavuso, and M. A. Boyd, "Dynamic fault-tree models for fault-tolerant computer systems," IEEE Transactions on Reliability, vol. 41, no. 3, pp. 363–377, 1992.

6.   L. Meshkat, J. B. Dugan, and J. D. Andrews, "Dependability analysis of systems with on-demand and active failure modes, using dynamic fault trees," IEEE Transactions on Reliability, vol. 51, no. 2, pp. 240–251, 2002.

7.   Y.-F. Li, H.-Z. Huang, Y. Liu, N. Xiao, and H. Li, "A new fault tree analysis method: fuzzy dynamic fault tree analysis," Eksploatacja i Niezawodnosc, vol. 14, no. 3, pp. 208–214, 2012. ·

8.   H. Boudali and J. B. Dugan, "A discrete-time Bayesian network reliability modeling and analysis framework," Reliability Engineering and System Safety, vol. 87, no. 3, pp. 337–349, 2005.

9.   Y.-F. Li, J. Mi, H.-Z. Huang, N.-C. Xiao, and S.-P. Zhu, "System reliability modeling and assessment for solar array drive assembly based on bayesian networks," Eksploatacja i Niezawodnosc, vol. 15, no. 2, pp. 117–122, 2013.

10. N. Khakzad, F. Khan, and P. Amyotte, "Risk-based design of process systems using discrete-time Bayesian networks," Reliability Engineering and System Safety, vol. 109, pp. 5–17, 2013.

11. S. Montani, L. Portinale, A. Bobbio, and D. Codetta-Raiteri, "Radyban: a tool for reliability analysis of dynamic fault trees through conversion into dynamic Bayesian networks," Reliability Engineering and System Safety, vol. 93, no. 7, pp. 922–932, 2008.

12. D. Huang, T. Chen, and M.-J. J. Wang, "A fuzzy set approach for event tree analysis," Fuzzy Sets and Systems, vol. 118, no. 1, pp. 153–165, 2001.

13. J. H. Purba, J. Lu, G. Zhang, and D. Ruan, "An area defuzzification technique to assess nuclear event reliability data from failure possibilities," International Journal of Computational Intelligence and Applications, vol. 11, no. 4, pp. 1–16, 2012.

14. T. Onisawa, "An approach to human reliability in man-machine systems using error possibility," Fuzzy Sets and Systems, vol. 27, no. 2, pp. 87–103, 1988.

15. Bobbio, L. Portinale, M. Minichino, and E. Ciancamerla, "Improving the analysis of dependable systems by mapping Fault Trees into Bayesian Networks," Reliability Engineering and System Safety, vol. 71, no. 3, pp. 249–260, 2001.

16. K. Wojtek Przytula and R. Milford, "An efficient framework for the conversion of fault trees to diagnostic bayesian network models," in Proceedings of IEEE Aerospace Conference, pp. 1–14, March 2006.

17. T. Yuge and S. Yanagi, "Dynamic fault tree analysis using bayesian networks and sequence probabilities," IEICE TRANSACTIONS on Fundamentals of Electronics, Communications and Computer Sciences, vol. 96, no. 5, pp. 953–962, 2013.

18. Y. Lu, Q. Li, and Z. Zhou, "Safety risk prediction of subway operation based on fuzzy Bayesian network," Journal of Southeast University, vol. 40, no. 5, pp. 1110–1114, 2010. ·

Chapter 8

3D ICS WITH OPTICAL INTERCONNECTIONS

Edward G. Kostsov[1], Sergey V. Piskunov[2] and Mike B. Ostapkevich[2]

[1]Institute of Automation and Electrometry, Russian Academy of Sciences, Russia

[2]Institute of Computational Mathematics and Mathematical Geophysics, Russian Academy of Sciences, Russia

INTRODUCTION

Today's microelectronics, whose main driving force of development has always been the needs in computing devices, has achieved exceptionally great results over the last decade.

The exponential growth in performance of microelectronic devices, predicted in 1965 by Gordon Moore and formulated as an empirical law, was sufficiently accurate for more than 45 years: the computing power of single-chip microprocessor-based systems increased almost by a factor of four every three years. Simultaneously with the improvement of parameters of ICs the performance of supercomputers has increased. The performance has increased both due to a reduced duration of a cycle and due to pipelining and parallelization of computations. Currently, petaflops computers are built on the basis of 10^5 - 10^6 single-chip processors with a clock frequency of 10 GHz.

At the same time, the analysis of the element base of microelectronics has shown that the main obstacle to improve the performance of computer devices is the problem of connections, both on the surface of individual substrates, and between them. The area occupied by conductors is about 70% of the overall area of a crystal itself in modern VLSI. The total length of conductors exceeds the linear dimensions of a crystal by more than a thousand times. The energy spent on charging the conductors is 60-70% of all energy losses. Placement of the conductors on the surface of a crystal requires significant technical efforts, while the limit value of capacitance of a conductor, which is 10^{-11} F/m., is attainable [1]. On schematic level, there is also a problem of RC propagation delay in the connections. And consequently, the velocity of signal propagation in ICs is much lower than that of light (5-30 times) depending on the degree of integration.

Thus, the main source of increasing the performance of computers at the previous period of their evolution, which is concluded in increasing the performance of a single gate and the degree of integration, is nearly exhausted. Thus, it can be stated that the clock frequency of computers is determined by circuit limitations, rather than by physical and technological ones.

No less complicated problems arise while exchanging information between processor chips. With an increase in their size the number of inter-chip connections increases as well. And this reduces the clock speed by almost an order of magnitude. With decreasing the cycle duration, the above-mentioned connections play an increasingly significant role.

From the stated-above it can be concluded that the solution to the problem of connections and thus further improvement of the performance of computer devices can be obtained by the following ways:

- increasing the degree of parallelization of information processing at all levels down to elementary operations (a maximum depth of parallelization) and increasing the decentralization of functions of storage, control, and data processing, and transition from long logical links within the surface of a chip to the local relations between the neighboring gates, i.e. by transition to a VLSI with a homogeneous structure;

- transition from connecting chips placed in a plane to 3D packs of VLSIs from these chips and to their vertical interconnections.

It follows from general physical considerations that a simple way of introducing the «third dimension» into the structure of communications is using optical links. Such channels have the following advantages:

- A high degree of parallelism of the information transfer from plane to plane makes possible to use highly parallel algorithms for processing (down to elementary operations), and thus to create high-performance computing.

- Capability of optical synchronization allowing delivering a synchronization signal to any point of a chip practically without delay, e.g. from one light source outside an IC.

- Optical communications channels are free from parasitic effects of the mutual influence because of the neutrality of photons. With increasing the clock frequencies (especially in the gigahertz frequency range), the advantages of optical signals grow as capacitive coupling between electronic conductors with increasing frequency grows as well.

There is always an interest in the use of optics for high-performance computer devices. Recently, a special attention of developers of such devices has been drawn to the development of a new element base, creation of optoelectronic integrated circuits (OEIC), for the organization of inter-IC and inter-processor connections in order to attain high throughput and low power consumption [1-4].

The optical communication bandwidth over short distances within a few millimeters already competes with electric conductors with sufficiently low energy consumption, up to 1 pJ/bit/m or less. The placement of optical communication channels directly on the chip surface can significantly reduce this bandwidth. The reports are already known about the first experimental data on providing the density of the bandwidth 37 Gbit/s/mm2 and higher by using a switched CMOS pair of a vertical-cavity surface-emitting laser (VCSEL) and a CMOS-compatible avalanche photodetector (CMOS-APD) system. The frequency of switching light sources VCSEL, electro-optic modulators MQW based on superlattices GaAs / GaAlAs, InGaAsP, etc. as well as that of their corresponding photodetectors has already reached values of $20 \div 50$ GHz, which in the short term, can grow up to $70 \div 80$ GHz. The level of technology of constructing field-programmable smart-pixel arrays and FP-SPA systems based on the use of optical channels in free space, is now able to provide the exchange of information between the two surfaces at a rate exceeding 10 Tbit/cm2.

Currently, however, optical communication channels are used only for the information transfer problems. Their application directly to the microelectronic structures and in the channels of information processing virtually does not evolve. The main reason is that the size of components of gates (modulators and light sources) due to the wave nature of the optical signal is much larger than that of the active elements of modern microelectronics. Entirely optical computers performing massively parallel computations typically contain rather large elements: lenses, shadowgrams, spatial light modulators, etc. and cannot be created using microelectronics technology.

At the same time, it may also be noted that with the development of nanophotonics, the advent of the light sources and receivers with nanometer dimensions (see, e.g. [5-7]), the optical channels could be used not only for the exchange of information between ICs, but also for the information processing, with the 3D logic devices. Thus, the construction of optoelectronic schemes, whose implementation fits well into the existing technology of building semiconductor circuits, is actual.

The *objectives* of this paper are: a) to consider the possibility of creating high performance and manufacturable multi-layer (3D) chips in which a processing of information and its mass exchange between the layers is performed using optical communication channels and whose parameters are compatible with those of microelectronic circuits, b) to represent the algorithmic and computer tools providing the design and research of simulation models of such 3D algorithms and structures with massive parallelism that are oriented to an optoelectronic implementation.

The principles of the gate, where logic signals are represented by the presence or absence of a light flux driven by an electric field are described in the paper. The possibility of the physical implementation of 3D logic circuits based on such elements is analyzed and its advantages are discussed. The methods and means of the algorithmic design of 3D ICs with a homogeneous structure for execution of algorithms with fine-grain parallelism are described. These tools include a formal model of fine-grained computing, Parallel Substitution Algorithm (PSA) and a simulation system (WinALT), which is based on PSA and provides the construction of models of devices with a 3D architecture. Features of the system are demonstrated by constructing models of 3D optoelectronic matrices for parallel data processing. Matrices can be characterized by a high performance, simplicity of cells, homogeneity and simplicity of the topology. Simulation models help in acquiring an objective view of the complexity and performance of the matrices and confirm the reasonability of the transition to 3D VLSI in order to overcome the problem of connections that arises in modern 2D VLSI. The WinALT system is available in Internet. The expediency of transforming WinALT to an online environment for supporting the simulation of 3D computational structures with the users› participation in the network (virtual) community is discussed.

THE PHYSICAL BASIS OF IMPLEMENTATION OF 3D OPTICAL LOGIC CIRCUITS

Optoelectronic Gate

Let us consider features of constructing a universal gate, which is based on the modulation of a light flux by an electric field [8-11], and when only a single transition of energy (between light and electric signal), the source of a light flux is shared by many gates. The scheme of such a gate is presented inFigure 1. It has the following main components: a light modulator LM controlled by the electric field, EC converter of a light signal to an electric signal (e.g. a photoelectric transducer), energy storage or a load element LE (capacitance for dynamic circuits). The element controls the intensity of the luminous flux transmitted by LM, and with the help of the light flux entering the optical input of EC.

The most optimal physical and technical solution, as follows from [8-11], is obtained when the light modulator is electro-optical, the energy converter is photovoltaic and the energy storage is electrostatic. Such a decision is mainly due to the consideration of energy. In this sense, the use of magneto-optical and acousto-optic light modulators is limited by a high energy consumption and by the complexity of the direct energy conversion (photomagnetic or photoacoustic).

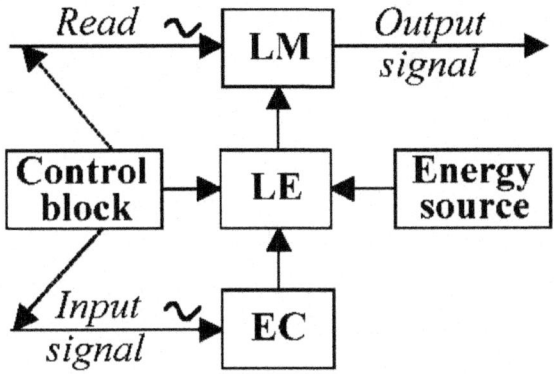

Figure 1: Flowchart of universal gate.

The state of a gate is determined by the ability of the light modulator to pass a light flux coming on its input (state "0" or "1"). This state is unambiguously related to the amount of energy stored, which, in turn, is determined by the intensity of the light flux entering the energy converter (entrance gate). According to Figure 1 a gate in the dynamic mode operates as follows:

- supply of energy from the energy source (voltage pulse) to LE at the time t_0 (operation "erasing information");
- supply of a light flux on the input of the transmitter PE at the time $t_0 + \Delta t$ (operation "information recording"), where Δt is the duration of a cycle;
- supply of a light flux on the input of LM at the time $t_0 + 2\Delta t$ (operation "read information").

Assume that the modulator lets the light flux pass when the amount of energy stored in LE, and does not let the light pass otherwise. Let us also assume that when EC receives a light flux composed of the streams X_1, X_2,..., X_n, such energy in the load element is released. Then the light flux on the output of the modulator Y (at the moment of reading) will be the Pierce logical function:

$$Y = \overline{X_1 \vee X_2 \vee \ldots \vee X_n}. \tag{1}$$

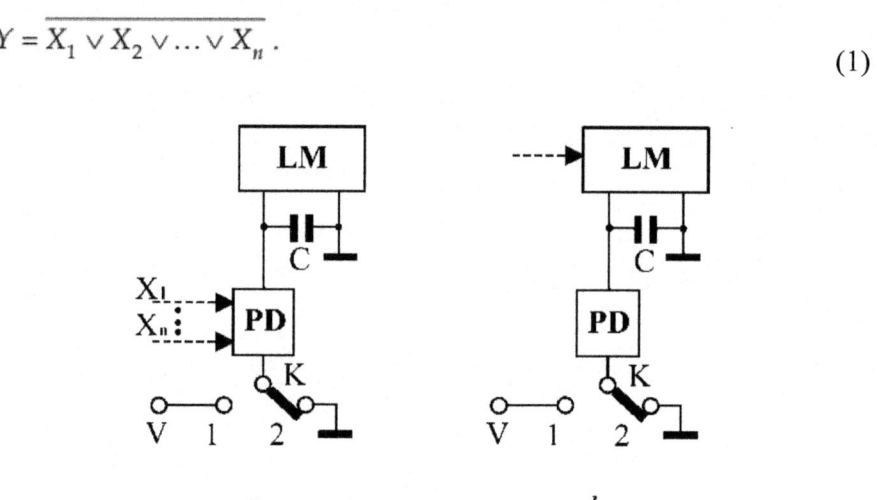

a b

Figure 2: Universal gate with optical connections.

This function forms a complete basis of Boolean functions, so the element proposed can be considered universal.

The electric functional scheme of the gate is shown in Figure 2, where K is the key, V is the power supply voltage. The gate implements the Pierce function and it operates by cycles according to the following general description.

cycle 1 - "erasing information", key K is in the position 1, and capacitance C is being charged by the source when the photodetector FP is illuminated.

cycle 2 - "storing" K is in position 2 (see Figure 2, a), the light flux with intensity J1 or J2 (depending on the level of illumination corresponding to signal «1» or «0», i.e. whether there is at least one of Xi not equal to «0») comes to the optical input of the gate (photodetector FP). According to the level of illumination the capacitance C is discharged with the time constants RTC, or RCC, and the $RTC \gg RCC$, where RT is dark resistance of FP, and RC is its resistance under illumination.

cycle 3 - "reading", key K is in position 2 (see Figure 2, b) and the light signal passes to the optical input of the modulator. The light flux corresponding to «0» light is obtained from the output if signal «1» arrived at the input in the second cycle. And vice versa, signal «1» with a high intensity of light is on the output of LM if there was signal «0» in the previous cycle.

The storage time of information received by the element is proportional to the value of RTC. However, we can create a dynamic memory by inclusion of two gates so that the output of one of them be connected to the input of another and vice versa, as shown in Figure 3. Its storage time is limited only by the time of maintaining the voltage on the power supply.

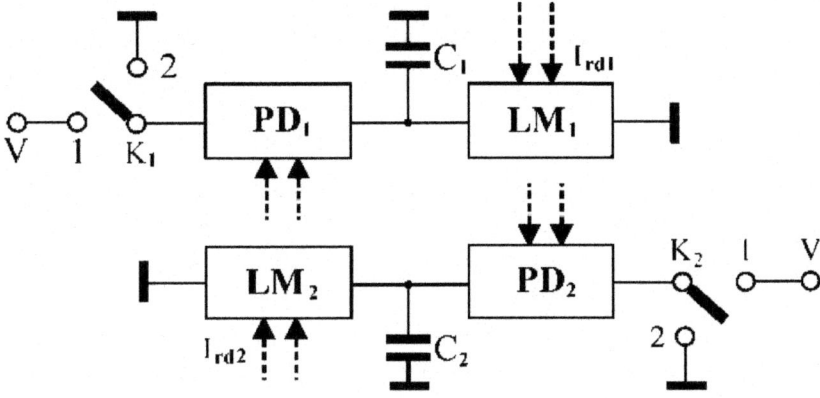

Figure 3: A dynamic memory cell.

When constructing digital devices based on these optical gates the interlayer connections are implemented by placing pairs of "modulator - photodetector" in different layers-planes so that the optical output light modulator of one gate is geometrically aligned with the photodetector optical input of another gate.

A Sample of Constructing a Dynamic Memory Cell

Such a cell is intended for storing information in the electric form on capacitors and transferring information between cells by pulsed optical signals, while a direct transfer of charge does not take place. The scheme of the cell is depicted in Figure 3.

Assume that a bit of information (e.g. "0") is written in the upper gate 1. The switch K_1 was shot to pin 2 and PD_1 was closed in order to do so.

Cycle 1. "Erasing" of information in lower gate 2 by charging the capacitance C2 under lit PD2 and the switch K2 shot to pin 1 of the power source with voltage V.

Cycle 2. The reading of information from element 1 by the light signal I_{rd1}. The switches K_1 and K_2 are in position 2. The output signal from LM_1 is the input writing signal for element 2.

Cycle 3. The "erasing" of information in element 1. The switch K_1 is in position 1 and the capacitance C_1 is connected to the power source with voltage V and is charging.

Cycle 4. The reading of information from element 2 by the light signal I_{rd2}. The switches K_1 and K_2 are in position 2. The output signal from LM_2 is the input writing signal for element 1.

The information in a cell is represented by a pair of light signals («0», «1») or by voltages taken from pins of capacitances. It can be kept as long as needed as a result of iterative repetition of cycles 1-4. If «1» was written to a cell in the initial state, it would have kept a couple of «1», «0».

Similarly, other schemes of optical elements can be built to implement the basic logic functions: repeater, inverter, AND, NOT-AND, OR, NOT-OR, sum by modulo 2, implication. The analysis of transfer characteristics for such elements is presented in [10].

The described principle of implementation of inter-layer communications allows creating functionally flexible devices by replacing electric logical communications by the optical ones and by using 3D structures of gates. A computational device would contain a minimum number of elements and electric connections if the following rules of its creation are followed: parallel electric circuits are used to supply power to elements and parallel logical links are done using optical channels.

An illustration of the functionality of a logic gate with optical connections is presented in [10, 11] using samples constructing 1D and 2D shift registers, switch and matrix processor for parallel image processing. The processor consists of a control block and a program controlled cellular automaton. The control block stores a program and fetches its instructions. The cellular automaton performs information storage and processing [11]. A cell in cellular automaton transforms information by changing its own state and those states of its neighbors. Any transformation is represented as a sequence of elementary transformations. Each elementary transformation is defined by the contents of a command that comes to a cell from the control unit. The transformation of information in a cellular automaton is performed simultaneously by all cells.

The following conclusions can be drawn from the analysis of the functioning of the described devices (primarily the matrix processor), the specific features of their design and comparison with microelectronic devices, capable of performing similar functions.

- A cell of matrix processor contains significantly less elementary components (estimated by two orders of magnitude) as compared to conventional chip with similar functional capabilities.
- The total number of cycles spent on execution of logical operations required for image processing is also smaller, at least by a factor of N

(where N is the number of lines in the image). The time of the complete image processing does not depend on its size.

In particular, a very small number of clock cycles is required to execute such operations as selection of a contour image, noise filtering, noise filtering with masking, extension of lines, etc., which are quite complex for electronic circuits. And the number of cycles remains unchanged when image dimensions increase.

- The area occupied by the connections is 7 - 8% of the total area of a substrate, as shown by the analysis made. The described devices have a higher volume density of placement of elements, the reliability of connections, noise immunity, and ultimate manufacturability. In addition, the design of an optical VLSI is considerably simpler, since there is no need to take a complex configuration of interconnections into consideration.

A breadboard construction of a cell of optic dynamic memory is created in order to demonstrate the possibility of practical implementation of optoelectronic gates [11]. It consists of two optoelectronic logic gates. Lithium niobate crystals are used as optoelectronic modulators.

The study of the dynamic memory model has made possible to draw the following conclusions:

- transfer of information from one element to another can be done without loss of information an unlimited number of times;

- the use of threshold photodetectors allows reducing of the operating voltage and of energy required for switching a gate down to acceptable levels for microelectronic implementation.

THE CELLULAR TECHNOLOGY CONSTRUCTING 3D LOGICAL STRUCTURES

Efforts to create optoelectronic circuits have given impetus to the development of algorithmic and software tools that aid in solution of the problem of design and study of 3D (multilayer) digital structures, focused on the use of optical interlayer connections.

These tools map a primordial parallel algorithm of a problem solution onto an architecture with a massive spatial-temporal parallelism. The orientation to constructing structures of models that consist of huge sets of rather simple and homogenous computing devices (cells), mainly with local links, and placed in a 3D space is their basic property.

The considered tools include:

- a formal model of fine-grain (cellular) computations, which is called the Algorithm of Parallel Substitutions (PSA) and which serves as the basis of a method for synthesis of parallel architectures;

- a simulation system of parallel computational processes, which is used for the construction, debugging of models of 3D logical structures as well as for the extraction of the characteristics from these models.

A detailed description of the cellular technology is given in [11].

PSA: A Generalized Model OF Fine-Grain Computations

Conceptually PSA unites in itself a substitutional character of Markov's algorithm [12] and spatial parallelism of a cellular automaton [13] basing on an associative mechanism of application of operations, which is common to both of them. PSA represents a "true parallelism" of computations, when all the allowable operations are executed at each step for all the available data. The main idea of PSA is concluded in the following three statements:

- The processed information is presented in the form of a cellular array, which is a set of cells. Each cell is data (a bit, a character, a number, etc.) with its name (its "location" within array, which is an element of a set of names M) in the array. A set of data belongs to a certain finite alphabet A.

- The algorithm is defined by a set of parallel substitutions. They have left-hand and right-hand sides (left and right parts). The expression in the left part generates cellular arrays, one for each cell name in the processed cellular array. If processed arrays contain one or more such arrays, then the substitution is applicable. Its execution means that a certain "base" part of the found array generated by its right part is replaced by array generated by its left part for the same cell name. Its execution means that a certain "base" part of the found array is replaced by the array generated by the right part of the parallel substitution for the same cell name.

- The process of computations is iterative. At each iteration, all the substitutions applicable to the processed cellular array are executed. The computation is over when no substitutions applicable to the array obtained at the previous iteration were found. This array is the result of work of PSA.

Executing the substitution in the form of replacement of one cellular array by another permits representation of such replacement as replacement of one spatial image by another. This is rather essential for a visual construction of computer-aided model of optoelectronic logical structure and visual representation of computational process of such a model, which is distributed in time and space.

A formal description of PSA is presented in [14]. Let us demonstrate the idea of PSA by a simple example. Let us consider PSS $\Phi\sigma$ for adding many non-negative binary numbers as a sample.

Let $A = \{0, 1\}$, $M = N^2$, $N = \{0, 1, 2,...\}$. A transformed cellular array is a 2D rectangular table whose cells are numbered according to left coordinate system (i is abscissa, j is ordinata). The name of cell m is a pair of values i, j. The digits of numbers are kept in the strings of the table. The least significant bits occupy the right-most column. PSA Φ_σ consists of two parallel substitutions:

$$
\Phi_\sigma = \begin{cases}
\Theta_1^\sigma : \{(1,\langle i,j\rangle)(1,\langle i,j+1\rangle)(0,\langle i+1,j\rangle)\} * \{(0,\langle i,j-1\rangle)(0,\langle i+1,j-1\rangle)\} \rightarrow \\
\qquad \{(0,\langle i,j\rangle)(0,\langle i,j+1\rangle)(1,\langle i+1,j\rangle)\}; \\
\Theta_2^\sigma : \{(1,\langle i, j\rangle)(0,\langle i,j+1\rangle)\} * \{(0,\langle i,j-1\rangle)\} \rightarrow \{(0,\langle i,j\rangle)(1,\langle i,j+1\rangle)\}.
\end{cases}
$$
(2)

The left part of the substitution is to the left of the arrow, while its right part is to the right of the arrow. Shift functions are written in angle brackets in the definitions of substitutions Φ_σ. These functions set the location of cells of the left and right parts of a substitution as related to each other. A cell description is confined in parentheses. Cell states are shown to the left of commas. The «base» parts of the left parts of substitutions are at the left of asterisks. When particular values of the pair i, j are substituted into the shift functions, the cellular arrays associated with this name are obtained.

The description of substitutions allows their geometric representation depicted in Figure 4, a. In this case, the left and right parts of commands are defined by templates, and the search of occurrences of the left parts of substitutions is done by shifting their templates above cells of the table along axes. One step of transformation of the source table containing numbers 9, 15, 5 is presented in Figure 4, b. After carrying out four steps the result is placed in the top string, while the rest of them contain zeros.

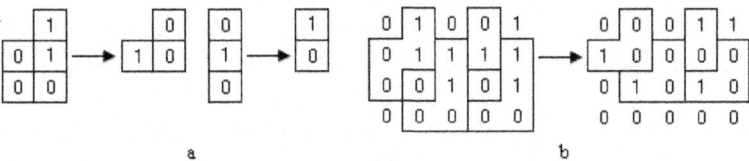

Figure 4: Cellular adder: a – graphic representation of PSA commands, b – one step of execution of algorithm.

The expressive capabilities of PSA augment if an alphabet A is extended by introduction of *variable* and *functional* symbols. An alphabet A serves as a domain for variables symbols and as a range for functional symbols. In the case of graphic representation of substitutions, variable symbols can be written into cells of templates of the left part of substitution, while the functional symbols can be in the cells of the right part. Such a substitution is called *functional* substitution. When an alteration of a state is performed for a cell located in the processed array under a template cell of the right part of a certain command containing a functional symbol, it is not specific data from the template cell, but a result of evaluation of a certain function that is written into this cell. The states of cells in the processed array that are below the cells of a template of the left part can be arguments of such functions. Using the functional substitutions helps: a) in reducing PSA notation to a considerably more concise form in theoretical issues, b) in practical issues to represent rather complicated devices such as ALU by a single cell.

A Simulation System of 3D Digital Structures

Overview of the System

A physical and technical rationale for the construction of multilayer optoelectronic structures is given In Section 2. Evidently the construction of real 3D structures requires a large amount of work associated in the first place with selection of an optimal kind of structure from the standpoint of technical parameters among a big number of possible variants. A manual solution of such a task is virtually impossible. Thus, a computer-aided tool must be available for the research and development of models of 3D structures. Such an instrument was built and it is called WinALT [15]. The user's interface of the system coincides with the standard user's interface in Windows applications. A simulation model is represented by a project that contains a number of sub-windows. Each sub-window can hold graphic or text objects of a model. Creation and editing of graphic objects are carried out by means of toolbars, menus and dialog windows. The system is freely distributed. System's site [16] contains a section "installation", which includes manuals on installation, uninstallation and system's distributive package. An open architecture of the system enables the user to participate in extension of the system functionality.

Adequacy of The System Language to The Problem Domain

The WinALT was developed simultaneously with computer-aided models of parallel algorithms and structures under their strong influence. This influence has manifested itself in the WinALT system by a wide employment of visualization tools both for supporting the construction of parallel algorithm descriptions (graphic representation of objects corresponding to cellular arrays, the left and right parts of substitutions) and for their simulation (capability to view the dynamics of application of each substitution). What is more, a special attention was given to the development of tools that provide the visualization of 3D (multilayer) objects and that allow viewing the transformation of cell states in any layer of a 3D cellular array. This is due to the orientation of the development of computer-aided models to fine-grain architectures that are promising for the implementation in the form of multilayer VLSI. The simulation language contains three parts.

The first part of the language is designed to describe parallel computations in the form of parallel substitutions. It is fully based on PSA.

The objective of the second part of the language is the description of sequential computations. This part is essentially based on Pascal. It provides statements for the description of simulation program structure, control operators, assignment operator and subroutine call by name or by reference. These statements can be used in a model program for the description of sequential control when needed. These means can be used in a model program for the description of sequential control, when it is needed, for the definition of functions that describe cell states and, also, for the construction of such service functions within a model program as menu definition, graph drawing or initial data input.

The third part of the language provides importing libraries into a model program. These are dll libraries written in C/C++ and embedded into the simulation system. This helps in enriching the functionality of simulation tools to suit the user.

Let us describe the first and the third parts in greater detail.

The Description of the Parallel Part of Simulation Language. This part has a clear division into graphic and analytical subparts. Graphic objects are cellular arrays and templates. An image of an object is composed of color cells located along horizontal and vertical axes and, also, along the axis that goes from the user to the screen. The origin in the template is called its *center.*

A color is used to visualize a cell's state. Its state can belong to any cell's data type supported by the library of data formats (see below). A name can be assigned to a cell in addition to a typed value as an additional property. It has to be unique within the scope of one cellular object. This makes possible to implement functional substitutions in WinALT. There is a special neutral void state of a cell (depicted by a diagonal cross on a color background).

A parallel substitution in WinALT simulation language is set by a bunch of operators in-at-do. The names of cellular arrays and templates can be used as parameters in these operators. In the case of one-block structure of a device (a single cellular array), a parallel substitution is described as follows. The parameter of in operator is the name of a processed cellular array. A name of template of the left part of substitution is a parameter in at operator, while that of its right part is a parameter in do operator. The execution of substitution is done in two phases. During the first one, the center of a template of the left part is moved along the axes in a processed cellular array, and all its occurrences in this array are marked accurate to the empty cells. At the second phase, the states of cells of the processed cellular array in all these occurrences are replaced by the states of cells from the template of the right part also accurate to the empty cells. Taking multi-block structure of the device into consideration means the following. The parameter in operator in is a list of names of cellular arrays placed in brackets and separated by commas. The lists in operators at and do are arranged similarly, but instead of names of cellular arrays the names of templates are used in them. Combining patterns in the list means that the movement of patterns in the images of their corresponding cellular arrays is coordinated by substitution of the same set of coordinates into the centers of all the templates of the operators at and do. The coordinates of cells of the cellular array that is in the head of the list in operator form these sets. Such substitutions are called vector substitutions and allow describing parallel transformations of information in compositions of cellular arrays.

A bunch of operators in-at-do and a description of a function serve as an analog of functional substitution. Operator at contains the names of templates, in which some cells are named. A function uses these names as input and output variables. The name of a function is used as a parameter in operator do. A functional substitution can also be a vector one.

A synchroblock exhaust-end (or shortly ex-end) is the main structure for definition of an algorithm of a device operation. This block implements an iterative procedure of PSA application for the composite operators describing parallel substitutions that it contains. In addition, there are two more kinds of synchroblocks that were introduced: clock – end (cl – end) and change – end (ch – end). The first one executes its substitutions a number of times specified as its parameter. The second one executes its body only once.

Remark. The combination of parallel and sequential parts of the language is attained by the possibility to use operators from the sequential part in synchroblocks. Let us also note that nested synchroblocks can be in WinALT simulation language. The described capabilities allow constructing any parallel-sequential compositions of synchronous transformations of cellular arrays.

Model program. The structure of a model program is quite conventional. It consists of a list of libraries imported to a program (using operators use, import and include), declarations of constants, variables and cellular objects, procedures, functions and the main operator block. The main operator block is placed in operator begin-end brackets and contains operators of the first and second parts of the language. A program may include comments, which can be placed in braces. A project can contain any number of simulation programs. They are capable of interacting with each other if necessary.

The third part of the language is based upon a set of WinALT libraries. The functionality of the system is extended by means of external modules. These modules are represented by Windows dll files. These external modules contain the interface functions that are used in versatile simulation models. The external modules form several groups that are called libraries. Some of them are briefly described below.

The library of data formats eliminates limitations of a data type that can be represented by cells in a cellular array. The library contains modules for representations of cellular arrays with integer cells (int8, int16, int32, uint8, uint16, uint32), bit cells (bit), float cells (float) and others. Some external formats are supported by the modules of library, such as bmp raster graphics format. The assignment ofdefault type for a cellular object means that any of its cells can have any of the above-mentioned formats. The latter can be used for the representation of heterogeneous cellular objects. In GUI, the type of a cellular object can be selected in a combo box within the dialog window of the new object creation.

The library of language functions provides the ability to use such functions in simulation programs as functions of object management (creation, deletion, modification or size alteration), GUI functions (construction of dialog windows and data input based upon them), mathematical functions (sin, cos,atan, cosh, log, j0), console I/O functions (WriteLn, ReadLn), file I/O functions (fopen, fgets, fread,feof, ...) and miscellaneous functions such as max, min, null, typeof, StringLength, Time. Operatoruse activates the modules of this library in a simulation program. A typical module of this library is adll written in C or C++.

The library of visual modes provides a customizable visualization of a cellular array and its cells. A cell state can be visualized e.g. by color, a directed arrow or by a number or by their certain combination. A 3D cellular array can be shown as a deck of layers or as layers unrolled in a line or in a grid.

New external modules can be added to any of these libraries. Such modules can be created, for example, using Microsoft Visual Studio. Their source texts can be either borrowed from a provided sample, written from the scratch, or taken from an existing library of functions (e.g. ANSI C runtime library).

Constructing Simulation Models of A Family Of Optoelectronic Matrices In Winalt

Mapping Digital Schemes onto A Customizable Cellular Automaton

The cellular automaton with Margolus (further CA) neighborhood [13], in which setting cells for execution of a certain set of elementary transformations of information, serves as a logical basis for the construction of a family of optoelectronic matrices. The CA is a double layer automaton. Its first layer functions in the same way as the Margolus automaton, being a rectangular matrix of cells. Let us split the matrix to blocks of 2x2 cells. Let us call it E-partition. Let us split the matrix again to blocks that are shifted as related to the blocks of E-partition by one cell along the vertical and horizontal axes. Let us call it O-partition. Each cell of the first layer can be in one of the three states: "white", "gray" or "black". Let us call this layer *informational*. The layer under the informational one is called *control*layer. Its E-partition coincides with that of the informational layer. This layer keeps the table of settings the control layer of CA. The size of a cell in the table coincides with that of a block. The CA is shown in figures as a sweep of two layers in a plane. The informational layer is on the left side and the control layer is on the right side. The blocks of E-partition are limited by solid lines. The blocks of O-partition are limited by dashed lines. An elementary transformation of information performed in the informational layer is a parallel substitution. Its left and right parts are matrices with 2x2 cells. These matrices are composed of white, gray and black cells. The elementary transformations performed in a CA are enumerated. Setting a CA consists in writing numbers of those elementary transformations that can be performed in the block of informational layer above a cell into that cell. Digital (combinatory) schemes of different kinds (adder, multiplier, etc.) can be implemented by setting a CA. A source combinatory scheme is initially transformed in a way that any of its gates has no more than two inputs and two outputs. All the cells of the informational layer are white in its initial state. The picture of a scheme to be imitated in CA, which is made by gray cells in informational layer of CA is called *image*of digital scheme. The signals transformed in a digital scheme are depicted by black cells in its image. The states of cells in each block of the information layer of a CA form a certain picture called *image* of a block.

The graphic images of commands of parallel substitutions are depicted in Figure 5, a. A number of a command is shown above an arrow. The transformations of information in Margolus cellular automata are performed by alternating E- and O-partitions [13]. The alternation of partitions is substituted by alternation of two groups of shifts of the image of simulated scheme as related to the fixed control layer in a CA. A group of shifts set into correspondence with E-partition is called *E-group*. A group of shifts set into correspondence with O-partition is called *O-group*. The introduction of a setting for CA allows reducing the number of parallel substitution commands that are executed by a single block to two. The command, whose number is specified first in a cell of the settings table of the control layer, can be executed when E-group is active, while the command, whose number is listed second, can be executed when O-group is active. A command is executed if its left part coincides with the image of a block in the informational layer. The execution of a command sets its right part as the image of a block.

Possible variants of signal transmission in the image of a digital scheme (horizontal and diagonal transmissions, branching and crossing of signals) and the indications to the functional elements (AND gate is denoted by the sign '&', OR gate is denoted by '1', addition by modulo 2 is denoted by '=1', half-adder" is depicted at end) are shown in Figure 5, b.

When E-group of shifts is active, the commands are executed for the source disposition of the image of a digital scheme, then for the image shifted by one row of blocks from E-partition up as related to the source image, and then one row of blocks down as related to the source image. The shifts are introduced in order to provide the execution of commands imitating the operation of gates and crossing of signals transmission channels in the image of a digital scheme. When O-group of shifts is turned on, commands are executed for the image obtained by shifting the source one by one row down and one column to the left.

The horizontal transmission of a signal is simulated by commands 1, 2. The diagonal transmission is performed by commands 3, 4. The signal branching is done by 5, 6 and signal crossing - by 8, 10, 15. The operation of AND gate is simulated by the two sets of commands: 9, 16, 17 and 12, 16, 17, because the result on the output of the gate can be recorded either in the right top cell or in the right bottom cell. of the block. Similarly, the simulation of OR gate is done by commands 7, 8, 9 and 10, 11, 12 and the simulation of addition - by modulo two is done by commands 7, 8, 18 and 10, 11, 18. A block of E-partition that does not contain white cells can simulate a gate operation not only with one output, but also with two outputs. Thus, the operation of half-adder is simulated by commands 7, 8, 12. The capability to simulate digital schemes with two inputs and two outputs allows performing both the signal transfer, e.g. by one of the diagonals, and a logical transformation of this signal and another signal. For example, a signal transition from the right bottom cell into the right top cell of a block along with performing AND operation with signals from the left column of cells in the block with writing the result to the right bottom cell of the block is simulated by commands 8, 15, 16.

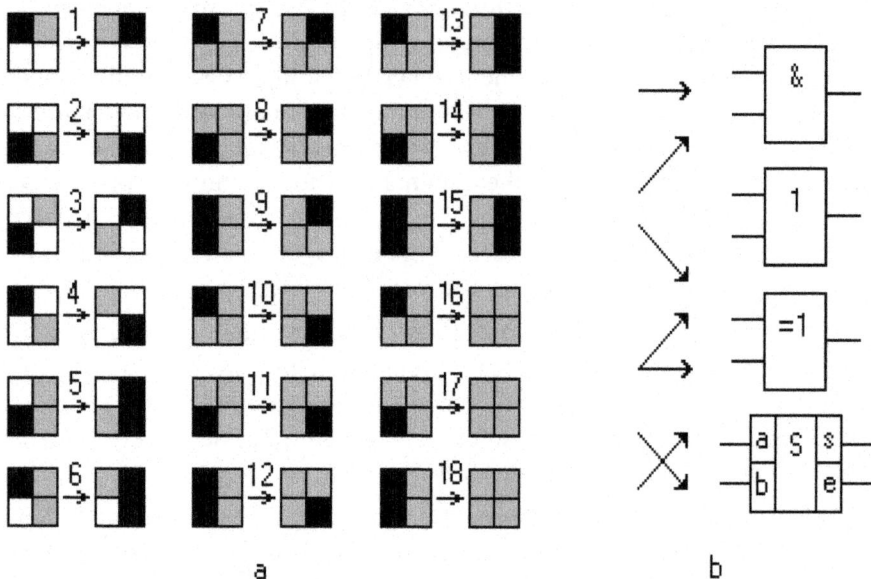

a   b

Figure 5: Graphic representation of parallel substitution commands, a - images of commands, b - symbols of signals and functional elements.

Let us demonstrate an image of a full adder in CA as a simple sample of what was stated above. It is presented in Figure 6.

Let us comment on this figure. The numbers of commands listed in Figure 5, a are used in the settings table. An image of a digital scheme with mnemonic specification of the functions of its composing blocks except those imitating the transfer of signals under O-partition is given in Figure 6, c in addition to the image in Figure 6, a, b. A sign of operation or an arrow in the image of scheme denotes a cell where result of transformation or data transfer is placed. Such a representation is rather easy to grasp and is introduced solely for the purpose of the reader's convenience. This helps in comparing a combinatory scheme and its image without the table of settings.

Usually, the signal transfers will be omitted when using such a kind of representation of a digital scheme unless that hampers the perception of the image of a scheme. Let us describe the resulting image of a digital scheme. Each of blocks (2, 2) and (3, 3) corresponds to a half-adder, which is a composition of AND gates and of addition by modulo two. Blocks (4, 2) correspond to OR gate.

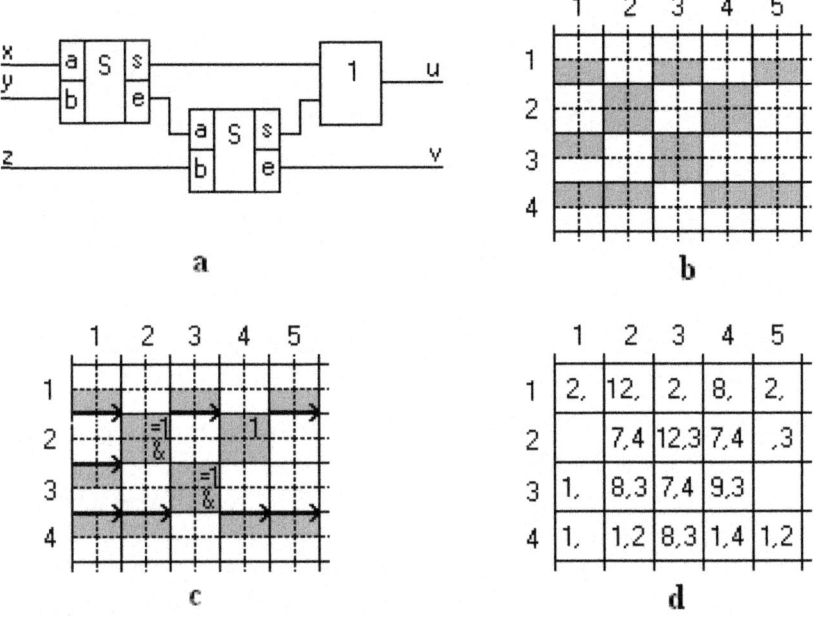

Figure 6: Full adder: a - a combinatory scheme, b - an image of scheme, c - a correspondence of blocks, d - the settings table.

Some or all of the gray cells in the first column of CA have to be altered to black ones in order to introduce input signals to a full adder.

A more complicated and realistic sample is presented in Figure 7. The following heuristic criteria are selected for the estimation of quality of a particular mapping. A mapping is optimal if there is at least one such chain from the inputs of a scheme to its outputs, that is composed only of blocks that imitate gates connected with other by corner cells. The image of an eight-bit pyramidal adder [17] transformed into a scheme, in which each gate has two inputs and one or two outputs is presented in Figure 7.

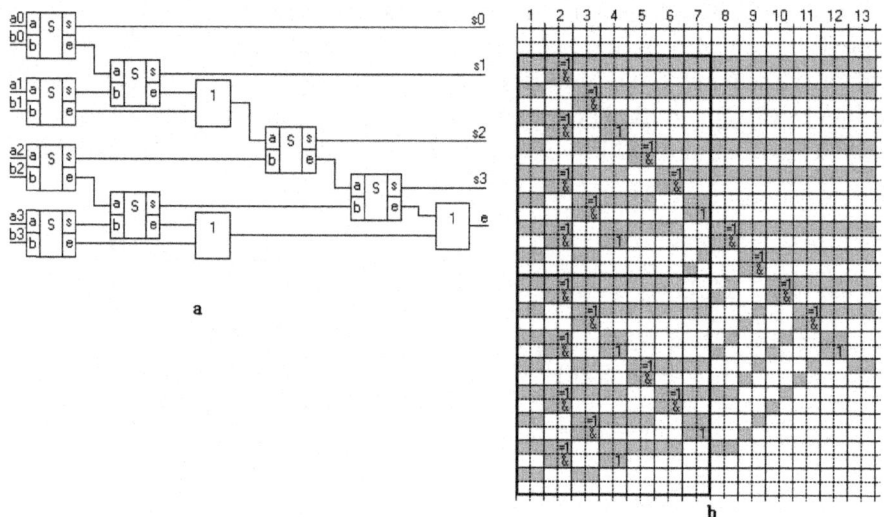

Figure 7: A pyramidal adder: a - the combinatory scheme of a four-bit adder, b - a concise image of eight-bit adder scheme

The image of a four-bit adder, which scheme is depicted in Figure 7, a, is a source fragment for its construction. The images of four-bit adders are contoured by a bold line in Figure 7, b. The rest of the scheme image provides daisy chaining of the selected images of adders. The image of an adder presented in Figure 8, b meets the heuristic criteria. The proposed way of constructing an image of a digital scheme with a specified bit width by connecting its homogeneous fragments of smaller bit width, for each of which the best image is already found, can be used for other kinds of schemes as well.

From a logic standpoint, a fine-grain structure (a micro-pipeline) is implemented by setting a CA up. This structure simulates a simultaneous operation of many copies of the same digital scheme. Each copy (let us call it a virtual digital scheme) performs a transformation of its own data set. A *stage* of a micro-pipeline forms a vertical column of blocks in E-partition of matrix. Each stage is a "cut" of image of a virtual scheme in a certain phase of transformation of the relevant information. This feature of matrix makes possible to obtain new results in each cycle. A cycle of micro-pipeline operation includes transformation and transfer of data from one stage to another. It means that a cycle includes an execution of the both groups of shifts. A model is constructed in WinALT in order to evaluate the correctness of an eight-bit adder implemented

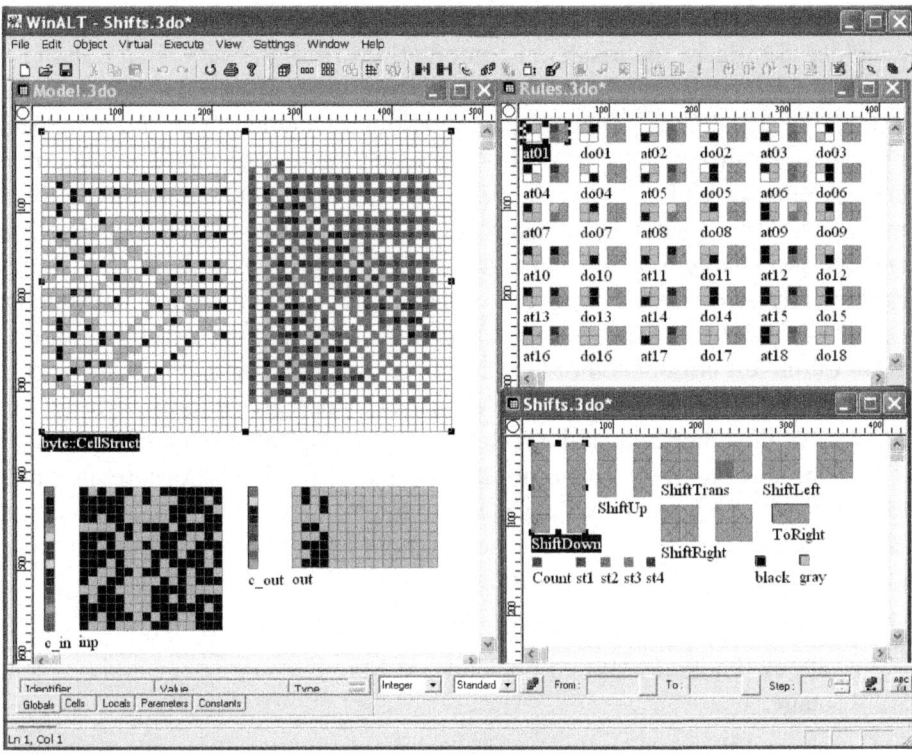

Figure 8: A project window of simulation model of CA that implements an eight-bit adder.

CA Simulation Model Implementing Eight-Bit Adder

The screenshot of a window of this project is presented in Figure 8. This window contains sheets with all the graphic objects of the model.

The sweep of a double layer cellular array byte::CellStruct is presented at sheet Model.3do. The upper layer containing the image of an eight-bit adder is presented at the left. The bottom layer is shown at the right. It contains a computer-aided representation of the settings table of an adder. The division of array layers into blocks is done by setting the cells of the bottom layer with both even coordinates into a state dedicated for this purpose. The numbers of substitution commands are represented by color in the settings table. The numbers of commands are distributed by the block cells as follows: the right and left top cells of block keep the numbers of the first and second commands respectively. The rest of cellular arrays shown in this window are auxiliary. Array inp keeps a set of input data. The columns of arrayinp are written in each cycle to the leftmost column of the informational layer of array byte::CellStructusing the array c_in and become the values of outputs of an adder. Similarly, the results of computations from the very right column of the informational layer of array byte::CellStruct are transferred to array out that stores the results of pipeline computations. Double layer templates are presented in sheet Rules.3do. The templates named ati and doi are put in correspondence with the left and right sides of substitution command with the number i. The upper layer of template ati coincides with the left part of the substitution number i, while that of template doi coincides with the right part of that substitution. The left upper cell of the bottom layer of template ati contains number i of substitution command. The number is represented by a color. A color of the left bottom cell of the same layer coincides with that of those cells in the bottom layer of array **byte::CellStruct**, which provides its partition into blocks. All the other cells of the both templates are empty (diagonally crossed). The templates used in substitution commands of the procedure imitating clocking of CA are presented in sheet Shifts.3do. A cycle of a model execution is divided into the four stages. At each of them parallel substitution commands kept in the settings table are executed:

- above the source image at the first stage;
- above the image shifted two cells up from the source one at the second stage;
- above the image shifted two cells down from the source one at the third stage;
- above the image shifted one cell down and one cell right from the source at the fourth stage (the command interleaving is done in the settings table during this stage: odd and even columns are swapped).

The operation of gates of a scheme is simulated at the first three stages of cycle. The data transfer between gates is simulated at the fourth stage. The cellular image and the settings table return to the initial («unshifted») state at the end of a cycle.

Let us briefly present main procedures of simulation program. The description of parallel substitution commands is kept in procedure mainProc, while the shifts performed at the each stage of a cycle are defined in procedure shiftImage. The body of procedure mainProc contains operator in with parameterbyte::CellStruct and eighteen bunches of at-do operators. The parameters of operators at and do in the i-th bunch are the templates named ati and doi respectively. The shifts of image of the digital scheme with respect to settings layer and the interleaving of substitution commands is done in procedureShiftImage by vector functional substitutions. Cellular arrays byte::CellStruct and Count listed in round brackets in the first operator in form a composition, which has the following meaning. The changes in array byte::CellStruct happen only when the unicellular array Count is in a certain state. For example, if the state of Count coincides with that of st2, the shift of the image of a digital scheme is done by one row of blocks down from its source placement. This shift is performed by the functionfShift that uses the values of variables x and y from template ShiftDown in its operator y:=x. The operators at-do in the second operator in alter the state of Count in order to set phases of cycle.

A Transformation of CA into Optoelectronic Matrix

Let us list those operations converting optical signals, which are to be carried out in an optoelectronic matrix to implement the above-mentioned CA. The operations have to provide the following for each block of CA: a) a comparison of the image of the left part of a command, whose number is written in the settings table, with the image of block in the image of a scheme, and b) if they coincide, replacement of the image of such block by the image of the right part of a command. Let us demonstrate that an execution of such operations is possible in a four layer matrix (called S). The images of left parts of substitution commands are kept in its first layer. The second layer contains the image of a digital scheme. The images of comparison schemes are in the third layer. The images of right parts of substitution commands are in the fourth layer. Matrix S is built using the basic gate rather schematically and without details of implementation. All the essential physical elements: modulators, photodetectors, memory cells are considered to have a size equal to one cell. The reasons for that are as follows: 1) matrix S built in such way can be easily turned into a data object when it is simulated, 2) the obtained matrix S has rather generalized form that makes possible to specify its versatile representations taking into consideration real sizes of its elements later, because these sizes are not known a priori and depend on physical principles of construction of the elements, technology, and the application domain.

The cells of informational layer of CA have three states. But the basic gate is binary. Thus, a transition must be done to binary encoding of the states of matrix S. For example, an encoding of white, gray and black cells presented in Figure 9 can be chosen.

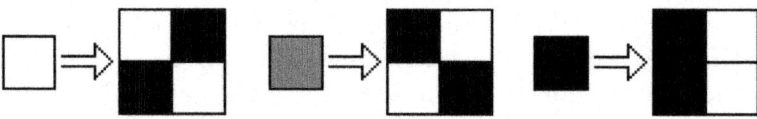

Figure 9: Binary encoding of CA states.

A white cell denotes a memory cell containing one, an opened modulator or an opened photodetector. A black cell denotes a memory cell containing zero, a closed modulator or a closed photodetector.

Let us commence the construction of matrix S using its second layer that contains an image of a digital scheme in a binary encoding. An image of each block in CA informational layer is composed of modulator states. Memory cells control the states of modulators (opened, closed). Assume that each cell has two outputs. Modulators are connected to both of them. Let us consider that each memory cell has two inputs, each of which is connected to a photodetector. Let us also assume that the signals from the inputs of a cell appear at its output with a fixed delay. A line of cells is composed of such elements. The first and second cells are modulators. The third cell is a memory cell. The fourth and fifth cells are photodetectors. A memory cell operates in two phases. If it contains one at the first phase (the phase of comparison), it opens the first modulator and closes the second one. And otherwise, it closes the first modulator and opens the second one. At the second phase (phase of recording) if the fourth photodetector is opened, and the fifth one is closed, one is written to memory cell. Otherwise, memory cell resets to zero.

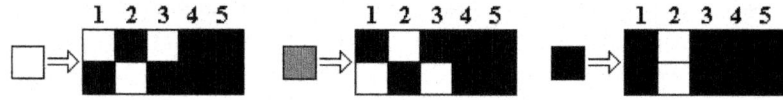

Figure 10: Binary encoding of cell states in the second layer of optoelectronic matrix.

An image of the second layer of matrix S when it is ready for the phase of comparison is obtained by replacing each cell in the information layer of CA by a group of cells according to the substitution rules in Figure 10. Thus, a block of optoelectronic matrix of size $10 \times 410 \times 4$ cells composed of four lines is put in correspondence to each block of the informational layer of CA.

Remark. Gray cells are also used in the transition to binary encoding, but they denote "void", absence of any hardware.

Let us perform a partition into blocks that have the same sizes and placement as in the rest of matrix layers.

Let us construct the first layer of matrix S. Unlike lines of cells form the second layer, the cell number four contains a memory cell and the fifth cell is gray. A memory cell is connected to modulators just as the memory cell occupying the third cell. The basis for setup of the first layer is the settings table of the control layer of CA. The setting of layer starts with encoding of its E-partition, i.e. the cells in columns 3 and 8 of the layer are set in such a way that the image formed by the states of modulators would coincide with the left part of that command, whose number is written first in the cell with exactly the same placement in the settings table of CA. The encoding for O-partition is embedded into the encoding of the layer that was just obtained. For this purpose binary codes set into correspondence to white, gray and black cells of the left part of the second command are written into the fourth column just as it is done for the cells of the third column in Figure 10.

The third layer of the matrix is constructed as follows. Black cells of a single block in the columns 1, 2, 6, 7 denote closed photodetectors. Black cells in the columns 4, 5, 9, 10 denote closed modulators. The rest of columns are empty. They are composed of gray cells. Pairs of photodetectors in rows 1, 2 and 3, 4 of the columns 1, 2 and 6, 7 are connected by OR scheme (parallel). Then all the OR schemes formed by pairs of photodetectors are connected sequentially forming AND scheme. A parallel assembly of all modulators of a block is connected as its load. The scheme obtained in the third layer allows to detect the situation when the images of blocks coded by the states of modulators located in the first and second layer one below another coincide, and in the case of full coincidence to prepare a substitution of the image of block of the second layer by the right part of command kept in the fourth layer.

The construction of the fourth layer can be done in two stages. First, a layer that is exactly the same as the first one for the right parts of commands is built. Then the columns in each block are transposed in the following order: 5, 4, 3, 1, 2, 10, 9, 8, 6, 7.

A fragment of matrix S presented in Figure 11 illustrates the procedure described above. The fragment is built for such a block of CA, whose settings table has a cell that contains the numbers of substitution commands 7, 4. The fragment is in such a state when E-shift is done and it appeared that the image of the block in the informational layer of CA coincides with the image of left part of substitution command number 7. A polarized light comes perpendicularly to the first and last layers of the matrix, respectively, from above and from below. The phase of comparison of images of blocks in the layers 1 and 2 of matrix S and detection of their coincidence is shown in Figure 11, a. The modulators in columns 4, 5 and 9, 10 are opened. The phase of writing new states of memory cells of the second layer using photodetectors connected to their inputs is depicted in Figure 11, b.

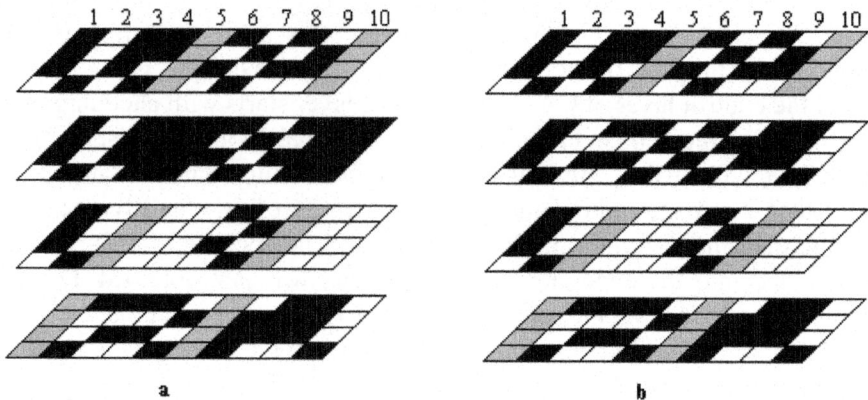

a   b

Figure 11: A fragment of the matrix S: a - detection of matching images, b - the phase of setting of new cell states in the second layer.

After a certain fixed delay, memory cells of the second layer set new states for modulators on their outputs. A fixed delay can be implemented, for example, by division of a memory cell into two elements: main and buffer. Modulators are connected to the outputs of the main buffer. Photodetectors are connected to the inputs of a buffer element. One cycle of state transition from a buffer element to a main one serves as a delay. Then the next shift of the image of a digital scheme is performed and the phase of comparison of image blocks can be done again.

It can be stated that CA turns into a simple homogeneous optoelectronic matrix S. It is a device that consists of four layers, which contain microscopic passive sources of light (optoelectronic modulators) and light detectors, forming a regular flat structure. The control of light sources and detectors is performed using electronic memory cells. Memory cells form a 2D shift register in order to provide setup of memory cells for storing the image of a scheme and alternation of commands. The obtained matrix is reconfigurable. It can be setup to imitate any combinatory scheme. The logical operations in the matrix (a comparison of the two codes and writing a code to memory) are preformed optically and their corresponding electronic gates simply do not exist. The proposed matrix has high performance. Indeed, after the pipeline has entered in the steady state, each next result is produced on the outputs of the matrix in each cycle.

The Family of Optoelectronic Matrices

Matrix S can serve as a basis for the construction of a family of similar matrices. Let us outline the possible ways of their construction. The circuitry of matrices may have rather different implementations. For example, one can construct a matrix of static memory elements (as in the previous section), but it is also possible to construct a matrix with dynamic memory elements. Another way of binary encoding of white, gray and black cells can be chosen. However, we consider more profound transformations of the original matrix. And here the following options are available. The first way. A number of layers can be reduced to two in a matrix. Let us first note that the need to alternate substitution commands in the control layer is induced by the necessity to execute commands in O-shifts. A modification of CA is proposed in [18] that eliminates the group of O-shifts and as a result the need to alternate commands in the settings layer. Let us select such a CA as a basis for the construction of a new optoelectronic matrix **S1**. The selection of such CA means that the memory cells in columns 4 and 2 can be erased in the lines of blocks of the first and fourth layers along with their modulators and photodetectors.

Let us introduce an additional limitation. Let us consider that a CA is setup once and only for implementation of a single digital scheme. Selecting a constant setting leads to the fact that there is no need to change the state of modulators in the first and fourth layers of the matrix S in the process of its operation. In its turn, this means that the memory cell can be removed, an open modulator can be replaced by a transparent plate, and a closed one by a dark plate. But then the first and fourth layers of the matrix S can be removed and the modulators in the second and third layers under dark plates can be masked or removed from the substrate. The execution of such a procedure can be demonstrated in greater details using masks in substitution command 1 (Figure 5) taken as a sample. A set of masks is presented in Figure 12.

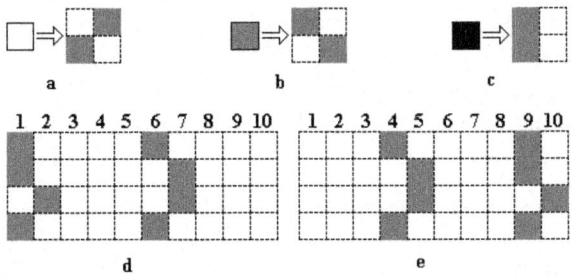

Figure 12: Masks: a - of a white cell, b - of a gray cell, c - of a black cell, d - the left part of substitution command 1, e - the right part of the same command.

Masks **d** and **e** are built using masks **a, b, c**. Mask **d** is superimposed on those blocks of the second layer, which are assigned to execute command 1. The modulators under gray cells are removed. Other elements of a block under white cells remain intact. Mask **e** is applied to such blocks of the third layer, that are located under similar blocks of the second layer, and the same kind of actions is performed. As a result a computer simulation model that implements a full adder for such a matrix was built.

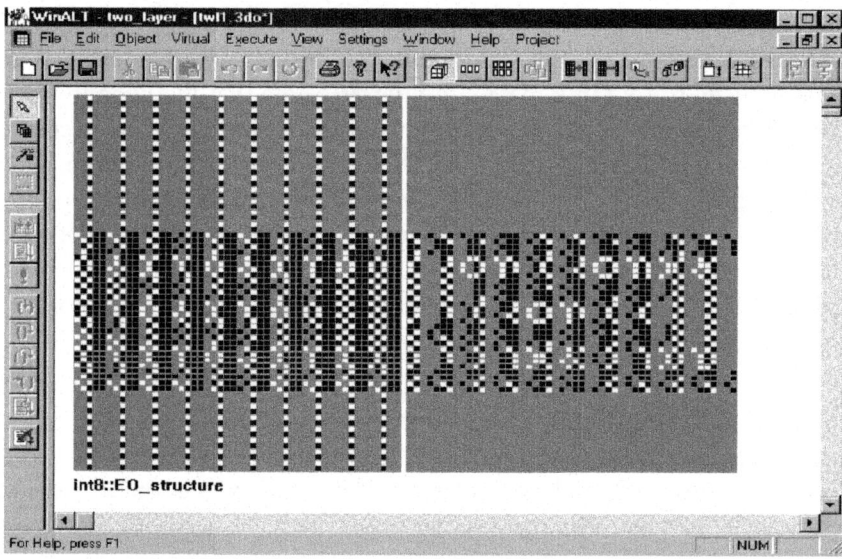

Figure 13: Computer image of a two layer optoelectronic matrix.

A certain step of computations performed in a double layer matrix is presented in Figure 13. The obtained model is rather realistic. All the optical and electronic components are shown in Figure 13. By analyzing the screenshot, one can easily draw a conclusion that the first layer of matrix is composed of logically isolated shift registers. Each of their digits has optical inputs and outputs. The second layer composed of isolated single comparison schemes. This way can be considered as one with the minimum fraction of electronic components in the matrix. The double layer matrix and its simulation are presented in details in [18].

However, it is clear that when considering a practical implementation of optoelectronic devices, one has to take into account the achievements of modern microelectronics, where the actual geometric dimensions of the transistors (approximately 30 nm) have become significantly smaller than the sources of light, which can not be less than 0.3 microns due to the physical limitations. They become a weak element in terms of achieving high-density packaging of elements in a chip. Therefore, when constructing such a family of matrices, it is useful to consider the maximum use of electronic components. "Optical" components keep only their main function, which is a three-dimensional organization of the logical connections and processing of cellular (logical) information. The electronic components are then used to perform simpler operations that support a computational process, such as the shift and storage of information, including settings, etc.

The second way. One can augment the electronic part of the hardware of matrix S and turn it into the matrix S2 with extended capabilities. Suppose that there are k (k > 1) four layer matrices and each i-th matrix is set to perform operation O_i. Assume that the sizes of the matrices and the location of inputs and outputs of simulated digital circuits are coordinated. Some of the considerations on how that can be obtained can be found in [19]. Let us combine all the electronic components of all these matrices in order to superpose them in a four layer matrix. It can be done the following way. A set of k memory cells is placed in layers 1, 2, 3 of the matrix instead of a single memory cell. Let us combine k_i cells into a 2D shift register set for execution of operation O_i. A decoder (named D1) is placed near each set of memory cells in the first and fourth layers. Similarly, two decoders are placed in the second layer (one D1 and one that is different from D1 is named D2). Decoder D1 has one group of outputs and two groups of inputs. The states of outputs can be 0 and 1. They are connected to modulators of layer the just the same way as the outputs of memory cells in the source matrix S. The first group of inputs of decoder D1 is for control. The codes of operations O_i (i = 1, ..., k) come to it. The second group of inputs has k pairs of inputs. The pair of outputs of the i-th memory cell of a set is connected to the i-th pair of inputs. If an operation code O_i comes to the control input of a decoder, the states of the i-th pair of the second input group are sent to the outputs of decoder D1. Decoder D2 is also controlled by operation codes. But its input group has two inputs, and it output group has k pairs of outputs. The inputs are connected to the outputs of photodetectors of the second layer the same way as the inputs of memory cells in the source matrix S. The i-th pair of outputs of decode D2 is connected to the inputs of the i-th memory cell of a set. If an operation code O_i came to the control input of decoder D2, the states of inputs are sent to the outputs of i-th pair of decoder D2. Each set of input data has to be accompanied by an operation code for such a matrix so as to specify what has to be done with these data. The transfer of operation codes from one micro-pipeline tier to another and their delivery to the decoders of the destination tier must be implemented. These actions can be performed for example by using additional shift registers that have to be inserted in the first, second and third layers of the matrix S2. As a result a matrix is built, in which one can change the image of a digital scheme and its setup table dynamically. That means that k different digital schemes can reside simultaneously in such a matrix. Thus from a logical standpoint the matrix is dynamically reconfigurable in the process of computations done by its pipeline. Matrix S2 has undoubted advantages in comparison with matrices S, S1. The delays of its pipeline are virtually eliminated, because its control device would always «find» a data set and an operation to perform and deliver those to the inputs of matrix S2.

The list of matrices is not limited by the samples presented above. Let us briefly outline some more possible ways of constructing new modifications of matrices. For example, a dynamically reconfigurable matrix can be built on the basis of a two layer matrix. In order to make it reconfigurable, one has to introduce electronic masking control first of all. Then an additional hardware is to be added just as it was shown in the previous section. A group of E-shifts of a digital image can be eliminated in all the proposed matrices by increasing the vertical sizes of blocks that are compared in layers at least by a factor of three. This technique helps in shortening the duration of a micro-pipeline cycle. An assessment of creating vertical assemblies built, for example, on the basis of two-layer matrices, seems to be rather interesting. Double-layer matrices are interleaved with layers of light sources in such kind of assemblies.

And finally, let us mark one more important advantage of optoelectronic configurable matrices. Their application replaces the design (logical and physical) and production of digital circuitry by a programming of an already existing matrix. Indeed, in order to get a digital circuit in a matrix, one need only to write specially selected long vectors to its layers, using the shift registers, which are formed by memory elements of the matrix layers.

CONCLUSION

Two most pressing problems appeared in constructing computing systems: 1) an increase in the number and length of the connections inside 2D ICs, 2) a massive parallelization of computations.

Using a 3D (multilayer) IC structure is proposed in this chapter in order to solve these problems. Each of the layers, forming such an IC, has a homogeneous cell structure. A data exchange between these cells is performed by the means of local electronic (intralayer) and parallel optical (interlayer) communication channels. Data exchange between layers is combined with their processing. Cellular 3D ICs are oriented to computations with fine grain parallelism.

The physical and computing aspects of creating such ICs were considered.

The physical aspects include constructing an optoelectronic gate that allows combining single-layer ICs in the two-layer structures and an experimental test of its operability. The assessment of potential of project of a customized device, a matrix processor for the image processing from the standpoint of its performance and of its manufacturability permits to draw the following conclusions about the prospects of the chosen direction of constructing optoelectronic circuits.

Despite the fact that the size of one of the components (light modulator) of the optoelectronic logic gate is much larger than the active components of the modern microelectronics because of the wave nature of the optical signal, the performance and manufacturability of specialized devices of this type can greatly exceed similar characteristics of a purely electronic device. This is due to the fact that the number of elementary components in an optoelectronic device is significantly less than that in a purely electronic device with the same functionality. Reducing the number of components results from a local structure of logical connections and from the unidirectional light propagation (in electronic conductors a required direction of motion of the electrons is obtained by using a set of gates). There are virtually no intersections in optical conductors. This greatly facilitates the design of ICs and lets them be cheaper, more reliable and easily manufacturable. High throughput between layers by the means of optical channels gives a hope that an essentially greater performance could be reachable in comparison with purely electronic schemes.

Computer aspects of the work are devoted to the foundations of the algorithmic design of 3D ICs based on the formal model of fine-grain parallel computing, Parallel Substitution Algorithm, and to WinALT simulation system, presented in [15, 16]. A family of optoelectronic matrices was developed using WinALT. Comparison of two graphic images in all matrices, which is a main logical operation, is carried out optically exactly the same way as in devices built on the basis of optoelectronic gate. The functions of information storage are carried out electrically. High homogeneity, simplicity of topology and a low complexity of cells are the features of the matrix. A wide range of matrices was built. They vary by a number of layers, by functionality and by ratio of optical and electronic components. The proposed matrices can serve as an ALU basis in a general purpose CPU. The selection of a particular choice depends primarily on the kind of physical parameters, on which a designer can be oriented to, and on the kind of operations that a matrix has to perform. And this requires interaction between experts in different domains: physics, algorithm, programming, etc. The above suggests the following conclusions about the directions in which algorithmic design tools are to be developed.

The system's site must evolve into a fully functional online resource providing the ability to create a network (virtual) team consisting of specialists in different domains, and let them actively participate in a joint development of a 3D IC using the Internet.

The mapping of only one version of a fine-grain structure, a reconfigurable CA, to an optoelectronic structure is presented in the paper. There are many fine-grain structures and algorithms for different purposes, which may be of interest for optoelectronic implementation. It is advisable to perform a constant replenishment of the collection of simulation models on the system's site both by its developers and by users. Also a collection of simulation models of optoelectronic implementations of fine grain structures and algorithms must be created. Let us also note the need to constantly replenish the system by new modules that extend its functionality with the emergence of new kinds of fine-grain structures and algorithms.

Constructing realistic optoelectronic structures requires constructing simulation models with huge amount of data and computations. It is expedient to launch such models on supercomputers. Thus, a WinALT subsystem must be developed for parallel execution of these models on clusters and supercomputers.

REFERENCES

1. Y. Jin-Sung, L. Myung-Jae, P. Kang-Yeob, C. 1. Woo-Young, 850-nm. C. M. O. S. O. E. I. C. Gb/s, with. a. Receiver, Avalanche. Silicon, I. E. E. Photodetector, IEEE journal of quantum electronics 2012482229236

2. I. A. Young, M. Edris, J. T. S. Liao, A. M. Kern, S. Palermo, B. A. Block, M. R. Reshotko, P. L. D. Chang, I. O. Optical, for. Technology, Computting. I. E. E. Tera-Scale, IEEE Journal of solid-state circuits 2009451235248

3. Beausoleil R G.Large-Scale Integrated Photonics for High-Performance Interconnects. ACM Journal on emerging technologies in computing systems 2011Special Issue SI N 6.

4. A. V. Krishnamoorthy, K. W. Goossen, J. William, et al.Progress in Low-Power Switched Optical Interconnects. IEEE Journal of selected topics in quantum electronics 2011172357376

5. K. Hoshino, K. Yamada, K. Matsumoto, I. Shimoyama, Creating a nano-sized light source by electrostatic trapping of nanoparticles in a nanogap. J. Micromech. Microeng. 2006161285

6. K. Nozaki, T. Baba, Laser characteristics with ultimate-small modal volume in photonic crystal slab point-shift nanolasers. Appl. Phys. Lett. 2006

7. K. Nozaki, S. Kita, T. Baba, Room temperature continuous wave operation and controlled spontaneous emission in ultrasmall photonic crystal nanolaser. Optics express. 2007151275067514

8. Egorov V M, Kostsov E G.The prospects for constructing high-performance optical digital computers. Optoelectronics, Instrumentation and Data Processing. 19851106

9. Egorov V M, Kostsov E G.Microelectronic optical digital computing device. Optoelectronics, Instrumentation and Data Processing. 1989361

10. V. M. Egorov, E. G. Kostsov, Integral optical digital computers. Appl. Opt. 199029811781185

11. E. G. Kostsov, S. V. . D. I. Piskunov, with. Cs, interconnections. optical, In: Tverdokhleb P.E. (ed.). 3D Laser Information Technologies. Novosibirsk; 2003168242

12. Markov A. A. Theory of Algorithms. Proc. Steklov. Moscow-Leningrad: USSR Academy of Sciences Publishing House. 1954;42.

13. T. Toffoli, N. Margolus, automata. Cellular, a. machines, environment. new, modeling. for, Cambridge, Massachusetts: MIT Press; 1987

14. S. Achasova, O. Bandman, V. Markova, S. Piskunov, Substitution. Parallel, Theory. Algorithm, Singapore. Application, World, Sci.; 1994

15. M. Ostapkevich, S. Piskunov, The Construction of Simulation Models of Algorithms and Structures with Fine-Grain Parallelism in WinALT. In: V. Malyshkin (ed.) PaCT 2011. LNCS. 6873. Heidelberg, Springer; 2011192203

16. WinALT site. http://winalt.sscc.ru/ (accessed 25Apr 2012

17. M. A. Kartsev, V. A. Brick, Computing systems and synchronous arithmetic. Moscow: Radio and Communications; 1981

18. E. G. Kostsov, S. V. Piskunov, Computer design of a two-layer computational array with optical interconnections. Optoelectronics Instrumentation and Data Processing C/C of Avtometriia. Allerton Press. 200033

19. Kostsov E G, Piskunov S V, Umrikhina E V.Configurable multilayer pipeline electro-optical structures. Optoelectronics Instrumentation and Data Processing C/C of Avtometriia. Allerton Press. 2005662

Chapter 9

WIRED/WIRELESS PHOTONIC COMMUNICATION SYSTEMS USING OPTICAL HETERODYNING

Alejandro García Juárez[1], Ignacio Enrique Zaldívar Huerta[2], Antonio Baylón Fuentes[3], María del Rocío Gómez Colín[4], Luis Arturo García Delgado[1], Ana Lilia Leal Cruz[1] and Alicia Vera Marquina[1]

[1]University of Sonora, Department of Physics Research, México

[2]National Institute of Astrophysics, Optics and Electronics. Department of Electronics, México

[3]Inst. FEMTO-ST, Université de Franche-Comté, Besançon, France

[4]University of Sonora, Department of Physics, México

INTRODUCTION

Over the past few years, there has been an increasing effort in researching new design of indoor wireless communications systems, due to connectivity that show in a room or in a building. Currently, several companies of telecommunications use purely omnidirectional antennas in their wireless routers to transmit data to laptops in close vicinity [1]. The properties of microstrip patch antennas and arrays with their planar configuration exhibit an attractive option for indoor communications where the gain is considerably enhanced. On the other hand, the generation of microwave and millimetre-wave (mm-wave) signals by using photonic techniques are being used in radio-over-fiber (RoF) systems, distribution antenna systems, broadband wireless access networks, and radar systems etc. In all these applications the microwave signals are generated at a remote central station and distributed transparently to several simplified antenna stations via optical fiber [1]. The main goal of these systems is to reduce infrastructure cost and to overcome the capacity bottleneck in wireless access networks, allowing, at the same time, flexible merging with conventional optical access networks. Thus, in order to design a reliable RoF-based access network infrastructure, RoF techniques must be capable of generating the microwave signals and allow a reliable microwave signals transmission over the optical link. For broadband wireless systems and distribution antenna systems operating at microwave and millimeter-wave carriers, several photonic techniques for generating microwave signals have been proposed. Among the most common used techniques are: optical heterodyning [2], optical injection locking [3], optical frequency/phase locked loops (OFLL/OPLL) [4], microwave generation using external modulation [5]. Optical injection locking and optical phase-locked loops (OPLL) are expensive in practice. The use of external intensity modulation generates frequency doubling or quadrupling of the driven radiofrequency (RF) sinusoid signal [6]. This method requires an external modulator which increases both loss and cost, and is more susceptible to bias drifting of the modulators, which can affect the output spectrum. The key advantage for generating microwave or millimeter-wave signals by optical means is that very high-frequency signals with very low phase noise and high purity can be generated. By using optical heterodyne technique it is very easy to tune frequencies with a spectral linewidth of a few ten MHz and over a wide range by simply tuning the wavelength of the two optical input signals; the obtained frequencies are limited only by the photodetector bandwidth [2]. Besides, the generated signals by using this technique can be generally used as both information carriers, and as a local oscillator for transmitting and receiving both analog and digital information signals by using not only RF schemes but also through an optical fiber. On the other hand, microwave photonics, which brings together radiofrequency engineering and optoelectronics, has attracted great interest in the field of telecommunications since it is an excellent alternative for the transmission of services such as high quality audio and video, e-mail, and Internet among others [7]. Furthermore, there has been an increase effort in researching new microwave photonics techniques for different interesting application that attracts interest in research is the filtering of microwave signals by using photonic techniques. The main feature of a photonic filter is that microwave signals are directly processed in the optical domain exploiting advantages inherent to photonics such as low loss, high bandwidth, immunity to electromagnetic interference, and tunability [8]. On the other hand, network architectures such as FTTx, where x can stand for home (H), building (B), neighborhood (N), or curb (C), are a communication architecture in which the final connection to the subscribers is optical fiber. Another important application of photonic telecommunications systems, which is very closely related to the FTTx systems, is the distribution of signals by integrating optical and wireless networks and passive optical networks (PONs). This particular type of scheme is referred as fiber-radio system [9]. Along with wavelength division multiplexing (WDM) technique, it would be more advantageous if RoF is integrated with a conventional PON where a base station (BS) plays the role of an optical network unit (ONU) to support both wired and wireless services. This integrated optical access system is capable not only to reduce the cost of multifunction BSs and the whole system but also meet the demands for bandwidth, mobility and connection options of users [10]. In this sense, the purpose of this chapter is to describe two an alternative optical communications systems. The first proposed system uses a couple microstrip antennas for distributing point to point analog TV with coherent demodulation based on optical heterodyne. In the proposed experimental setup, two optical waves at different wavelengths are mixed and applied to a photodetector. Then a beat signal with a frequency equivalent to the spacing of the two wavelengths is obtained at the output of the photodetector. This signal corresponds to a microwave signal located at 2.8 GHz, which it is used as a microwave carrier in the transmitter and as a local oscillator in the receiver of our optical communication system. The feasibility of this technique is to demonstrate the transmission of a TV signal located at 66-72 MHz. The second system deals with the experimental transmission of analog TV signal in a fiber-radio scheme using a microwave photonic filter (MPF). For that purpose, filtering of a microwave band-pass window located at 2.8 GHz is obtained by the interaction of an externally modulated multimode laser diode emitting at 1.5 μm associated to the chromatic dispersion parameter of an optical fiber. Transmission of TV signal coded on the microwave band-pass window is achieved over an optical link of 20.70 km. Demodulated signal is transmitted via radiofrequency using printed antennas.

OPTICAL HETERODYNE TECHNIQUE

The basic principle for generating microwave carriers is based on optical heterodyne technique, it represents a physical process called optical beating or frequency beating, where two phase-locked optical sources with angular frequencies ω_1 and ω_2 are superimposed and injected into a high frequency photodetector that permits to obtain a photocurrent at a frequency $\omega_2 - \omega_1$. To explain this in more detail, let us consider the relation between the generated electrical output signal and the two superimposed optical input waves from a more physical point of view. For simplicity, we assume that the two optical input waves are linearly polarized monochromatic plane waves in the infrared which propagate in the +z direction. Let

$$E_1 = \hat{E}_1 \exp\left[i\left(\omega_1 t - k_1 z + \varphi_1\right)\right]e_1,$$

(1)

and

$$E_2 = \hat{E}_2 \exp\left[i\left(\omega_2 t - k_2 z + \varphi_2\right)\right]e_2,$$

$$(2)$$

be the complex electrical field vectors of the two optical waves, with field amplitudes $\hat{E}_1$ and $\hat{E}_2$, angular frequencies ω_1 and ω_2 and wave numbers k_1 and k_1 φ_1 and φ_1 and e_1 and e_2 are the unit vectors determining the orientation of the electrical field vector of the linearly polarized optical input waves. The intensities of the constituent waves are given by the magnitude of their Poynting vectors and are therefore given by [11]

$$I_1 = \frac{1}{2}\left(\frac{\varepsilon_r \varepsilon_o}{\mu_o}\right)^{1/2} |E_1|^2.$$

$$(3)$$

$$I_2 = \frac{1}{2}\left(\frac{\varepsilon_r \varepsilon_o}{\mu_o}\right)^{1/2} |E_2|^2.$$

$$(4)$$

If the two incident optical waves are perfect plane waves and have precisely the same polarization (

$e_1 = e_2$, the resulting electrical field E_o of the optical interference signal is the sum of the two constituent input fields and hence we can write $E_o = E_1 + E_2$. Taking the squared absolute value of the optical interference signal we obtain

$$|E_o|^2 = |E_1 + E_2|^2 = |E_1|^2 + |E_2|^2 + E_1 E_2^* + E_1^* E_2$$
$$= |E_1|^2 + |E_2|^2 + 2|E_1||E_2|\cos\left((\omega_2 - \omega_1)t - (\varphi_2 - \varphi_1)\right).$$

$$(5)$$

From equation (5) and by using equations (3) and (4), it follows that the intensity of the interference signal I_o is given by [11]

$$I_o = I_1 + I_2 + 2(I_1 I_2)^{1/2} \cos\left((\omega_2 - \omega_1)t - (\varphi_2 - \varphi_1)\right).$$

$$(6)$$

By launching this optical interference signal into a photodetector, a photocurrent i is generated which can be expressed as [11]

$$i = \frac{\eta_o q}{hf_1}P_1 + \frac{\eta_o q}{hf_2}P_2 + 2\frac{\eta_{f_c} q}{h}\left(\frac{P_1 P_2}{f_1 f_2}\right)^{1/2}\cos\left((\omega_2 - \omega_1)t - (\varphi_2 - \varphi_1)\right),$$

$$(7)$$

where

q is the electron charge and P_1 and P_2 denote the optical power levels of the two constituent optical input waves. The photodetector's DC and high-frequency quantum efficiencies are represented by η_o and η_{f_c}

It is of course important to consider that the detector's quantum efficiency is not independent of the frequency. Several intrinsic and extrinsic effects such as transit time limitations or microwave losses will eventually limit the high-frequency performance of the detector and thus the detector's DC responsivity η_o is typically much larger than its high-frequency responsivity η_{f_c}. In our case, we can further simplify the photocurrent equation (Eq. (7)) by considering the fact that the two optical input waves are close in frequency $(f_1 \approx f_2)$ whereas the difference frequency f_c is by far smaller $(f_c = | f_2 - f_1 | << f_1, f_2)$. If we further assume for simplicity that the power levels of the two optical input waves are equal ($P_{opt} \approx P_1 \approx P_2$), Eq. (7) becomes [11]

$$i = 2s_o P_{opt} + 2s_{f_c} P_{opt} \cos\left(2\pi f_c t + \Delta\varphi\right).$$

(8)

Where $\Delta\varphi = \varphi_2 - \varphi_1$. Here $s_o = \dfrac{\eta_o q}{hf}$ and $s_{f_c} = \dfrac{\eta_{f_c} q}{hf}$ are the photodetector's DC and high frequency responsivities given in A/W. Eq. (8) is the fundamental equation describing optical heterodyning in a photodetector. The first term is the DC photocurrent generated by the constituent optical input waves and the second term is the desired high-frequency signal oscillating at the difference frequency f_c (down-converter) or intermediate frequency (IF) [1]. In our case it represents the microwave signal that we will use as both information carriers, and as a local oscillator for transmitting and receiving TV signals in a wireless communication system.

EXPERIMENTAL SCHEME FOR GENERATING MICROWAVE SIGNALS

The heterodyne technique for generating microwave signals has been done using the experimental setup shown in Figure 1. In this experiment, two laser diodes emitting at different wavelengths are used. One of them is a tunable laser (New Focus, model TLB-3902) which can be tuned over the C band with a channel spacing of 25 GHz, and the other one is a fiber coupled distributed feedback (DFB) laser source (Thorlabs, model S3FC1550) with a central wavelength at 1550 nm. For the generation of the microwave signals, the outputs of both lasers are coupled to optical isolators to avoid a feedback into the lasers and consequently instabilities to the system. A pair of polarization controllers is used to minimize the angle between the polarization directions of both optical sources. Thus, the polarization of the light issued from each optical source is matched and therefore, there is not degradation of the power levels in the microwave signals generated from the photodetector. The output of each controller is launched to a 3 dB coupler to combine both optical spectrums.

After that, an optical output signal is received by a fast photodetector (MITEQ model SCMR-50K6G-10-20-10) with a typical gain of 25 dB, and –3 dB bandwidth of 6 GHz, The resulting photocurrent from the photodetector corresponds to the microwave beat signal which is analyzed with an Electrical Spectrum Analyzer (ESA), (Agilent model E4407B). The other optical output resulting from optical coupler is applied to an Optical Spectrum Analyzer (OSA) (Anritsu model MS9710C), for monitoring the wavelength of the two beams.

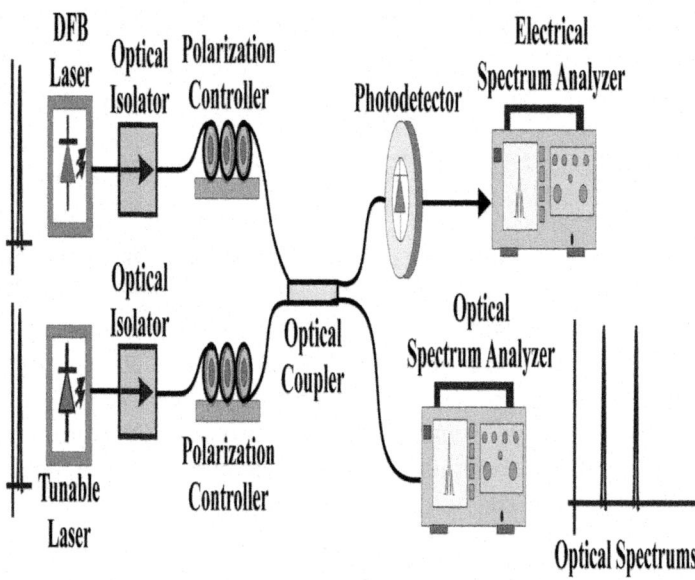

Figure 1: Experimental setup for generating microwave signals by using optical heterodyne technique.

DFB laser can be used to control not only the output power of the fiber coupled laser diode, but also the precise control of the temperature at which the laser is operating. Both controls can be used to tune the fiber coupled laser diode to an optimum operating point, providing a stable output. In this way, it is possible to observe that the wavelength of the DFB laser is shifting, by varying its temperature with a scale of 1 °C. Consequently, the beat signal frequency is continuously over the band of photodetector. On the other hand, the frequency difference from both lasers can be expressed by [1]

$$\Delta f = \frac{c}{\lambda_1} - \frac{c}{\lambda_2} = \frac{c(\lambda_2 - \lambda_1)}{\lambda_1 \lambda_2} \approx \frac{c}{\lambda^2}|\Delta\lambda|,$$

(9)

where λ_1 and λ_2 are the wavelengths of the two beams, respectively, and Δ_λ is the difference between the two wavelengths. To obtain a microwave signal, in a first step, the tunable laser is biased and its optical spectrum is displayed on the OSA screen. In a second step, the DFB laser is also biased, fixing an optical power of 2.2 mW and its central wavelength is settled as near as possible to the central wavelength of the tunable laser. As can be seen from Figure 2, the value of Δ_λ = 0.023739 nm is the wavelength difference between both lasers and it corresponds to the beat signal frequency of 2.8 GHz.

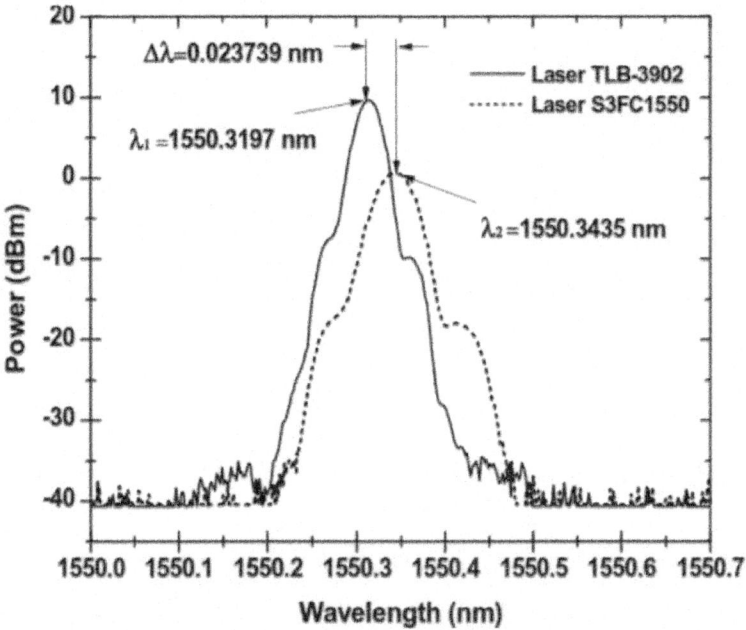

Figure 2: Optical spectrum corresponding to the mixed optical sources. The peaks located at λ_1=1550.3197 nm and λ_2=1550.3435 nm corresponds to the tunable and DFB lasers, respectively.

A precise control of the difference, between the two central wavelengths and by consequence over the frequency difference, is obtained by tuning the DFB. The wavelength variation of the laser source is obtained by changing the junction temperature between 22.8 °C, 23.2 °C and 23.7 °C corresponding to the frequency range of 0 to 5.0 GHz. Figure 3 illustrates the electrical spectrums of four generated microwave signals by using optical heterodyne. These signals are located at f_1=1.0, f_2=2.0, f_3=2.8 and f_4=4.0 GHz respectively. It can be seen, that the microwave signals are in good agreement with theoretical value given by Eq. (9). Therefore, when one laser source is operating at a fixed wavelength and the other is being continuously tuned, the beat frequency will shift correspondingly. In particular, the frequency of the microwave drive signal is set at 2.8 GHz.

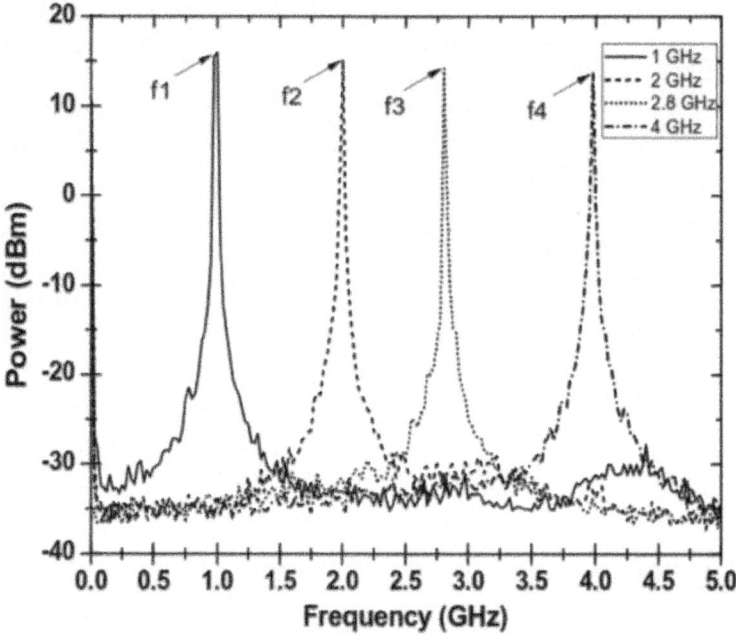

Figure 3: Spectrum for the microwave signal generated by using optical heterodyne.

MODULATION AND DEMODULATION

Some form of modulation is always needed in an RF system to translate a baseband signal (e.g., audio, video, data) from its original frequency bandwidth to a specified RF frequency spectrum. There are many modulation techniques, for example, amplitude modulation (AM), frequency modulation (FM), amplitude shift keying (ASK), frequency shift keying (FSK), phase shift keying (PSK), biphase shift keying (BPSK), quadriphase shift keying (QPSK), 8-phase shift keying (8-PSK), 16-phase shift keying (16-PSK), minimum shift keying (MSK), and quadrature amplitude modulation (QAM). AM and FM are classified as analog modulation techniques, and the others are digital modulation techniques [12]. In this section we describe the AM modulation and demodulation due to it was used in our proposed wireless communication system.

Amplitude Modulation

Analog modulation uses the baseband signal (modulating signal) to vary one of three variables: amplitude

A_c, electrical frequency $(\omega_1-\omega_2)=\omega_c=2\pi f_c$; or phase $(\phi_1-\phi_2)=\Delta\phi$. According toEq. (8), the obtained carrier signal by using optical heterodyne technique can be written by

$$p(t) = A_c \cos\big((\omega_1 - \omega_2)t + \phi_1 - \phi_2\big) = A_c \cos\big(2\pi f_c t + \Delta\phi\big).$$
(10)

Where $A_c=2s_{f_c}P_{opt}$ In amplitude modulation, if we assume that s(t) is the information signal, and considering $A_c=1$, $\Delta\phi=0$, then a modulated signal can be written by

$$g(t) = s(t)\cos 2\pi f_c t.$$
(11)

Applying the modulation property of the Fourier transform to Eq. (11), we can find the density spectral of

g(t) is

$$G(f) = \frac{1}{2}S(f - f_c) + \frac{1}{2}S(f + f_c).$$
(12)

Amplitude modulation therefore translates the frequency spectrum of a signal by $\pm$ f $_c$ hertz, but leaves the spectral shape unaltered. This type of amplitude modulation is called sup- pressed-carrier because the spectral density of g(t) has no identifiable carrier in it, although the spectrum is centered at the frequency f$_c$.

Amplitude Demodulation

Recovery the signal information s(t) from the signal p(t) requires another translation in frequency to shift the spectrum to its original position. This process is called demodulation or detection. Because the modulation property of the Fourier transform proved useful in translating spectra for modulation, we try it again for demodulation. Assuming that $g(t)=s(t)\cos 2\pi f_c t$ is the transmitted signal, we have

$$g(t)\cos 2\pi f_c t = s(t)\cos^2 2\pi f_c t = \frac{1}{2}s(t) + \frac{1}{2}\cos 4\pi f_c t.$$
(13)

Taking the Fourier transform of both sides of Eq. (13) and using the modulation property, we get

$$\Im\big[g(t)\cos 2\pi f_c t\big] = \frac{1}{2}S(f) + \frac{1}{4}S(f + 2f_c) + \frac{1}{4}S(f - 2f_c).$$
(14)

The mathematical process described in this section can be obtained by convolving the spectrum of the received signal g(t) with that of $\cos 2\pi f_c t$ (i.e., with impulses at $\pm f_c$). A low-pass filter is required to separate out the double frequency terms from the original spectral components. Obviously we need a filter with a cut frequency $f_{cut} > 2f_m$ for proper signal recovery. In this case f_m represents the information frequency.

Effects in Frequency and Phase Variations

When the local oscillator at the receiver, has a small frequency error Δ_f and a phase error Δ_θ, then this signal can be written as

$$p_L(t) = \cos\left[2\pi\left(f_c + \Delta f\right)t + \Delta\theta\right].$$
(15)

Assuming again that $g(t) = s(t)\cos 2\pi f_c t$ is the transmitted signal; then we have that at the receiver, the recovered signal can be written by

$$g(t)\cos\left[2\pi(f_c + \Delta f)t + \Delta\theta\right] = s(t)\cos(2\pi f_c t)\cos\left[2\pi(f_c + \Delta f)t + \Delta\theta\right]$$

$$= s(t)\left(\frac{\cos(2\pi\Delta ft + \Delta\theta)}{2} + \frac{\cos\left[2\pi(2f_c + \Delta f)t + \Delta\theta\right]}{2}\right).$$
(16)

The second term on the right hand side of Eq. (16) is centered at $\pm 2f_c + \Delta f$ and can be filtered out by using a low pass filter. The output of this filter $s_F(t)$ will then be given by the remaining term in Eq. (16).

$$s_F(t) = \left[\frac{s(t)}{2}\left(\cos 2\pi(\Delta f)t\cos(\Delta\theta) - sen 2\pi(\Delta f)t sen(\Delta\theta)\right)\right].$$
(17)

As can been from equation (17), the output signal is not $\frac{s(t)}{2}$, unless both Δ_f and Δ_θ are zero. The effects of both frequency errors and random phase errors render this demodulation of the signal unsatisfactory. It is necessary, therefore, to have synchronization in both frequency and phase between the transmitter and the receiver when amplitude modulation is used. The synchronization of the carrier signals presents no major problem when the transmitter and the receiver are in close proximity. Recovering the original signal s(t) from the modulated signal g(t) using a synchronized oscillator is called coherent demodulation. In our case we take advantage of proposed optical heterodyne technique permits to obtain microwave carrier and local oscillator simultaneously in the transmitter and receiver respectively.

DESIGN OF A PATCH ANTENNA AT 2.8 GHZ

The microstrip patch antenna is a popular printed resonant antenna for narrow-band microwave wireless links that require semihemispherical coverage. Due to its planar configuration and ease of integration with microstrip technology, the microstrip patch antenna has been studied heavily and is often used as an element for an array. Common microstrip antenna shapes are square, rectangular, circular, ring, equilateral triangular, and elliptical, but any continuous shape is possible [13]. Furthermore, a patch antenna is an excellent device due to its small size, low cost, and good performance [14-16]. In this chapter, a rectangular printed patch antenna is proposed. Simulation results have been obtained by using Advanced Design System (ADS) that is a computer-aided-engineering software tool. The radiating structure consists of a patch and a microstrip inset-feed line, allowing that the characteristic impedance (Z_o) to be improved. Figure 4 shows the geometry and configuration of the top layer. The proposed antenna in this work was designed to operate in the band S of telecommunications (2.8 GHz). FR4 is used as a dielectric substrate exhibiting a thickness h=1.524mm, and relative dielectric constant ε_r=4.2 . In a first step, the width (W) of the patch is computed by using [17]:

$$W = \frac{c_o}{2f_c}\sqrt{\frac{2}{\varepsilon_r+1}},$$

$$\text{(18)}$$

where c_o is the light velocity in the free space, and f_o is the operation frequency. Next, the value of the effective dielectric constant ε_{eff} is evaluated considering W/h>1

$$\varepsilon_{eff} = \frac{\varepsilon_r+1}{2} + \frac{\varepsilon_r-1}{2}\left(1+12\frac{h}{W}\right)^{-1/2}$$

$$\text{(19)}$$

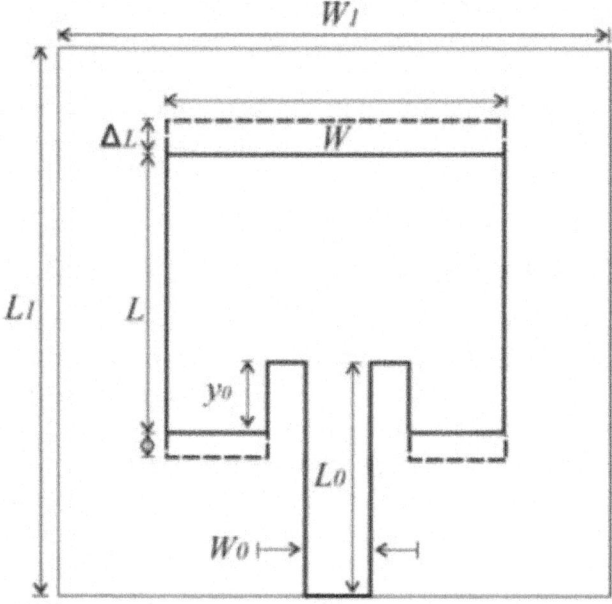

Figure 4: Layout of the patch antenna.

Border effects [17] must to be considered in the design of the antenna. For this reason, Δ_L from Figure 4 can be evaluated as:

$$\frac{\Delta L}{h} = 0.412\frac{\left(\varepsilon_{eff}+0.3\right)\left(\dfrac{W}{h}+0.264\right)}{\left(\varepsilon_{eff}-0.258\right)\left(\dfrac{W}{h}+0.8\right)}$$

(20)

This allows that the length (LL) of the patch to be evaluated as:

$$L = \frac{c_o}{2f_o\sqrt{\varepsilon_r}} - 2\Delta L$$

(21)

Considering the values previously obtained, the effective dimensions (L_{eff} and W_{eff}) can be calculated, respectively as:

$$L_{eff} = L + 2\Delta L$$

(22)

$$W_{eff} = W + \frac{t}{\pi}\left(1 + \ln\left(\frac{2h}{t}\right)\right)$$

(23)

From Eq. (23), t is the conductor thickness and $W/h > 1/2\pi$ must to be considered. The ground plane dimensions are computed as:

$$L_1 = 6h + L_{eff}$$
$$W_1 = 6h + W_{eff}$$

$$(24)$$

The best dimensions which assure a good matching between the impedances ($R_{in} = Z_0 = 50\Omega$) of the antenna and generator can be calculated by the use of LineCalc tool from ADS and by the next expression:

$$R_{in}(y = y_o) = \frac{1}{2(G_1 \pm G_{12})} \cos^2\left(\frac{\pi}{L} y_o\right)$$

$$(25)$$

where G_1 and G_{12} are the conductance values obtained by the cavity method. Finally, Table 1 shows a summary of the dimensions for the patch and the ground plane.

Table 1: Dimensions of the fabricated antenna

Operation Frequency(Antenna)	Dimensions (cm)						
	$_W0$	$_L0$	W	L	$_W1$	$_L1$	$_v0$
2.8 GHz	0.13	3.08	3.32	2.56	10	10	0.93

Figure 5(a) shows a picture of the fabricated patch antenna where a SubMiniature version A (SMA) connector is added. Figure 5(b) illustrates simulation and experimental results corresponding to the S_{11} parameter. Electrical measurements are obtained by using a Vector Network Analyzer (VNA) (Agilent Technologies model: E8361A). It is clearly observable that experimental result is in good agreement with the simulation.

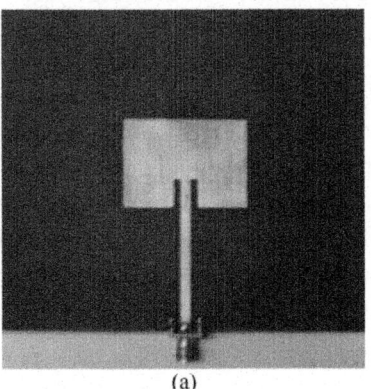

(a)

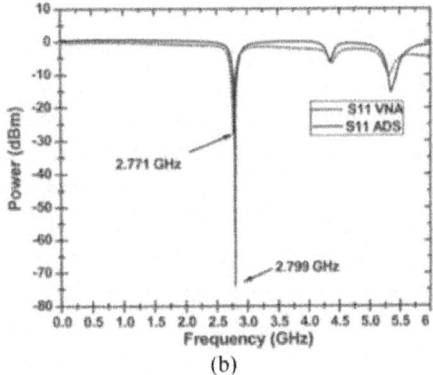

(b)

Figure 5: Fabricated antenna (a), Experimental and simulation return loss curve for the antenna (b).

TRANSMISSION OF TV SIGNALS BY USING HETERODYNE TECHNIQUE

In order to show a potential application of optical heterodyne technique in the field of the wireless communications, we have proposed a coherent wired/ wireless photonic communication system as shown in Figure 6. This system is not a truly wireless communication system, since an optical fiber is required to deliver both microwave carrier and local oscillator for transmitting and receiving information of TV signals as an approximation to point to point indoor wireless communications systems. From the photodetector 1 in the transmitter, a microwave signal located at 2.8 GHz is obtained and mixed with an analog TV signal located at 62.25 MHz. Then the resulting signal is amplified before being applied to our fabricated microstrip antenna. After that, the obtained modulated signal as shown in Figure 7, is transmitted through a point to point wireless link by using the microstrip antenna. Finally in the receiver, another microstrip antenna is used to receive the transmitted information, which it is processed using optical heterodyne technique again to recover in this case the TV signal (66-72 MHz). From the photodetector 2 in the receiver, a local oscillator that is synchronized, in frequency as well as in phase with to that obtained from the photodetector 1, is mixed with the received signal. Then the resulting signal is filtered and the power spectral density obtained is displayed on an electrical spectrum analyzer, where it is analyzed to measure the power level of recovered information.

Figure 8 shows the frequency spectrum of an analog National Television System Committee (NTSC) TV signal at the input of the transmitter located at 67.25 MHz (before being applied to frequency mixer). In the same figure we can see the obtained analog NTSC TV signal at the output of the receiver. In order to measure the quality of the received signal, it is necessary to quantify the parameter of signal-to-noise ratio (SNR), in this case it is approximately 45 dB. The analog information is successfully transmitted from the transmitter to the receiver, and the received signal is satisfactorily reproduced on TV monitor. The differential gain and differential phase were not measured Nevertheless we demonstrated that the generated microwave signal by using optical heterodyning can be used as carrier information in a traditional communication system and we have used a TV signal of test to verify it.

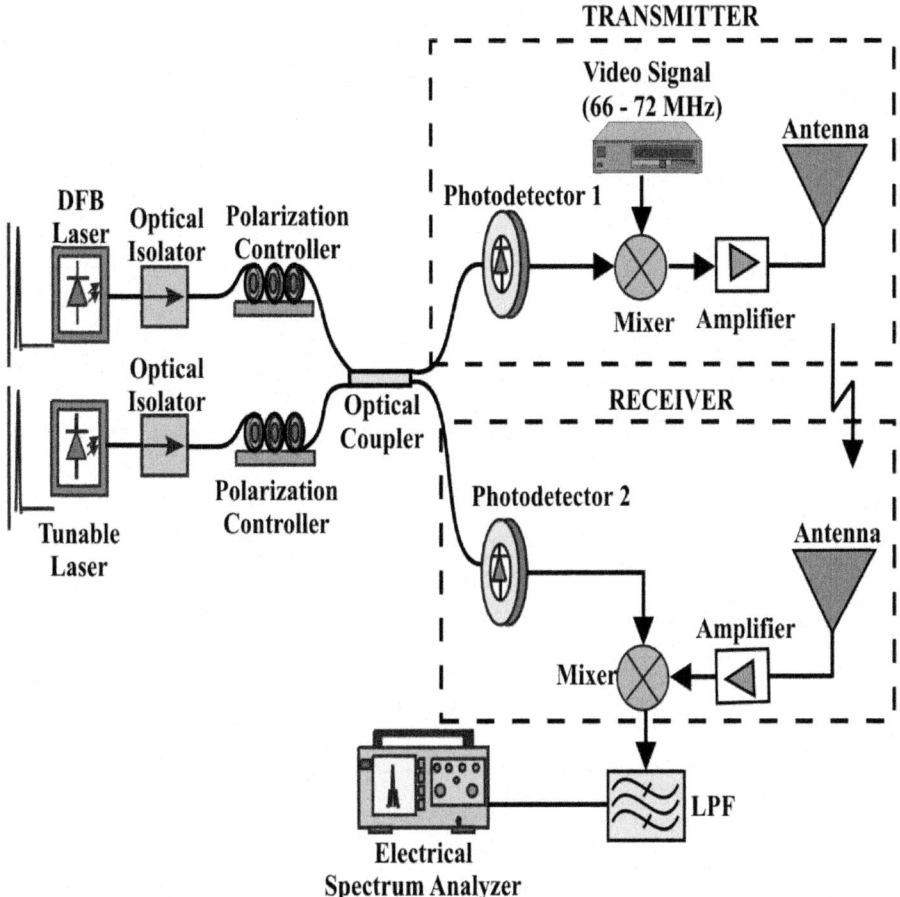

Figure 6: Wired/wireless photonic communication system for transmitting and receiving TV signals.

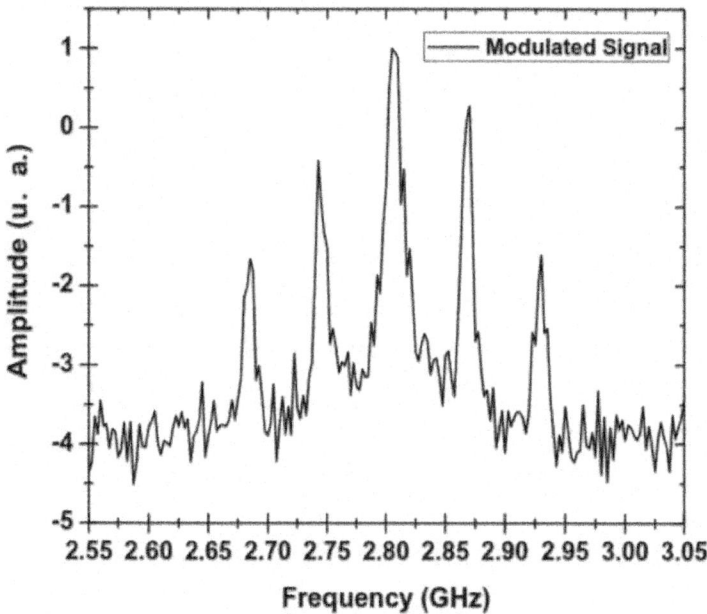

Figure 7: Electrical spectrum of the modulated signal.

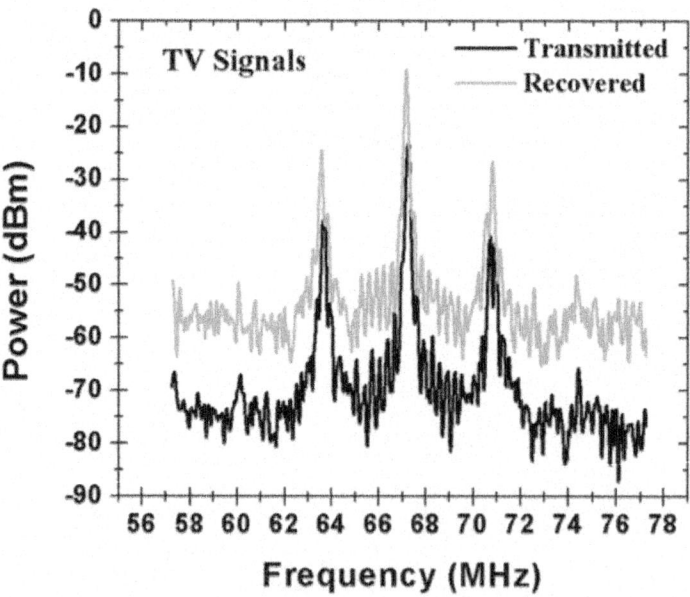

Figure 8: TV signals at 67.25 MHz, transmitted and recovered.

ANALYTICAL MODEL OF THE MICROWAVE PHOTONIC FILTER

The scheme of the MPF is illustrated in Figure 9. Consider that the optical signal of a poly- chromatic source with spectrum P(ω), centered at an optical frequency ω_n, is launched into the input of the Mach-Zehnder intensity modulator (MZ-IM). A single spectral component of such an optical signal can be modeled by a stochastic process e(t)= E_o(t)exp(jω_n t), where E_o(t) is the complex amplitude and ω_n is the optical angular frequency. If the intensity of such optical signal is externally modulated by an electrical signal V_m =1 + 2mcos(ω_m t), where m is the modulation index and ω_m is the angular frequency of external modulation, then the optical field at the input of the optical fiber can be expressed by Eq. (26). The modulation index m is related to the electrical input signal amplitude, V_m, as: $2m=\pi(V_m / V_\pi)$, where Vπ is the half wave voltage of the MZ-IM [18].

$$e_i(t) = e(t)s(t)$$

The optical fiber can be considered as a linear time invariant (LTI) system. If, for simplicity, the attenuation is ignored, then the transfer function of the optical link, for a given length L, is $H(j\omega) = \exp(-j\beta L)$, where β is the propagation constant. Thus, the optical field at the end of the link is given by

$$e_L(t) = e_i(t)\exp(-j\beta L) \tag{27}$$

Substituting e(t) and s(t) in Eq. (26), and then replacing this in Eq. (27), it becomes:

$$e_L(t) = E_o(t)\exp\left(j(\omega_m t - \beta L)\right) + E_o(t)m\exp\left(j\left[(\omega_m - \omega_n)t - \beta L\right]\right)$$
$$+ E_o(t)m\exp\left(j\left[(\omega_m + \omega_n)t - \beta L\right]\right) \tag{28}$$

In the frequency domain Eq. (28) can be expressed as:

$$E_L(\omega) = E_o(\omega - \omega_n)\exp(-j\beta L) + E_o(\omega - (\omega_n - \omega_m))\exp(-j\beta L)$$
$$+ E_o(\omega - (\omega_n + \omega_m))\exp(-j\beta L) \tag{29}$$

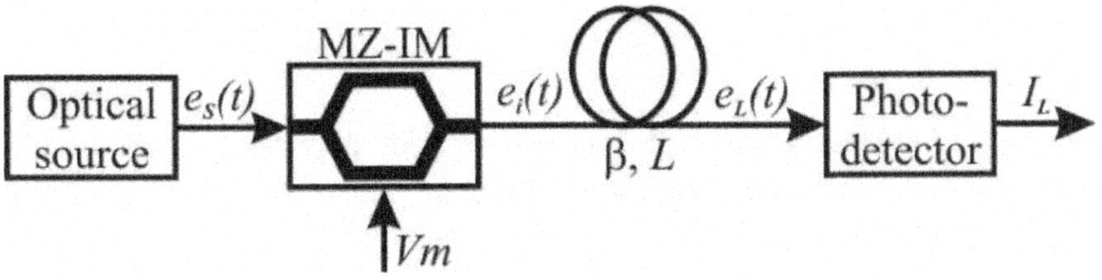

Figure 9: Scheme of the microwave photonic filter.

There are three spectral components. In the presence of chromatic dispersion, there is a propagation constant associated to each one of them, i.e. $\beta(\omega-\omega_n)$, $\beta(\omega-(\omega_n+\omega_m))$, and $\beta(\omega-(\omega_n-\omega_m))$. By denoting $W=\omega-\omega_n$, Eq. (29) . By denoting , Eq. (29) then becomes:

$$E_L(\omega) = E_o(W)\exp(-j\beta(W)L) + E_o(W+\omega_m)\exp(-j\beta(W+\omega_m)L)$$
$$+E_o(W-\omega_m)\exp(-j\beta(W-\omega_m)L) \tag{30}$$

Assuming that within the frequency range $\omega_n-\omega_m$ to $\omega_n+\omega_m$, centered at ω_n the propagation constant varies only slightly and gradually with ω , it can be approximated by the first three terms of a Taylor series expansion, and it can be shown that

$$\beta(W\pm\omega_m) = \beta(W)\pm\beta_1\omega_m + \beta_2\left[\frac{1}{2}\omega_m^2 \pm \omega_m(\omega-\omega_n)\right] \tag{31}$$

where $\beta_i = [d^i\beta(\omega)/d\omega^i]_{(\omega=\omega_n)}$

The optical intensity, I , is obtained by integrating the power spectral density over all the frequency range, i.e.

$$I = \int_{-\infty}^{\infty}|E_L(\omega)|^2 d\omega \tag{32}$$

Considering that the MZ-IM is operating on its linear region, it is valid to note that m 2 p_0 . On the other hand, if $\omega_n \gg \omega_m$ then E_o (W)pE_o (W + ω_m) pE_o (W 3ω_m) . Furthermore, in the fre□ quency domain, the spectrum of the source is defined as $P(\omega)=E_o(\omega)E_o^*(\omega)$. Thus, developing the product | EL (ω)| 2 in Eq. (32) and replacing Eq. (31), it is possible to demonstrate that the intensity at the end of the optical fiber is given by Eq. (31), it is possible to demonstrate that the intensity at the end of the optical fiber is given by:

$$I = \int_{-\infty}^{\infty} P(W)dW + 4m\cos\left(\beta_2\frac{\omega_m^2}{2}L\right)\cos(\beta_1\omega_m L)\Re\left\{\int_{-\infty}^{\infty} P(W)\exp(-j2\pi ZW)dW\right\} \tag{33}$$

where $Z = \beta_2\omega_m L /2\pi$, $W=\omega-\omega_n$ and its derivative, $dW = d\omega$. The total average intensity is

$I_o = \int_{-\infty}^{\infty} P(W)dW$, and the integral $\Re\left\{\int_{-\infty}^{\infty} P(W)\exp(-j2\pi ZW)dW\right\}$ corresponds to the real part of the Fourier transform of the spectrum of the optical source. This means that the optical intensity which reaches the surface of the photodetector is proportional to:

$$F(W) = \Re\{FT\{P(W)\}\} \tag{34}$$

A spectrum with Gaussian shape can be modeled by an analytical expression as:

$$P(\omega) = \frac{2P_o}{\Delta\omega\sqrt{\pi}} \exp\left(-\frac{4(\omega - \omega_m)^2}{\Delta\omega^2}\right)$$

(35)

where ω is the angular frequency, ω is the central angular frequency, ω is the maximum power emission and $\Delta\omega$ is the full width at half maximum (FWHM) of the optical source. If the emission spectrum of the optical source has a Gaussian shape, as defined in Eq. (35), then the Eq. (34) becomes:

$$F(\omega) = \exp\left(-\left(\frac{\beta_2\omega_m L\Delta\omega}{4}\right)^2\right)$$

(36)

In such case the FWHM of the frequency response can be determined equating $F(\omega)=0.5F(\omega)=0.5$
, which implies:

$$\left(\frac{\beta_2\omega_m L\Delta\omega}{4}\right)^2 = \ln(2)$$

(37)

For finding the value of the frequency f m that yields that condition, it is necessary to express ωm in terms of f m , i.e. $\omega_m = 2_\pi f_m$. But this, in turn, yields an expression that can be reduced by expressing $\Delta\omega$ in terms of Δ_λ and β_2 in terms of dispersion D . For Δ_ω this is done as follows: given $d\omega / d\lambda = 3(_2\pi_c / \lambda^2)$, where c is the speed of light in the free space and λ is the wave- length of the optical signal, it is possible to establish the following correspondence:

$$d\omega = -\frac{2\pi c}{\lambda^2} d\lambda \Leftrightarrow \Delta\omega = -\frac{2\pi c}{\lambda^2}\Delta\lambda$$

(38)

Now, for the factor β_2 , given that the group velocity, $v_g = L / \tau_g$ where τ_g is the group delay, is related to $\beta(\omega_n)$ as $\tau_g / L = d\beta(\omega_n)/d\omega$, and its derivative is $(d\tau_g / d\omega)/L = d^2\beta(\omega_n)/d\omega^2 = \beta_2$, then $(1/L)(d\tau_g) = d\omega\beta_2$. Thus, the derivative of this expression by $d\lambda$ is $(1/L)(d\tau_g / d\lambda) = (d\omega / d\lambda)\beta_2$. Furthermore, the dispersion, as a function of the wavelength is defined as $D = (1/L)(d\tau_g / d\lambda)$. This means that $\beta_2 = -D(\lambda^2 / 2\pi c)$. Finally, by substituting $\omega_m = 2\pi f_m$, $\Delta\omega$, in Eq. (38), and the expression for β_2 in Eq. (37), the frequency f_m , which corresponds to the low-pass bandwidth Δf_{lp} , can be expressed as:

$$\Delta f_{lp} = \frac{2\sqrt{\ln(2)}}{\pi DL\Delta\lambda}$$

(39)

where the dispersion D has units of ps nm-1 km-1, length L is given in km, and the FWHM of the optical source, $\Delta\lambda$, in nm. This means that in the presence of an optical source, like a super luminescent light-emitting diode (LED), the frequency response of the system is low-pass, and its bandwidth is given by

Eq. (39). In the context of this chapter, the optical source is an multimode laser diode (MLD). The emission spectrum of this type of optical sources can be modeled by means of an analytical expression as expressed in Eq. (40):

$$P(\omega) = \frac{2P_o}{\Delta\omega\sqrt{\pi}} \exp\left(-\frac{4(\omega-\omega_n)^2}{\Delta\omega^2}\right)\left[\frac{2P_o}{\sigma\omega\sqrt{\pi}} \exp\left(-\frac{4(\omega-\omega_n)^2}{\sigma\omega^2}\right) * \sum_{n=-\infty}^{\infty} \delta(\omega-n\delta\omega)\right]$$

(40)

where ω is the angular frequency, ω_n is the central angular frequency, Po is the maximum power emission, $\Delta\omega$ is the FWHM of the optical source, σ_ω is the FWHM of each emission mode and $\delta\omega$ is the free spectral range (FSR) between the emission modes. By using variables Z and W , as defined earlier, and substituting Eq. (40) in Eq. (34), it can be expressed as:

$$F(\omega) = \exp\left(-\left(\frac{\beta_2\omega_m L\Delta\omega}{4}\right)^2\right) * \left[\exp\left(-\left(\frac{\beta_2\omega_m L\sigma\omega}{4}\right)^2\right)\frac{1}{\delta\omega}\sum_{n=-\infty}^{\infty}\delta\left(\frac{\beta_2\omega_m L}{2\pi}-\frac{n}{\delta\omega}\right)\right]$$

(41)

The term between crochets indicates the presence of a periodic pattern. The frequency of the first maximum can be determined by equating:

$$\frac{\beta_2\omega_1 L}{2\pi} = \frac{1}{\delta\omega}$$

(42)

For finding the value of the frequency f_1 that yields that condition, it is necessary to express δ_ω in terms of f_1 . In a similar way as in Eq. (38), it is possible to establish the following correspondence::

$$d\omega = -\frac{2\pi c}{\lambda^2}d\lambda \Leftrightarrow \delta\omega = -\frac{2\pi c}{\lambda^2}\delta\lambda$$

(43)

thus, substituting $\delta\omega$ in Eq. (42), expressing $\omega 1$ in terms of f 1 and using $\beta 2 = -D(\lambda^2 / 2\pi c)$ then the frequency f_1 can be expressed as:

$$f_1 = \frac{1}{DL\delta\lambda}$$

(44)

and, in general, the central frequency of the n-th band-pass lobe is given by

$$f_n = \frac{n}{DL\delta\lambda}$$

(45)

where n is a positive integer, dispersion D is given in ps nm^{-1} km^{-1}, length L in km, and the FSR δ_λ in nm. The bandwidth of each of these band-pass lobes is equal to:

$$\Delta f_{bp} = \frac{4\sqrt{\ln(2)}}{\pi DL\Delta\lambda}$$

(46)

which is twice Eq. (39). The periodic pattern in the frequency response of the system will appear only when an MLD is used in the system. This behavior will allow that microwave signals to be filtered and transmitted over a wide range of frequencies.

EXPERIMENTAL SETUP OF OPTICAL AND WIRELESS TRANSMISSION

In a first step, the MLD used in this experiment (OKI OL5200N-5) is optically characterized by means of an optical spectrum analyzer (Agilent, model 86143B). Figure 10 corresponds to the measured optical spectrum obtaining λ_0=1553.53nm, $\Delta\lambda$=5.65nm, and $\delta\lambda$=1.00nm for a driver current of 25 mA. The use of a laser diode temperature-controller (Thorlabs, model LTC100-C) allows us to guarantee the stability of the optical parameters to thermal fluctuations.In a second step, considering a length L=20.70 km of single-mode-standard-fiber (SM-SF) exhibiting a chromatic fiber-dispersion parameter of D=16.67 ps/nm km. Eq. (45) allows us to determine the value of the central frequency corresponding to the first filtered microwave or first band-pass as

$$f_1 = \frac{1}{DL\delta\lambda} = \frac{1}{(16.67\mathrm{x}10^{-12}\mathrm{seg/nm} \cdot \mathrm{km}) \cdot (20.70 \ \mathrm{km}) \cdot (1.0 \ \mathrm{nm})} = 2.8 \ \mathrm{GHz}$$

Eq. (39) permits us to determine the value of the low-pass band as

$$\Delta f_{lp} = \frac{2\sqrt{\ln(2)}}{\pi DL\Delta\lambda} = \frac{2\sqrt{\ln 2}}{(\pi) \cdot (16.67\mathrm{x}10^{-12}\mathrm{seg/nm} \cdot \mathrm{km}) \cdot (20.70 \ \mathrm{km}) \cdot (5.65 \ \mathrm{nm})} = 271.85 \ \mathrm{MHz}$$

Finally, according to Eq. (46), the corresponding bandwidth of the band-pass window is Δf_{bp}=543.70 MHz

.At this point, it is well worth highlighting the advantageous use of the chromatic dispersion parameter to obtain the filtered microwave signal. Once the main parameters are known, the topology illustrated in Figure 11 is assembled in order to evaluate the frequency response of the MPF.

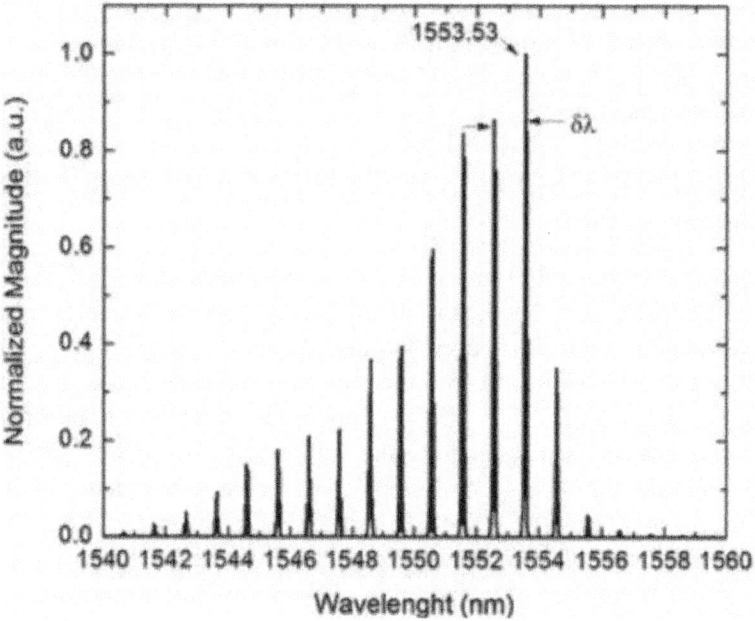

Figure 10: Optical spectrum for the MLD used in the experiment.

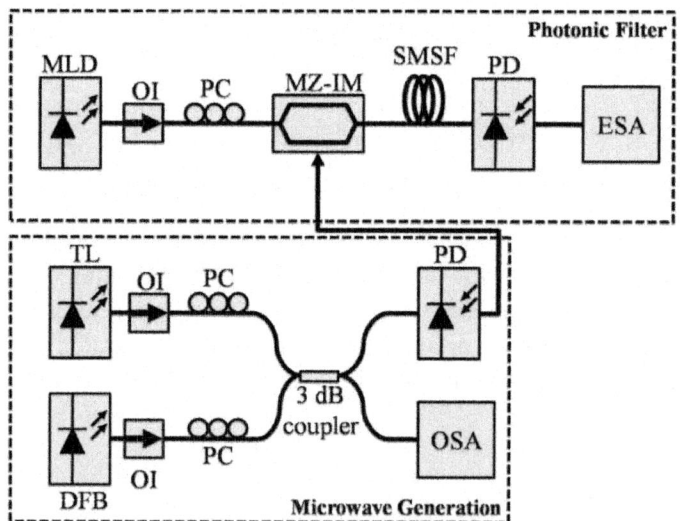

Figure 11: Experimental microwave photonic filter.

At the output of the MLD, an optical isolator (OI) is placed in order to avoid reflections to the optical source. Since the MZ-IM (Photline MX-LN-10) is polarization-sensitive, a polarization controller (PC) is used to maximize the modulator output power. The optical signal is launched into the MZ-IM. The microwave electrical signal (RF) for modulating the optical intensity is supplied by using optical heterodyne as described in Figure 1. The registered frequency response is located from 0.01 to 4 GHz at 0 dBm. The intensity-modulated optical signal is then coupled into a 20.70 km of SM-SF coil. The length of the optical fiber is corroborated by using an optical time domain reflectometer, OTDR (EXFO, model FTB-7300E). At the end of the link, the optical signal is applied to a fast Photo-Detector (PD, Miteq DR-125G-A), and its output connected to an electrical spectrum analyzer (Anritsu, model MS2830A-044), in order to measure the frequency response of the MPF. Figure 12corresponds to the measured experimental frequency response where a low-pass band centered at zero frequency and the presence of a band-pass band centered at 2.8 GHz are clearly appreciable.

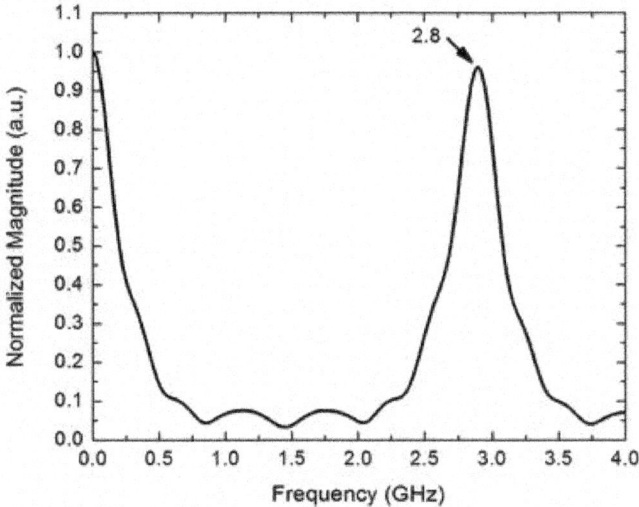

Figure 12: Experimental frequency response of the filter.

The bandwidth of 543.70 MHz associated to the band-pass window centered at 2.8 GHz allows us to guarantee enough bandwidth in case of fluctuations (in the order of nanometers) between mode spacing. On the other hand, a considerable increase on the length of the optical fiber due to thermal expansion is practically impossible. These considerations permit us to guarantee a good stability for the microwave photonic filter. Once the frequency response of the MPF is determined, the setup illustrated in Figure 13 is assembled for carrying out the fiber-radio transmission.

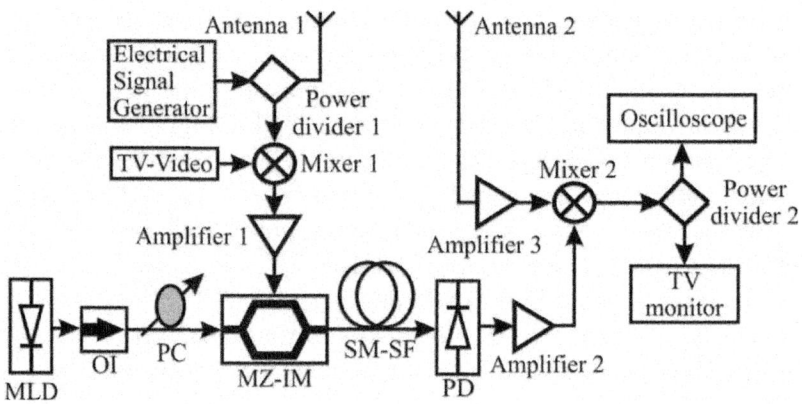

Figure 13: Experimental setup for optical and wireless transmission.

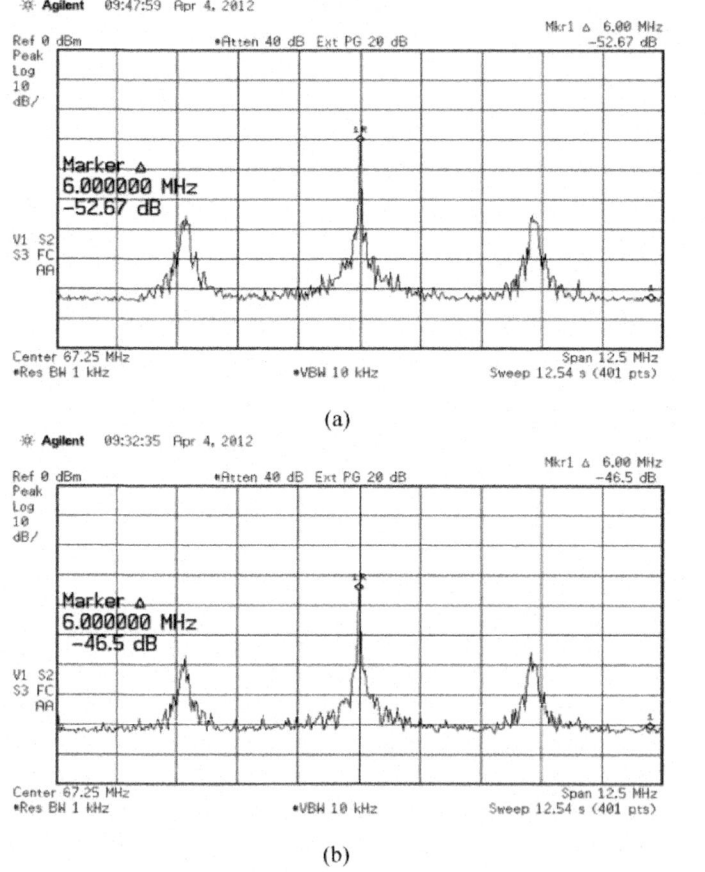

Figure 14: Electrical Spectrums for (a) Transmitted and (b) recovered TV signal.

Now, the electrical signal generator provides a signal of 2.8 GHz at 0 dBm that is used as the electrical carrier and demodulated signal. This signal is separated by using a power divider. Part of this signal is transmitted via radio frequency by the fabricated microstip antenna shown in Figure 5, and the rest is mixed with an analog NTSC TV signal of 67.25 MHz. The resulting mixed electrical signal is then applied to the electrodes of the MZ-IM for modulating the light emitted by the MLD. The modulated light is coupled into the 20.70 km SM-SF coil. At the end of the optical link, the signal is injected to a fast photo-detector (PD), and its electrical output is then amplified and launched to an electrical mixer. Another microstrip patch antenna placed at a distance of 10 meters is connected to a port of the mixer in order to recuperate the microwave signal that plays the role of the demodulated signal. Finally, by using another power divider, recovered analog TV signal can be launched to a digital oscilloscope or to the electrical spectrum analyzer in order to evaluate the quality of the recovered signal and at the same time display the TV signal on a TV monitor. Figure 14 (a) shows the measured electrical spectrum (Agilent, E4407B) corresponding to the transmitted TV signal where the SNR is 52.67 dB, whereas Figure (b) corresponds to the recovered TV signal with a SNR of 46.5 dB.

Finally, Figure 15 corresponds to a photograph of the screen of the oscilloscope where upper and lower traces are the waveforms of the transmitted and recuperated signal, respectively.

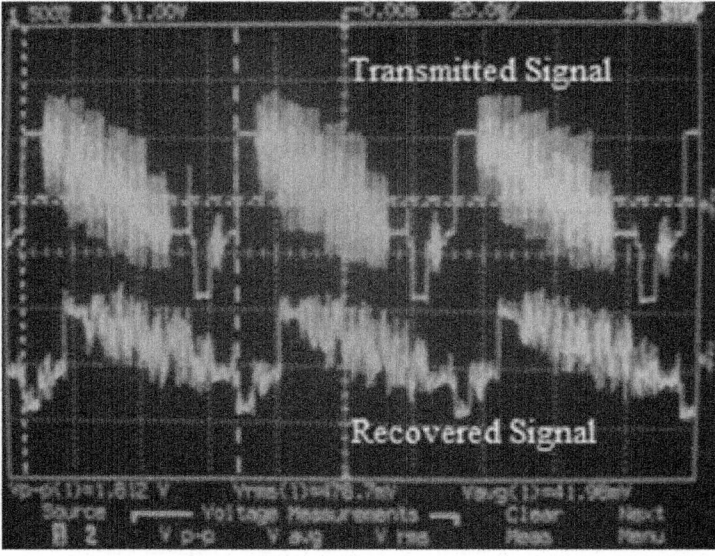

Figure 15: Transmitted and recovered TV signal.

CONCLUSIONS

Wireless communication systems require compact sources for the generation of mm-wave signals, that must have high spectral purity (linewidth < 100 kHz, phase noise < 100 dBc @100 kHz offset), tuneability, low power consumption and low cost, and although optical heterodyne of two DFB lasers has phase noise of −75 dBc/Hz even at an offset frequency of 100 MHz and it does not very compact, we have demonstrated in this chapter that by using optical heterodyne technique, a TV signal was transmitted and received satisfactory as a result of our proposed communication system generates a microwave carrier and a local oscillator simultaneously ensuring synchronization in frequency as well as in phase between microwave carrier and a local oscillator and avoiding in this case the use of an analog phase locked loop in the receiver to recover the TV information. The authors consider that the first proposed scheme in this chapter is not a truly wireless communication system, since an optical fiber is required to deliver the local oscillator in the receiver, however in order to obtain a wireless communication systems by using optical heterodyne technique, it is necessary to have collimated beams from optical fiber to photodetectors. On the other hand, due to the fact that the distribution of TV over microwave signals in the electrical domain presents loss associated with electrical distribution lines, the authors consider that the optical fiber is an ideal solution to fulfill this task because of its extremely broad bandwidth and low loss. In that case the distribution of TV over microwave can be directly by using optical fiber. In this way the second proposed experiment in this chapter represents a novel fiber-radio scheme to transmit an analog NTSC TV signal coded on a microwave band-pass located at 2.8 GHz. Filtering of microwave signal was achieved through the appropriate use of the chromatic fiber dispersion parameter, the physical length of the optical fiber, and the free spectral value of the multimode laser. Transmission of a TV signal was achieved over an optical link of 20.70 km, whereas a demodulated signal was transmitted via radiofrequency using the fabricated microstrip patch antennas. Although the distance between antennas was short, this distance can be lengthened if an array of antennas is used. Besides, a mathematical analysis corresponding to the microwave photonic filter was described demonstrating that the frequency response of the microwave photonic filter is proportional to the Fourier transform of the spectrum of the optical source used. The proposed microwave photonic filter represents an interesting technological alternative for transmitting information by using optoelectronic techniques. The results obtained in this work ensure that as an interesting alternative, several modulation schemes can be used for transmitting not only analog information but also digital information. Besides as optical heterodyne technique described here can generate microwaves continually tuned, we can use this feature to transmit several TV signals using frequency division multiplexing schemes FDM [19] and wavelength division multiplexing WDM techniques, not only point to point but also with bidirectional schemes by using simultaneous wired and wireless systems.

ACKNOWLEDGEMENTS

This work was supported by CONACyT (grants No 102046 and 154691).

REFERENCES

1. Li An, Shieh W., and Tucker R.S. Wavelet packet transform-based OFDM for optical communications. Journal of Lightwave Technology, 2010; 28(24), 3519-3528.

2. Ben Ezra Y., Lembrikov B.I., Zadok Avi, Ran Halifa R., and Brodeski D. All-optical Signal Processing for High Spectral Efficiency (SE) Optical Communication. In: Narottam Das (ed.) Optical Communication. Rijeka: InTech; 2012. p343-366.

3. Schmogrow R., Bouziane R., Meyer M., Milder P.A., Schindler P.C., Killey R.I., Bayvel P., Koos C., Freude W., and Leuthold J. Real-time OFDM or Nyquist pulse generation – which performs better with limited resources? Optics Express 2012; 20(26) B543-B551.

4. Bosco G., Carena A., Curri V., Poggiolini P., and Forghieri F. Performance limits of Nyquist-WDM and CO-OFDM in high-speed PM-QPSK systems. IEEE Photonic Technology Letters, 2010; 22(15) 1129-1131.

5. Li An, Shieh W., and Tucker R.S. Impact of polarization-mode dispersion on wavelet transform based optical OFDM systems, In: proceedings of National Fiber Optic Engineers Conference, San-Diego, California, USA March 21-25, 2010, JThA5, pp.1-3.

6. Shieh W., Yi X., Ma Y., and Tang Y. Theoretical and experimental study on PMD-supported transmission using polarization diversity in coherent optical OFDM systems. Optics Express 2007; 15(16) 9936-9947.

7. Kingsbury N. Complex wavelets for shift invariant analysis and filtering of signals Journal of Applied and Computational Analysis 2001 10 (3) 234-253.

8. Bayram I. and Selesnick I.W. On the dual-tree wavelet packet and M-band transforms. IEEE Transactions on Signal Processing 2008; 56(6) 2298-2310.

9. Shieh W, Djordjevic I. Orthogonal Frequency Division Multiplexing for

Optical Communications. London: Academic Press; 2010.

10. Hillerkuss, D.et al. Simple all-optical FFT scheme enabling Tbit/s real-time signal processing, Optics Express, April 2010; 18(9) 9324-9340.

11. Wang X.; Ho P., and Wu Y. Robust Channel Estimation and ISI Cancellation for OFDM Systems with Suppressed Features, IEEE Journal on Selected Areas in Communications 2005; 23(5) 963-972.

12. Shieh W., Bao H., and Tang Y. Coherent optical OFDM: theory and design. Optics Express 2008; 16(2) 841-859.

13. Armstrong J. OFDM for Optical Communications, IEEE Journal of Lightwave Technology, February 2009; 27(3) 189-204.

14. Kikuchi K. Coherent optical communication systems, In: Kaminov, I. P.; Li, T. & Willner, A. E. (Eds.) Optical Fiber Telecommunications VB: Systems and Networks, Academic Press, Amsterdam, London, New York: Academic Press; 2008. p91-129.

15. Da Silva E., Pataca D.M., Ranzini S. M., de Carvalho L.H.H., Juriollo A.A., da Silva M.L., Oliveira J.C.R.F. Transmission of 1.15 Tb/s NGI-CO-OFDM DP-QPSK superchannel over 4520 km of PSCF with EDFA-only amplification. Journal of Microwaves, Optoelectronics and Electromagnetic Applications, 2013; 12(SI-2) 96-103.

16. Cincotti G., Moreolo M.S. and Neri A. Optical Wavelet Signals Processing and Multiplexing, EURASIP Journal on Applied Signal Processing, 2005; 10, 1574-1583.

17. Rao R.M.& Bopardikar A. S. Wavelet Transforms. Introduction to Theory and Applications. Reading, Massachusetts: Addison-Wesley; 1998.

18. Daubechies I. Ten Lectures on Wavelets. Philadelphia, Pennsylvania: Society for Industrial and Applied Mathematics; 2006.

19. Sarkar K.T., Salazar-Palma M., Wicks M.C. Wavelet Applications in Engineering Electromagnetics. Boston: Artech House; 2002.

20. Nerma M. H. M., Kamel N. S., Jeoti V. An OFDM based on dual tree complex wavelet transform (DT-CWT), Signal Processing: An International Journal (SPIJ), 2009; 3 (2) 14-26.

21. Ben-Ezra Y., Brodeski D., Lembrikov B.I. High Spectral Efficiency OFDM Based on Complex Wavelet Packets. In: ICTON 2014: Proceedings of the 16th International Conference on Transparent Optical Networks, 6-10 July 2014, Graz, Austria, We.A1.4 p1-3.

22. Kuschnerov M., Hauske F. N., Piyawanno K., Spinnler, B., Alfiad, M.S. Napoli, A., and Lankl, B. DSP for coherent single-carrier receivers, Journal of Lightwave Technology, 2009; 27(16) 3614-3622.

23. Kingsbury N. Design of Q-shift complex wavelets for image processing using frequency domain energy minimization. In: Image Processing 2003. ICIP 2003. Proceedings of International Conference on Image Processing, 14-18 September 2003, Barcelona, Catalonia, Spain, 1, 1-1013-16.

24. Kingsbury N. A dual-tree complex wavelet transform with improved orthogonality and symmetry properties. In: Image Processing 2000. ICIP 2000. Proceedings of International Conference on Image Processing, 10-13 September 2000, Vancouver, BC, Canada, 375-378.

25. Selesnik I. W. Hilbert transform pairs of wavelet bases, IEEE Signal Processing Letters, 2001; 8(6) 170-173.

26. Ma Y., Yang Q., Tang Y., Chen S., and Shieh W. 1-Tb/s single-channel coherent optical OFDM transmission over 600-km SSMF fiber with subwavelength bandwidth access. Optics Express 2009; 17(11) 9421-9427.

27. Ip, Ezra, Lau, A.P.T., Barros, D.J.F., Kahn, J.M. Coherent detection in optical fiber systems. Optics Express, January 2008; 16(2) 753-791.

Chapter 10

APPLICATION OF COMPLEX WAVELET PACKET TRANSFORM (CWPT) IN COHERENT OPTICAL OFDM (CO-OFDM) COMMUNICATION SYSTEMS

Y. Ben-Ezra[1] and B.I. Lembrikov[1]

[1]Department of Electrical Engineering, Holon Institute of Technology, Holon, Israel

INTRODUCTION

Coherent optical orthogonal frequency division multiplexing (CO-OFDM) is a modulation format that attracted wide interest due to its high spectral efficiency (SE) and robustness against chromatic dispersion (CD) and polarization mode dispersion (PMD). CO-OFDM communication systems with coherent detection combine high SE and high receiver sensitivity [1], [2]. CO-OFDM transmission at 1 Tb/s can be realized [1]. OFDM is implemented using fast Fourier transform (FFT) which results in inter carrier interference (ICI) and inter symbol interference (ISI) [1], [2]. Usually, in order to avoid ICI and ISI, a so-called cyclic prefix (CP) is inserted into OFDM symbols. The addition of CP requires an increase of a bandwidth and sampling rate of analog-to-digital converter (ADC) and digital-to-analog converter (DAC) decreasing SE [1], [2]. The need for CP can be avoided if the wavelet packet transform (WPT) is used in CO-OFDM systems instead of Discrete Fourier Transform (DFT) and inverse DFT (IDFT) [2]. In such a case, a signal can be expanded in an orthogonal set of so-called wavelets [1], [2]. WPTs provide orthogonality between OFDM subcarriers based on the wavelets instead of sinusoids. Wavelets have finite length. For this reason, wavelet transforms (WTs) have both frequency and time localization. It has been shown that WPT-OFDM single-polarization system can mitigate CD of 3380 ps/nm at bit rate of 112 Gb/s [1].

An alternative real-time multiplexing technique characterized by a high SE is the Nyquist wavelength division multiplexing (N-WDM) [3]. N-WDM is made up of temporal sinc-pulses [3]. Electrically generated Nyquist pulses are shaped with finite duration impulse response (FIR) filters, have nearly rectangular spectra, and are transmitted in independent WDM channels [3]. For the higher order filters the spectrum approaches to a rectangle, and the inter channel guard bands can be kept small without introducing ICI [3]. N-WDM has been investigated both theoretically and experimentally [3], [4]. For the polarization-multiplexed quadrature phase-shift keying (PM-QPSK) transmitter based on two QPSK integrated modulators in nonreturn-to-zero (NRZ) modulation regime, and standard single mode fiber (SSMF) with the typical values of parameters such as bit rate of 111 Gb/s per channel, loss 0.22dB/km, dispersion 16.7 ps/nm/km, noise-figure 5 dB, BER=4x10⁻³ *at* the maximum distance of 2300 km, performance of CO-OFDM and N-WDM techniques appeared to be similar [3],[4].

Recently, dual-polarization transmission has been investigated as a promising technique for future high bit rate communication systems [1], [5]. In such a case, the optical fiber two-input two-output channel model is described by a two-element Jones vector for any OFDM symbol on a subcarrier [6]. In Fourier transform (FT) based optical OFDM (FTO-OFDM), individual subcarriers are single-sideband, and the phase dispersion caused by CD and PMD can be compensated. In WPT-OFDM systems, on the contrary, the modulated signals are double-sideband. Unlike CD, PMD does not possess the phase symmetry [5]. Consequently, the Jones matrixes for the positive and negative sidebands are not equal, the dispersions of the two sidebands are different, and their addition does not represent the initial real WT basis. As a result, the orthogonality is broken, and inter-packet-interference occurs [7].

We proposed a novel type of CO-OFDM based on the dual-tree complex WPT (DT-CWPT) recently developed by N. Kingsbury [7]. DT-CWPT utilizes two wavelet filter banks (FBs) [8].

- A first wavelet FB is used in the discrete WT (DWT).
- A second wavelet FB is constructed in such a way that its impulse responses are approximately the discrete Hilbert transforms of the first wavelet FB.

The first and the second FBs are the real and imaginary parts of the DT-CWPT, respectively. We have shown that PMD can be compensated by digital signal processing using a DT-CWPT. SE and the transmission distance are substantially increased. Numerical simulations show that the 1 Tb/s single-channel CO-OFDM transmission with high SE of 7.88 bit/s/Hz over the distance of 1800 km can be realized.

The proposed Chapter is constructed as follows. In Section 2, we discuss the fundamentals of CO-OFDM systems; in Section 3, the WPT based OFDM systems are considered; PMD influence on the CO-OFDM dual-polarization transmission is reviewed in Section 4. In Sections 5 and 6, the structure and the properties of CO-OFDM system based on DT-CWT and DT-CWPT are described, respectively. In Section 7, the original results are presented for the DT-CWPT based single-channel CO-OFDM transmission system including a laser-diode and Mach-Zehnder external modulator. We describe the methods of the CD, PMD, and nonlinearity mitigation and present the numerical simulations results. Conclusions are presented in Section 8.

FUNDAMENTALS OF CO-OFDM SYSTEMS

In this Section, we consider the fundamentals of CO-OFDM communication systems OFDM is a kind of multicarrier modulation (MCM) where the data information is carried over many lower rate subcarriers [9]. The block diagram of an OFDM system is shown in Fig. 1 [9].

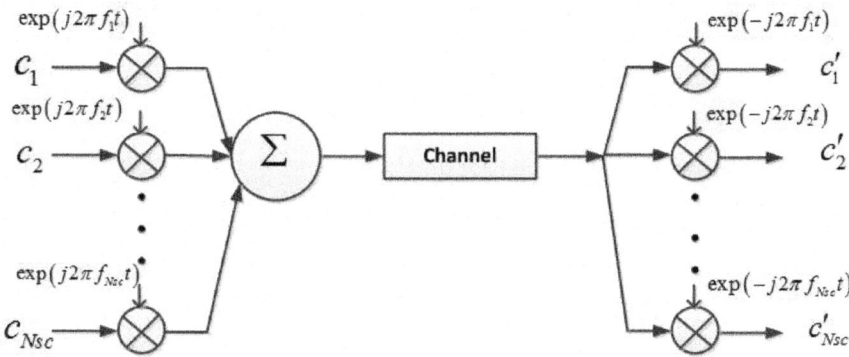

Figure 1: A diagram of OFDM system.

The fundamental advantages of OFDM are its robustness against channel dispersion and the ease of phase and channel estimation in a time-varying environment; the essential disadvantages of OFDM are high peak-to-average power ratio (PAPR) and sensitivity to frequency and phase noise [9]. The MCM transmitted signal is represented as follows [9].

$$s(t) = \sum_{i=-\infty}^{\infty} \sum_{k=1}^{N_{sc}} c_{ki} s_k(t - iT_s)$$

$$(1)$$

$$s_k(t) = \Pi(t)\exp(j2\pi f_k t)$$

(2)

$$\Pi(t) = \begin{cases} 1, & (0 < t \le T_s) \\ 0, & (t \le 0, t > T_s) \end{cases}$$

(3)

where c_{ki} is the ii-th information symbol at the kth subcarrier, s_k is the waveform for the k-th subcarrier, N_{sc} is the number of subcarriers, f_k is the frequency of the k-th subcarrier, T_s is the symbol period, and $\Pi(t)$ is the pulse shaping function. The detected information symbol c'_{ki} is given by [9].

$$c'_{ki} = \frac{1}{T_s} \int_0^{T_s} r(t - iT_s)\exp(-j2\pi f_k t)dt$$

(4)

where r(t)r(t) is the received time domain signal. In the case of the classical MCM nonoverlapped bandlimited signals are used which results in the excessive bandwidth and decrease of SE. In the case of OFDM, the overlapped orthogonal signal sets are determined by the following condition for any two orthogonal subcarriers [9].

$$\frac{1}{T_s} \int_0^{T_s} s_k s_l^* dt = \delta_{kl}$$

(5)

Substituting expression (2) into (5) and carrying out the integration we obtain.

$$\exp\left[j\pi(f_k - f_l)T_s\right] \frac{\sin\left[\pi(f_k - f_l)T_s\right]}{\left[\pi(f_k - f_l)T_s\right]} = \delta_{kl}$$

(6)

The orthogonality condition (6) is satisfied for the following relationship between the subcarrier frequencies

$$f_k - f_l = m\frac{1}{T_s}; m = 1, 2, \ldots$$

(7)

In such a case, the orthogonal subcarrier sets can be recovered with the matched filters without ICI even for the strong signal overlapping [9]. OFDM modulation and demodulation can be implemented by using IDFT/DFT where the mm-th samples s_m and r_m of the MCM transmitted and received signals s(t) and r(t), respectively, are given by [9].

$$s_m = \sum_{k=1}^{N} c_k \exp\left[j2\pi \frac{(k-1)(m-1)}{N} \right] = F^{-1}\{c_k\}; c'_k = F\{r_m\}$$

$$(8)$$

Here $F\{rm\}$ is the Fourier transform,

$m \in [1, N]$, $f_k = (k-1)/T_s$; $k \in [k_{min}+1, k_{min}+N]$, k_{min} i k_{min} is an arbitrary integer, and Ts/NTs/N is the sampling interval. Expressions (8) show that the discrete value of the transmitted signal s(t) is an NN-point IDFT of the information symbol c_k, and the received information symbol c'_k is an N-point DFT of the received sampled signal rmrm [9]. The PAPR of the OFDM signal is given by the following expression [9]:

$$PAPR = \max\left\{ \left| s(t) \right|^2 \right\} / E\left\{ \left| s(t) \right|^2 \right\}, t \in [0, T_s]$$

$$(9)$$

where $\max\{|s(t)|^2\}$ and $E\{|s(t)|^2\}$ are the maximum value and the expectance of $|s(t)|^2$.

In order to implement IDFT/DFT DAC and ADC are needed. IDFT/DFT is characterized by two important advantages [9].

- The computation complexity measured with the number of complex multiplications is reduced to $(N/2)\log_2(N)$ which is almost linear with respect to the subcarriers number N.

- The OFDM system architecture is relatively simple. It does not require complex radio frequency (RF) oscillators and filters.

OFDM is characterized by the ISI and ICI as it was mentioned above. These phenomena are caused by a large number of subcarriers [9]. CP, i.e. the cyclic extension of the OFDM waveform into the guard interval (GI) ΔG, has been proposed in order to prevent ISI and ICI [9]. If the GI is long enough to contain the intersymbol transition, then the remaining part of the OFDM symbol satisfies the orthogonality conditions (6), (7) and receiver cross-talk occurs only within GI [10]. The addition of CP requires an increase of a bandwidth and sampling rate of ADC and DAC. CP is an easily recognizable feature of an OFDM system which results in the signal vulnerability to interception by surveillance receiver [11]. Hence, the elimination of CP reduces the probability of interception and improves SE [11]. CP can be inserted implicitly into the definition of the MCM transmitted signal s(t) (1)-(3) by modification of equation (3) which takes the form [9].

$$\Pi(t) = \begin{cases} 1, & \left(-\Delta_G < t \le T_s\right) \\ 0, & \left(t \le -\Delta_G, t > T_s\right) \end{cases}$$

(10)

OFDM is characterized by the dispersion robustness, high SE, possibility of linear and nonlinear impairment mitigation [9]. For this reason, OFDM appeared to be an attractive long-haul transmission format for the optical communication systems [9], [12], [13].

The optical communication systems can be divided into two groups. The first group is based on optical wireless, multimode fiber (MMF) systems and plastic optical fiber (POF) systems where the OFDM signal is represented by the intensity of the optical signal; the second group is based on the single mode fiber (SMF) techniques where the OFDM signal is represented by the optical field [13]. The first group uses the direct detection technique, while the second group uses the coherent detection technique [9]. The optical OFDM systems based on the coherent detection are called CO-OFDM [9], [12]. A generic CO-OFDM system consists of five functional blocks: the RF OFDM transmitter, the RF-to-optical (RTO) up-converter, the optical channel, the optical-to-RF (OTR) down-converter, and the RF OFDM receiver with the coherent detector [12]. CO-OFDM manifests higher receiver sensitivity, SE, and robustness against polarization dispersion [9]. CO-OFDM can achieve high SE by overlapping subcarrier spectrum and at the same time avoiding interference by using coherent detection and signal set orthogonality [9], [12]. CO-OFDM communication systems are characterized by the linearity in RTO up-conversion and optical-to-RF (OTR) down-conversion [9], [12]. The electrical bandwidth requirements for the CO-OFDM transceiver can be greatly reduced by using direct up/down conversion which results in the low cost of the high-speed electronic circuits [9], [12]. SE $\eta\eta$ of the CO-OFDM system is given by [9], [12].

$$\eta = \frac{2R}{B_{OFDM}} \approx 2\alpha; \alpha = \frac{t_s}{T_s}$$

(11)

where $R = N_{sc}/T_s$ is the total symbol rate, $B_{OFDM} = (2/T_s) + (N_{sc}-1)/t_s$ is the OFDM bandwidth, $N_{sc} \gg 1$, the factor of 2 accounts for two light wave polarizations in the fiber, and $t_s t_s$ is the observation period.

CO-OFDM communication systems possess the following advantages as compared to the intensity modulation with direct detection (IM/DD) systems [14].

• The shot-noise limited receiver can be achieved with a sufficient local oscillator (LO) power.

- The frequency resolution at the intermediate frequency (IF) or baseband stage is high enough in order to separate close wavelength-division multiplexed (WDM) channels in the electric domain.

- The phase detection improves the receiver sensitivity compared with IM/DD systems.

- The advanced multilevel modulation formats can be introduced into optical communications by using phase modulation.

CO-OFDM communication systems can achieve high SE transmission through higher order modulation such as 64-QAM in single polarization and 16-QAM in dual polarization due to the DACs at the transmitter [9]. As a result, CO-OFDM techniques can be applied to the optical long-haul transmission at 100 Gb/s and beyond [9]. Recently, the 1.15 Tb/s no-guard-interval CO-OFDM (NGI-CO-OFDM) DP-QPSK superchannel transmission over the distance of 4520 km with SE=3.75b/s/Hz with BERs below the stringent forward error correction (FEC) limit has been demonstrated experimentally [15]. The pure silica core fiber and the erbium doped fiber amplifier (EDFA) have been used [15].

WPT BASED CO-OFDM SYSTEMS

Conventional CO-OFDM systems based on DFT and IDFT can be replaced with WPT based CO-OFDM [1]. The sinusoidal functions are infinitely long in the time domain while wavelets have finite length being localized in time and in frequency domains [1]. Wavelet signal analysis can be a base for an effective computational algorithm which is faster and simpler than FFT [16]. Wavelets have been used in optical communications for time-frequency multiplexing and ultrafast image transmission [16]. A signal may be expanded in an orthogonal set of wavelet packets (WPs) as the basis functions, each channel occupies a wavelet packet (WP), and IDWPT/ DWPT are used at the transmitter and receiver, respectively [1], [16]. The theory and possible applications of continuous wavelet transform (CWT) and DWT are presented in [1], [16]-[19].

In particular, CO-OFDM system SE can be improved by avoiding CP. The problem of the CP elimination can be solved by using WPT-OFDM [16]. A mitigation of CD of 3380 ps/nm at 112 Gb/s without CP in the WPT-OFDM system has been demonstrated [1]. Typically, WPT-OFDM optical communication systems are single-polarization. However, the dual-polarization transmission is a promising technique for 1 Tb/s Ethernet transport [1]. Here we present some definitions and relationships essential for DWT applications.

CWT $W_T(a,\tau)$ of a given function f(t) with respect to a mother wavelet (MW) $\psi(t)$ is defined as follows [17], [18]:

$$W_T(a,\tau) = \frac{1}{\sqrt{|a|}} \int_{-\infty}^{\infty} \psi^* \left(\frac{t-\tau}{a} \right) f(t)dt$$

(12)

where the real numbers aa and ττ are the scaling and shifting, or translation parameters, respectively, and asterisk means complex conjugation. Note that in many practically important cases MW $\psi(t)$ is real. The functions $\psi^{a,\tau}(s) = |a|^{-1/2}\psi((s-\tau)/a)$ are called wavelets [18]. The set of wavelets is orthogonal and can be used as a basis instead of sinusoidal functions [16]. It is possible to localize the events described by f(t)f(t) in time and frequency domains simultaneously by means of W_T choosing the appropriate values of the parameters aa and ττ [17].

DWT is given by [17], [18]

$$W_T^{m,n}(a,\tau) = a_0^{-m/2} \int_{-\infty}^{\infty} \psi^* \left(a_0^{-m}t - n\tau_0 \right) f(t)dt$$

(13)

where $m, n \in Z, Z$ is the set of all integers, and the constants $a_0 > 1, \tau_0 > 1$. . Comparison of eqs. (12)and (13) shows that $a = a_0^m$ and $\tau = n\tau_0 a_0^m$ [17]. The orthogonal wavelet series expansions can be successfully used in DSP and multiplexing when the scaling and translation parameters are discrete [17]. In such a case, a signal $s(t) \in V_0$ can be represented by a smooth approximation at resolution 2_M, obtained by combining translated versions of the basic scaling function $\varphi(t)$, and M details at the dyadic scales $a = 2^l$, $(l = 1, 2, ..., M-1)$ obtained by combining shifted and dilated versions of the MW $\psi(t)$ as follows [17].

$$s(t) = \sum_k 2^{-M/2} c_M[k] \phi \left(2^{-M}t - k\Delta\tau \right) + \sum_{l=1}^{M} \sum_k 2^{-l/2} d_l[k] \psi \left(2^{-l}t - k\Delta\tau \right)$$

(14)

Here a subspace $V_0 \in L^2(R)$, $L^2(R)$ is a the linear vector space of square integrable functions, $2^{-l/2}\phi(2^{-l}t - k\Delta\tau)$ and $2^{-l/2}\psi(2^{-l}t - k\Delta\tau)$ are the orthonormal bases for the subspaces $V_l \in L^2(R)$ and $W_l \in L^2(R)$, respectively, $V_l \perp W_l$, $(l, k) \in Z$, $c_l[k]$ and $d_l[k]$ are the scaling and detail coefficients, respectively, at resolution 2^l, $\Delta\tau$ is the time interval coinciding with the inverse of the free spectral range (FSR).

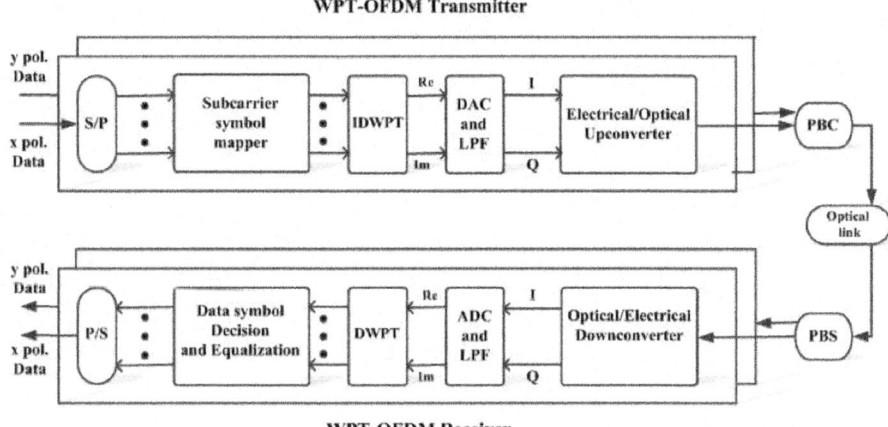

Figure 2: Block diagram of WPT-OFDM communication system. PBC/PBS: polarization beam Combiner/Splitter.

The block diagram of the WPT-OFDM communication system is shown in Fig. 2 [1]. In WPT-OFDM system each channel occupies a wavelet packet which plays a role of a subcarrier in wavelet domain. At the transmitter, IDWPT reconstructs the time domain signal from wavelet packets [1]. At the receiver, DWPT is used for the decomposition of the time domain signal into wavelet packets. The decomposition is carried out by means of successive low-pass and high-pass filtering in the time domain [1]. For WPT-OFDM, the basis function wavelets are finite in time, and the inter-symbol orthogonality in WT is caused by the shift orthogonal property of the waveforms [1], [17], [18]. In W_T, symbols are overlapped in time domain [1]. As a result, the symbol duration increases. The symbol duration increase, in turn, provides the CD tolerance and eliminates the need for CP [1].

Real-valued wavelets are used for the real signal processing [1]. However for fiber optic channels, the up-conversion to the optical domain generates two spectral sidebands: a positive sideband and a negative one [1], [5]. In such a case, PMD influence on WT-OFDM is essential. It will be considered in Section 4.

PMD INFLUENCE ON THE CO-OFDM COMMUNICATION SYSTEM PERFORMANCE

In this Section we consider PMD in CO-OFDM communication systems and its influence on their performance [1], [5], [6], [9]. CO-OFDM transmission system can be defined as 2×22×2 multiple-input multiple-output (MIMO)-OFDM channel in the presence of polarization dispersion effects [7]. SMF supports two degenerate modes of orthogonal polarization [9]. However, the degeneracy may be broken due to the asymmetry in fiber geometry caused by the manufacture process or mechanical stress [9]. In such a case, a fiber birefrigence $B=|n_x-n_y|$ occurs where nx,ynx,y are the refractive indices for x and y polarization, respectively [9]. An optical signal with the wavelength λ propagating through the SMF is coupled from one polarization to the other with the coupling period LB=λ/B [9]. Typical values of these parameters are B≈10−7,LB≈10m at λ=1.5μm [9]. Then, the transmitted OFDM time-domain signal s(t)s(t) (1) and the ii-th information symbol at the kk-th subcarrier ckicki (2) should be replaced with the Jones vectors for the two polarizations given by [6], [9].

$$s(t) = \begin{pmatrix} s_x \\ s_y \end{pmatrix}; \vec{c}_{ki} = \begin{pmatrix} c_{ki}^x \\ c_{ki}^y \end{pmatrix}$$

(17)

GI ΔGΔG in such a case must be long enough in order to handle the fiber CD and PMD. It takes the form [6], [9].

$$\Delta_G \geq \frac{c}{f^2}|D_t|N_{sc}\Delta f + (DGD)_{max}$$

(16)

where c is the speed of light, f is the optical carrier frequency, Dt is the total accumulated CD in units of ps/pm, Δf is the subcarrier channel spacing, and (DGD)max(DGD)max is the maximum budgeted differential-group-delay (DGD) which is typically chosen to be of 3.5 times of the mean PMD in order to satisfy inequality (16). The Jones vector of the received symbol $\vec{c}'_{ki}=(c'_{ki}{}^x, c'_{ki}{}^y)^t$ with a sufficiently long symbol period for the k-th subcarrier in the i-th OFDM symbol is given by [6], [9].

$$\vec{c}'_{ki} = \exp(j\phi_i)\exp(j\Phi_D(f_k))T_k\vec{c}_{ki} + \vec{n}_{ki}$$

(17)

where $\vec{n}_{ki}=(n_{ki}^x, n_{ki}^y)^t$ is the noise including two polarization components, φiφii is the OFDM symbol phase noise, or common phase error (CPE) caused by the phase noises from the lasers and RF LOs at the transmitter and receiver, $\Phi_D(f_k)=\pi c D_t f_k^2/f^2$ is the phase dispersion caused by the fiber CD, T(k)T(k) is the Jones matrix for the fiber link given by [6], [9].

$$T(k) = \prod_{l=1}^N \exp\left\{\left(-\frac{1}{2}j\vec{\beta}_l f_k - \frac{1}{2}\vec{\alpha}_l\right)\cdot\vec{\sigma}\right\}$$

(18)

NN is the number of PMD and polarization dependent loss (PDL) cascading elements, $\beta l \rightarrow \beta l \rightarrow$ is the l-th element birefringence vector, $\alpha^{\rightarrow} l \alpha \rightarrow l$ is the l-th element PDL vector, and $\sigma^{\rightarrow} \sigma \rightarrow$ is the Pauli matrix vector.

The CO-OFDM system architecture in the fiber optic channel can be divided into four groups related to the number of the transmitters and receivers used in the polarization dimension [6], [9].

- Single-input single-output (SISO) where one optical OFDM transmitter and one optical OFDM receiver are used for CO-OFDM transmission. The SISO configuration is susceptible to the polarization mode coupling in a fiber. It requires a polarization controller (PC) before the receiver in order to align the input signal polarization with the LO polarization. However, in the case of large PMD, PC cannot prevent the polarization rotation between subcarriers. As a result, SISO CO-OFDM is susceptible to polarization-induced fading and should not be implemented in applications.

- Single-input two-output (SITO) where one optical OFDM transmitter and two optical OFDM receivers are used, one for each polarization. In such a system, PC is not needed. SITO CO-OFDM system is resilient to PMD when the polarization diversity receiver is used. The PMD in the fiber link may even improve the system margin against PDL-induced fading.

- Two-input single-output (TISO) where two optical OFDM transmitters and one optical OFDM receiver are used. This time, each transmitter is used for one specified polarization. The PC is not needed at the transmit end. The polarization diversity transmitter can achieve PMD resilience. In the TISO scheme, the same information symbol is repeated in two consecutive symbols which results in the electrical and optical efficiency reducing by half and the two times increase of the optical signal-to-noise ratio (OSNR) requirement as compared to SITO scheme.

- Two-input two-output (TITO) where a polarization diveristy optical OFDM transmitter and a polarization diveristy optical OFDM receiver are used. In such a scheme, the transmitted OFDM information symbol cikcik is considered as polarization modulation or polarization multiplexing. Hence, the channel capacity is doubled as compared to the SITO scheme. The PMD rotates subcarrier polarization and does not influence the channel capacity doubling. In the case of the TITO scheme, the polarization diversity receiver permits to avoid the polarization tracking at the receiver.

Comparison of the properties of the different CO-MIMO-OFDM architectures shows that SITO-and TITO-OFDM are the preferred configurations [9].

Consider the PMD influence on the transmission in a 2x2 MIMO CO-OFDM system. In such a case, the Jones matrix (18) for the fiber link on the k-th subcarrier in the first-order PMD approximation is given by [1], [5].

$$T(k) = M^{-1} \begin{bmatrix} \exp(-j\pi f_k \tau_{link}) & 0 \\ 0 & \exp(j\pi f_k \tau_{link}) \end{bmatrix} M$$

(21)

where τlinkτlink is the DGD of the link, and the matrix M has the form.

$$M = \begin{bmatrix} \cos\left(\frac{\theta}{2}\right)\exp\left(-j\frac{\varphi}{2}\right) & -\sin\left(\frac{\theta}{2}\right)\exp\left(-j\frac{\varphi}{2}\right) \\ \sin\left(\frac{\theta}{2}\right)\exp\left(j\frac{\varphi}{2}\right) & \cos\left(\frac{\theta}{2}\right)\exp\left(j\frac{\varphi}{2}\right) \end{bmatrix}$$

(20)

Here $\theta,\varphi\theta,\varphi$ are the polar and azimuth angle of the principle state of polarization. In FTO-OFDM systems, the individual subcarrier is single-sideband by nature, and $\Phi_D(f_k)$, $T(k)$ can be estimated and compensated [5]. In WT based OFDM systems the modulated signals are double-sideband as it was mentioned above. CD influence on such a system is insignificant since the two sidebands have the same phase dispersion: $\Phi_D(f_k) = \Phi_D(-f_k)$. On the contrary, PMD does not have phase symmetry: $T(k) \neq T(-k)$. As a result, at the receiver two sidebands experience two different dispersions, and their addition does not reproduce the real wavelet basis, the orthogonality is broken, and the inter-packet-interference occurs [1], [5]. The WPT-OFDM systems are more susceptible to PMD then conventional FT-OFDM systems [1], [5].

DUAL-TREE COMPLEX WAVELET TRANSFORM (DT-CWT)

DWT has two main disadvantages [7], [8].

- It is not shift-invariant. For this reason, the small shift in the input signal can cause major variations in the energy distribution between DWT coefficients at different scales.

- It is characterized by a low directional selectivity for the diagonal elements since the wavelet filters are separable and real.

The DT-CWT has been proposed recently [7], [8]. DT-CWT consists of two W_Ts operating in parallel on an input signal [8]. In order to construct a DT-CWT, each sub band should be repeatedly decomposed using low-pass/high-pass PR FBs [8]. The response of each branch of the second WP FB is the discrete Hilbert transform of the corresponding branch of the first WP FB [8].

These WTs correspond to two FBs. The block diagram of DT-CWT is shown in Fig. 3 [8].

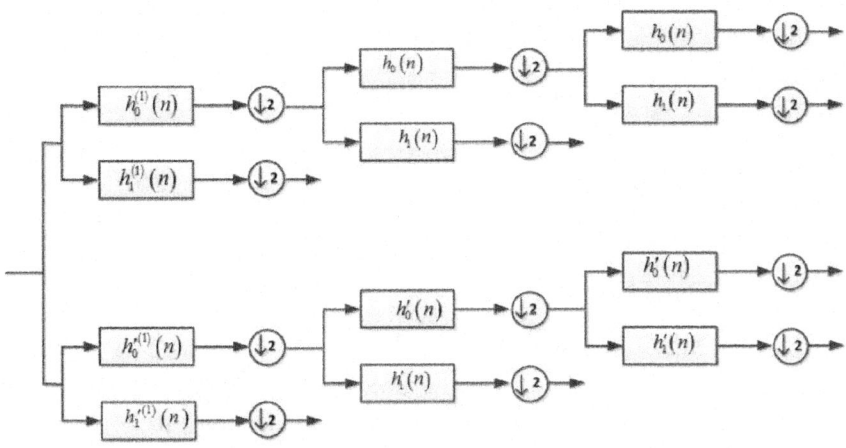

Figure 3: The DT-CWT consists of two wavelet FBs operating in parallel.

The first wavelet FB is determined by DWTs, the second FB contains the discrete Hilbert transforms of the first wavelet FB [8]. The first FB is the real part and the second FB is the imaginary part of the CWPT [8].

The wavelet function ψ(t) associated with the first FB and the corresponding scaling function φ(t)have the form [8].

$$\psi(t) = \sqrt{2}\sum_n h_1(n)\phi(2t-n); \phi(t) = \sqrt{2}\sum_n h_0(n)\phi(2t-n) \qquad (21)$$

For an orthonormal wavelet basis, the LPFs and high-pass filters (HPFs) have the following relationships $H_1(e^{j\omega}) = -e^{-j\omega d}H_0^*(e^{j(\omega-\pi)})$, or equivalently $h_1(n) = (-1)^n h_0(d-n)$ where d is an odd integer [8]. The wavelet function $\psi'(t)$ and the filters $h'_0(n)$, $h_1'(n)$ for the DT-CWPT imaginary part are defined in a similar way [6]. The filters $\{h_0(n), h_1(n)\}$ and $\{h'_0(n), h'_1(n)\}$ are assumed to be finite impulse response (FIR) conjugate quadrature filter (CQF) pairs [8]. In the case of the ideal DT-CWT we have.

$$\psi_c(t) = H_{Hilbert}\{\psi(t)\} \qquad (22)$$

where $H_{Hilbert}\{\psi(t)\}$ is the Hilbert transform of ψ(t) and the complex wavelet $\psi(t) = \psi(t) + j\psi'(t)$ is approximately analytic [20]. In order to satisfy the perfect reconstruction (PR) conditions, the filters are designed in such way that the two low pass filters (LPFs) should satisfy the half-sample delay condition: $h'_0(n) \approx h0(n-0.5)$ [20]. If each wavelet $\psi(t)$, $\psi'(t)$ are orthogonal to their integer translates, then the Hilbert relation between them is satisfied under the condition [8]:

$$H'_0\left(\exp\left(j\omega\right)\right) = \exp\left(-j0.5\omega\right)H_0\left(\exp\left(j\omega\right)\right), |\omega| < \pi$$

(23)

Then, for the HPFs we have [8]

$$H'_1\left(\exp\left(j\omega\right)\right) = -j\,\mathrm{sgn}\left(\omega\right)\exp\left(j0.5\omega\right)H_1\left(\exp\left(j\omega\right)\right), |\omega| < \pi$$

(24)

Here sgnsgn is the signum function, $H'_{0,1}(\exp(j\omega))$, $H_{0,1}(\exp(j\omega))$ are the z-transforms of $h'_{0,1}(n)$, $h_{0,1}(n)$ on the unit circle, respectively [8]. It can be shown that the first stage of the DT-CWT must be different from the following stages [8]. The equivalent response $H^{(k)}(\exp(j\omega))$ of the kth stage of the first FB terminated with $H_1(\exp(j\omega))$ for $k>1$ is given by:

$$H^{(k)}\left(\exp\left(j\omega\right)\right) = H_1\left(\exp\left(j2^{k-1}\omega\right)\right)\prod_{m=0}^{k-2}H_0\left(\exp\left(j2^m\omega\right)\right), |\omega| < \pi$$

The equivalent response

$H'^{(k)}(\exp(j\omega))$ of the second FB's corresponding branch can be obtained by replacing H_i with H'_i given by [8]:

$$H'^{(k)}\left(\exp\left(j\omega\right)\right) = -j\,\mathrm{sgn}\left(\omega\right)\exp\left(j0.5\omega\right)H^{(k)}\left(\exp\left(j\omega\right)\right), |\omega| < \pi$$

(26)

which is equivalent to the relationship

$$H'^{(k)}(\exp(j\omega)) = -\exp(j0.5\omega)H_{Hilbert}\{H^{(k)}(\exp(j\omega))\}$$

DT-CWT has the following properties [7].

- Approximately shift-invariance.
- Good directional selectivity in 2 dimensions.
- Perfect reconstruction (PR) using short linear-phase filters.
- Limited redundancy.
- Efficient order-N computation.

DUAL-TREE COMPLEX WAVELET PACKET TRANSFORM (DT-CWPT)

DT-CWT can be generalized to DT-CWPT by using the bases of discrete wavelet packet transforms (DWPTs). DT-CWPT is approximately shift-invariant, provides a geometrically oriented signal analysis in multiple dimensions and permits the transformation of a double-sideband spectrum into an approximately single-sideband one [7], [8]. The invert DT-CWPT is provided by the inversion of its real and imaginary parts [8].

The construction of the DT-CWPT requires the repeated decomposition of the sub bands by using low-pass/high-pass PR FBs [8]. The PR FBs should be chosen in such a way that the response of each branch of the second wavelet packet FB is the discrete Hilbert transform of the corresponding branch of the first wavelet packet FB [8]. Under this condition, each sub band of the DT-CWPT will be analytic [8]. Taking into account that the discrete Hilbert transform is a linear time-invariant (LTI) system we can write the following relationships for the discrete Hilbert transform pair of the filters g(n),h(n) g(n),h(n) [8].

$$G\left(\exp\left(j\omega\right)\right) = j\,\mathrm{sgn}\left(\omega\right)H\left(\exp\left(j\omega\right)\right), |\omega| < \pi \qquad (27)$$

$$G\left(\exp\left(j\omega\right)\right)C\left(\exp\left(j\omega\right)\right) = j\,\mathrm{sgn}\left(\omega\right)H\left(\exp\left(j\omega\right)\right)C\left(\exp\left(j\omega\right)\right), |\omega| < \pi \qquad (28)$$

Then, the convolutions $g(n)*c(n), h(n)*c(n)$ are also a discrete Hilbert transform pair [8].

It has been shown that a DT-CWPT consisting of two wavelet packet FBs operating in parallel can be produced where some filters in the second wavelet packet FB are the same as those in the first wavelet packet FB [8].

The first of these two wavelet FBs for a four-stage DT-CWPT is shown in Fig. 4 [8]. The second wavelet packet FB is obtained by replacing the first stage filters $h_i^{(1)}(n)$ with $h_i^{(1)}(n-1)$ and by replacing $h_i(n)$ with $h_i'(n)$ for $i \in \{0, 1\}$ [8]. The filters FiFi are unchanged in the second wavelet packet FB [8].

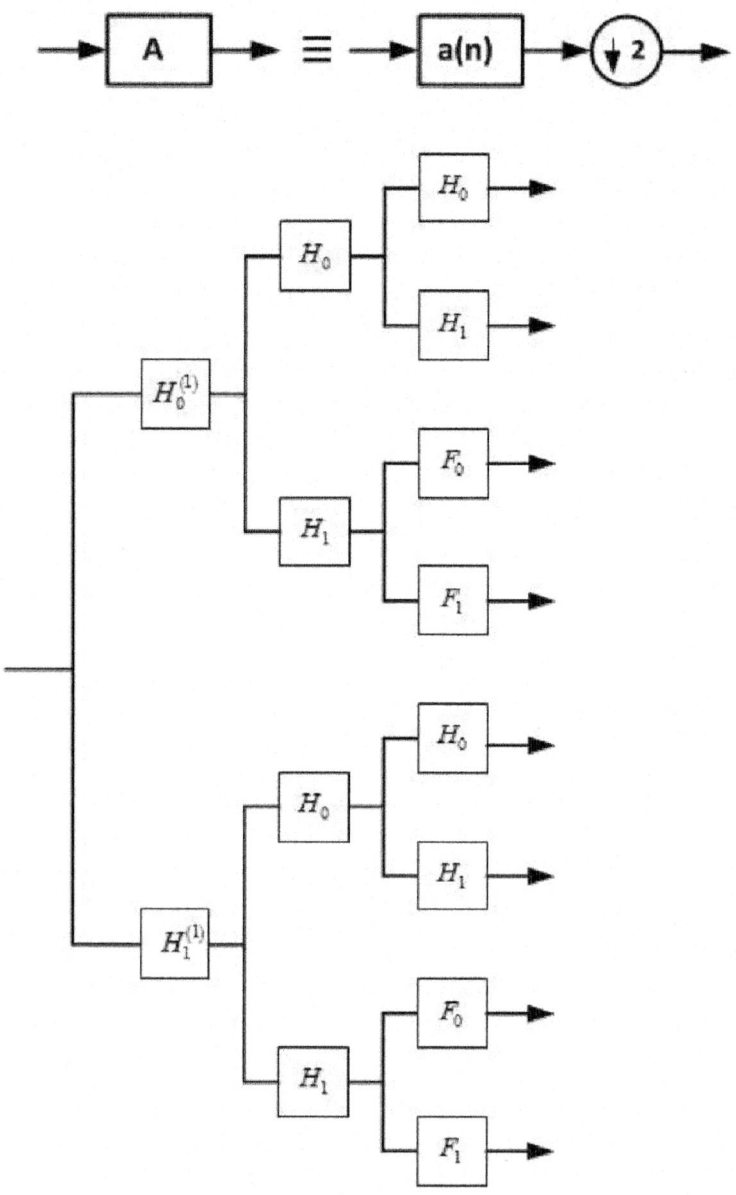

Figure 4: First wavelet packet FB of a four-stage DT-CWPT.

In order to demonstrate the development of the M-band DT-CWT based on the Hilbert pairs of M-band wavelets, consider the construction of the four-band DT-CWT (M=4) from a given two-band DT-CWT which is defined by a two-channel orthonormal FB $\{h_0^{(2)}(n), \tilde{h}_1^{(2)}(n)\}$ associated with the scaling function $\phi^{(2)}(t)$ and the wavelet $\psi^{(2)}(t)$ [8]. Their Fourier transforms (FTs) have the form [8].

$$\Phi^{(2)}(\omega) = \prod_{i=1}^{\infty}\left[\frac{1}{\sqrt{2}}H_0^{(2)}\left(\frac{\omega}{2^i}\right)\right]; \Psi^{(2)}(\omega) = \frac{1}{\sqrt{2}}H_1^{(2)}\left(\frac{\omega}{2}\right)\Phi^{(2)}\left(\frac{\omega}{2}\right)$$

(29)

A two-band DT-CWT can be created if we have a second two-channel orthonormal FB $\{h_0'^{(2)}(n), h_1'^{(2)}(n)\}$ with the associated scaling function $\phi'^{(2)}(t)$ and the wavelet $\psi'^{(2)}(t) = H_{Hilbert}\{\psi^{(2)}(t)\}$, where the complex wavelet $\psi^{(2)}(t) + j\psi'^{(2)}(t)$ is analytic [8]. In order to construct a four-band CWT, another two-channel orthonormal FB $\{f_0(n), f_1(n)\}$ is necessary [8]. The corresponding four-channel orthonormal FB, i.e a DWPT is shown in Fig. 5 [8].

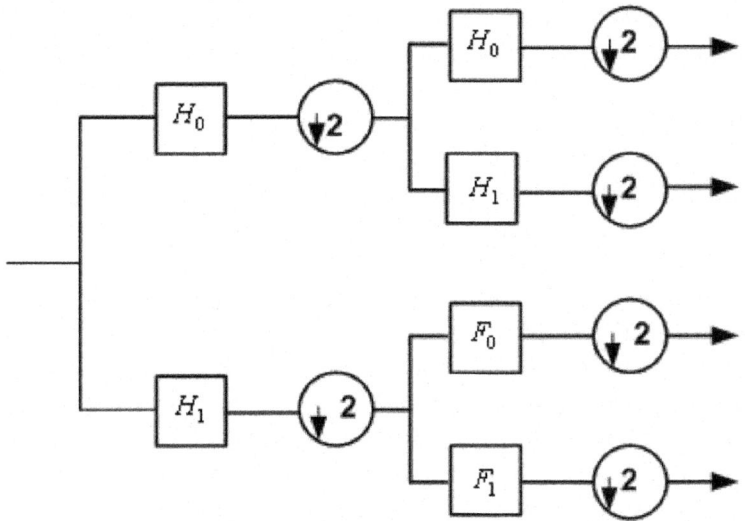

Figure 5: Discrete wavelet packet transform (DWPT).

A second WPT such that the wavelets associated with the two WPTs form the Hilbert transform pairs is given by [8].

$$H_0^{(4)}(\exp(j\omega)) = H_0^{(2)}(\exp(j\omega))H_0^{(2)}(\exp(j2\omega))$$

(30)

$$H_1^{(4)}(\exp(j\omega)) = H_0^{(2)}(\exp(j\omega))H_1^{(2)}(\exp(j2\omega))$$

(31)

$$H_2^{(4)}\left(\exp(j\omega)\right) = H_1^{(2)}\left(\exp(j\omega)\right)F_0\left(\exp(j2\omega)\right)$$

(32)

$$H_3^{(4)}\left(\exp(j\omega)\right) = H_1^{(2)}\left(\exp(j\omega)\right)F_1\left(\exp(j2\omega)\right)$$

(33)

It can be shown that the FB determined by eqs. (30)-(33) is equivalent to the FB shown in Fig. 5 [8].

SIMULATION RESULTS FOR THE SINGLE CHANNEL DT-CWPT BASED CO-OFDM SYSTEM

We have carried out the numerical simulations based on DT-CWPT for the single-channel CO-OFDM transmission system including the optical front elements such as a laser-diode, and Mach-Zehnder external modulator [21]. The nonlinear effects are neglected since Hilbert transform is LTI system [8]. We used the 16 quadrature amplitude modulation (QAM) with $128=2^7$ sub bands, 7 decomposition levels, bit rate of 100 Gb/s and the bandwidth of 25 GHz. The typical values of transmission process are following: the input optical signal power is equal to 10 mW; CP is 1%; the phase noise frequency in the case of the coherent detection is equal to 500 KHz; the overall noise amplification level of the link is 50 dB; the pre-amplifier gain at the receiver input is 25 dB; the optical fiber dispersion constant is 17 *ps* / (*nm·km*); the PMD constant is 0.1 ps/km$^{1/2}$. The 8 bit DAC and ADC have been used. In order to mitigate the group velocity dispersion and PMD we used the Least Mean Square (LMS) and Viterbi-Viterbi Digital Signal Processing (DSP) algorithms [22]. We used in our simulations the complex wavelets based on the Q-shift filters [7], [8], [23]. The design of pairs of wavelet bases where the wavelets form a Hilbert transform pair was also discussed in detail in [24], [25]. The numerical simulation results are presented in Figs. 6-11. The QAM-16 modulated signal constellations for the transmission distance of 100 km, 700 km and 1700 km in the case of the DT-CWPT based CO-OFDM system are shown in Figs.6-8, respectively.

The QAM-16 modulated signal constellations for the transmission distance of 1200 km and 1300 km in the case of the conventional CO-OFDM system are shown in Figs.9-10, respectively. In both cases CP=1% has been used.

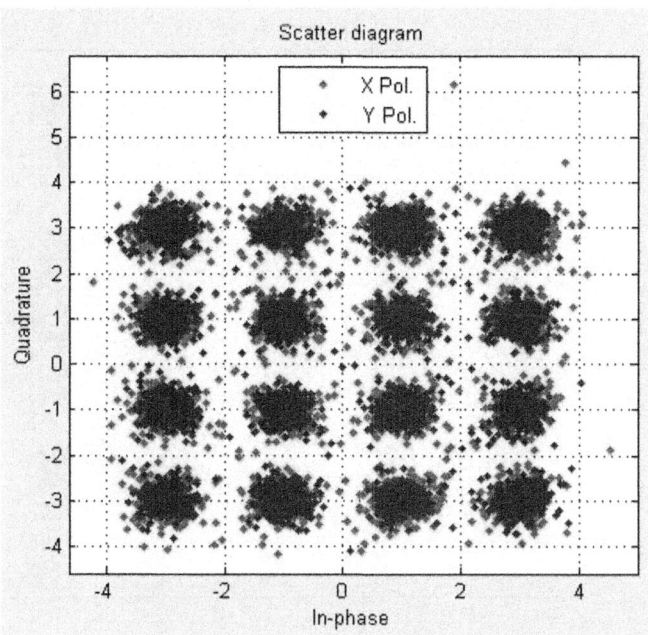

Figure 6: The QAM-16 modulated signal constellation for the transmission distance of 100 km, CP 1%, BER=$10^{-2.56}$, DT-CWPT based CO-OFDM system.

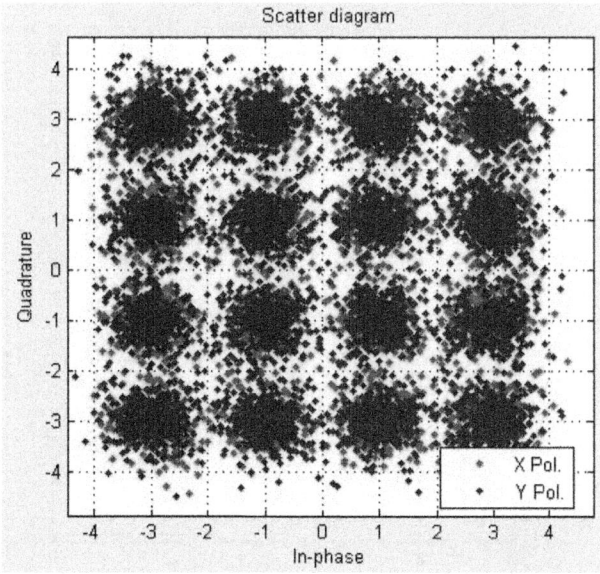

Figure 7: The QAM-16 modulated signal constellation for the transmission distance of 700 km, CP 1%, BER=$10^{-2.13}$, DT-CWPT based CO-OFDM system.

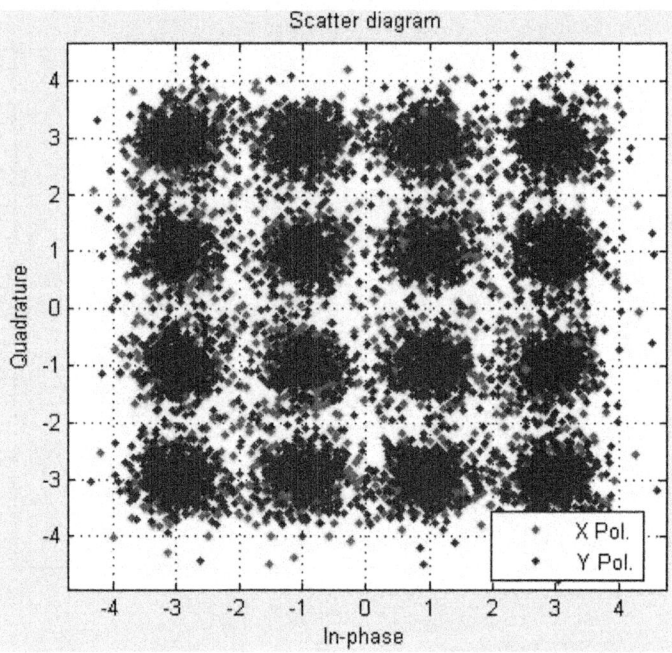

Figure 8: The QAM-16 modulated signal constellation for the transmission distance of 1700 km, CP 1%, BER=$10^{-2.08}$, DT-CWPT based CO-OFDM system.

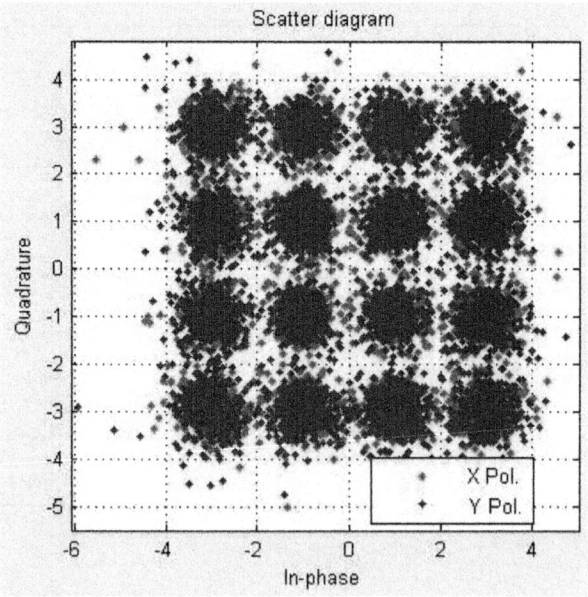

Figure 9: The QAM 16 modulated signal constellation for the transmission distance of 1200 km, BER=$10^{-2.08}$, CP 1% Conventional CO-OFDM system.

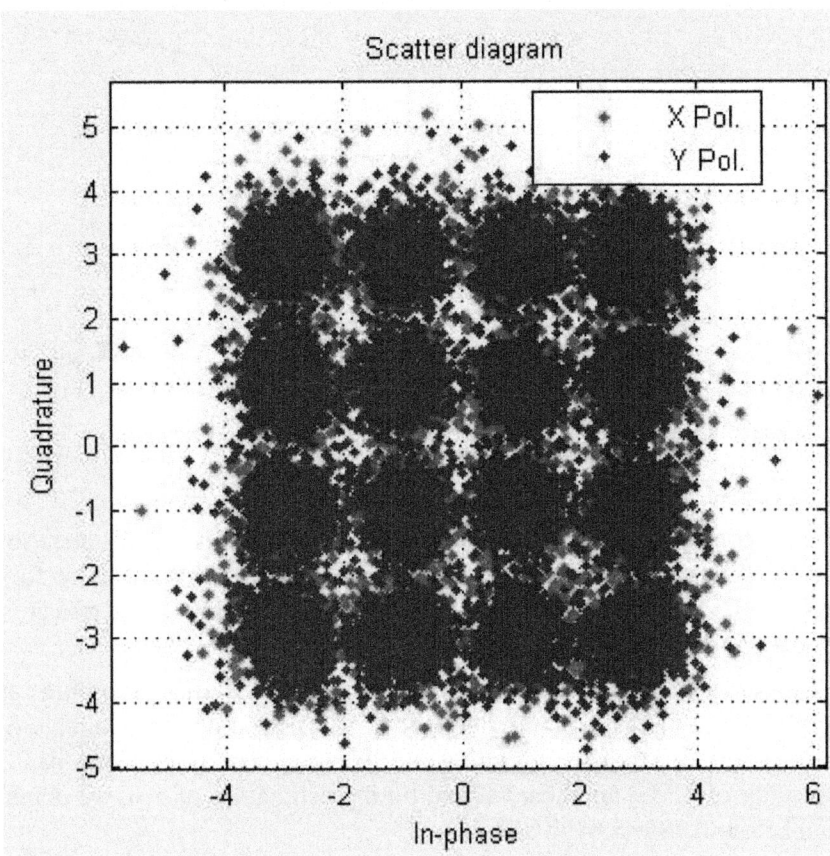

Figure 10: The QAM 16 modulated signal constellation for the transmission distance of 1300 km, BER=$10^{-1.83}$, CP 1% Conventional CO-OFDM system.

The QAM-16 modulated signal constellations for the transmission distance of 1200 km and 1300 km in the case of the conventional CO-OFDM system are shown in Figs.9-10, respectively. In both cases CP=1% has been used.

The comparison of Figs. 8 and 10 clearly shows that the performance of the DT-CWPT based OFDM communication system is better than the performance of the conventional CO-OFDM communication system. Indeed, for the same modulation format and CP, BER=$10-1.83=1.48\times10-2$ at the distance of 1300 km for the conventional CO-OFDM as compared to BER=$10-2.08=8.32\times10-3$ at the distance of 1700 km for the DT-CWPT based OFDM system.

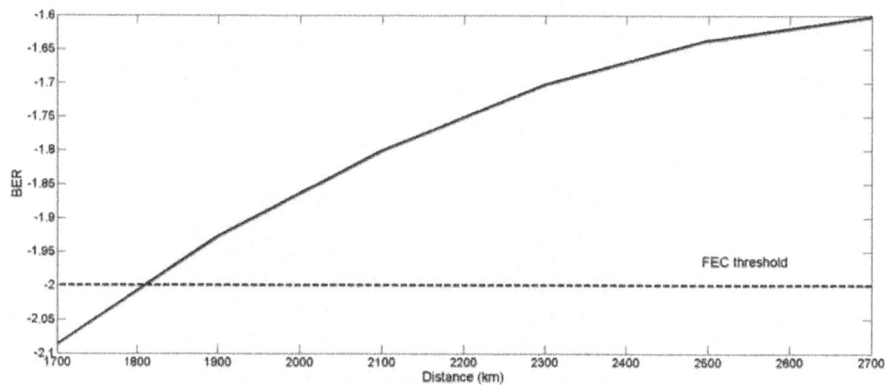

Figure 11: The BER dependence on the transmission distance for the DT-CWPT based CO-OFDM system.

The bit error rate (BER) dependence on the transmission distance for the DT-CWPT based CO-OFDM system is presented in Fig. 11. The forward error correction (FEC) threshold of BER=10^{-2} corresponds to the transmission distance of about 1800 km.

We have evaluated SE for the dual polarization signal, sampling rate of 26.6 GHz, and QAM 16 modulation format. The maximum available SE in such a case is 8bit/s/Hz. Assuming the FEC penalty of 20% per band, the CP penalty of 1.5%, and channel and timing estimation of 5%, we obtain the actual SE=8×0.735=5.84bit/s/Hz.

CONCLUSIONS

CO-OFDM communication systems are characterized by high SE, possibility of the advanced modulation formats applications due to the coherent detection, and the high receiver sensitivity [1], [9], [13]. Conventional CO-OFDM systems are based on FTs [1], [9]. In such systems, low rate subcarriers are orthogonally transformed into time domain signals [1]. The condition of subcarrier orthogonality prevents spectral overlapping and ICI [1], [9]. The basis functions of the conventional CO-OFDM system are the infinitely long in time sinusoids. The 1-Tb/s CO-OFDM single-channel signal transmission over 600 km SSMF with SE of 3.3 bit/s/Hz and without Raman amplification and dispersion compensation has been reported [26]. In CO-OFDM systems, the CD and PMD impairments can be compensated by choosing an appropriate length of CP [1], [9]. However, the CP insertion decreases the CO-OFDM communication system SE [1], [9].

Recently, the CO-OFDM systems based on WPTs have been proposed [1], [2]. Wavelets used as the basis functions in these advanced systems have finite length in the time domain. For this reason, WPT-OFDM systems do not need CP which results in a higher SE [1], [2]. The WPT-OFDM can mitigate a CD of 3380 ps/nm at 112 Gb/s rate without CP [1]. However, the high performance has been demonstrated for a CO-OFDM system based on a single-polarization [5]. The most promising technique for achieving high SE and maximizing power efficiency is coherent detection with polarization multiplexing where the in-phase (I) and quadrature (Q) signals are used in two field polarizations [27]. Information can be encoded in all available degrees of freedom, while the the compensation of transmission impairments can be provided by DSP [22], [27]. Advanced FEC coding can be also implemented [27]. Taking into account the importance of the polarization multiplexing, it is necessary to investigate the PMD influence on the dual-polarization transmission [5], [6].

In WPT-OFDM communication systems the modulated signals are double-sideband [5]. It has been shown that the CD influence on WPT-OFDM is compensated automatically due to the phase symmetry of the both sidebands while the PMD does not possess the phase symmetry [5]. As a result, the two sidebands have two different dispersions, their addition does not reproduce the real wavelet basis, orthogonality condition is broken, and inter-packet-interference occurs [5].

In this chapter, we considered the fundamentals of WPT based CO-OFDM communication systems and the PMD influence on such systems. It appeared to be that DT-CWPT can generate the single-sideband wavelets in frequency domain [5], [7], [8], [20], [21], [23]-[25]. We briefly discussed the properties of complex wavelets. The DT-CWPT consists of two real wavelet FBs where the second wavelet FB is the Hilbert transform of the first one and it represents an imaginary part of the CWT. The DT-CWPT approach to CO-OFDM communication systems mitigates the PMD impact on the system perfomance.

We have carried out the numerical simulations for the DT-CWPT based CO-OFDM system and QAM-16 modulation format. The constellation and BER dependence on the distance clearly show that the transmission over the distance of 1800 km with high actual SE of 5.85 bit/s/Hz taking into account the FEC and CP penalties, channel and timing estimation can be achieved due to the PMD mitigation provided by the DT-CWPT based CO-OFDM system.

We compared the performance of the conventional CO-OFDM and the DT-CWPT based CO-OFDM communication systems. The performance of the DT-CWPT based OFDM communication system is better than the performnace of the conventional CO-OFDM communication system: for the same modulation format and CP, BER at the distance of 1300 km for the conventional CO-

OFDM is larger than BER at the distance of 1700 km for the DT-CWPT based OFDM system. We used in our simulations the already developed complex wavelets [23]-[25]. In future theoretical investigations, we suppose to find more appropriate wavelet bases using the so-called best-basis algorithms [8].

REFERENCES

1. Li An, Shieh W., and Tucker R.S. Wavelet packet transform-based OFDM for optical communications. Journal of Lightwave Technology, 2010; 28(24), 3519-3528.

2. Ben Ezra Y., Lembrikov B.I., Zadok Avi, Ran Halifa R., and Brodeski D. All-optical Signal Processing for High Spectral Efficiency (SE) Optical Communication. In: Narottam Das (ed.) Optical Communication. Rijeka: InTech; 2012. p343-366.

3. Schmogrow R., Bouziane R., Meyer M., Milder P.A., Schindler P.C., Killey R.I., Bayvel P., Koos C., Freude W., and Leuthold J. Real-time OFDM or Nyquist pulse generation – which performs better with limited resources? Optics Express 2012; 20(26) B543-B551.

4. Bosco G., Carena A., Curri V., Poggiolini P., and Forghieri F. Performance limits of Nyquist-WDM and CO-OFDM in high-speed PM-QPSK systems. IEEE Photonic Technology Letters, 2010; 22(15) 1129-1131.

5. Li An, Shieh W., and Tucker R.S. Impact of polarization-mode dispersion on wavelet transform based optical OFDM systems, In: proceedings of National Fiber Optic Engineers Conference, San-Diego, California, USA March 21-25, 2010, JThA5, pp.1-3.

6. Shieh W., Yi X., Ma Y., and Tang Y. Theoretical and experimental study on PMD-supported transmission using polarization diversity in coherent optical OFDM systems. Optics Express 2007; 15(16) 9936-9947.

7. Kingsbury N. Complex wavelets for shift invariant analysis and filtering of signals Journal of Applied and Computational Analysis 2001 10 (3) 234-253.

8. Bayram I. and Selesnick I.W. On the dual-tree wavelet packet and M-band transforms. IEEE Transactions on Signal Processing 2008; 56(6) 2298-2310.

9. Shieh W, Djordjevic I. Orthogonal Frequency Division Multiplexing for Optical Communications. London: Academic Press; 2010.

10. Hillerkuss, D.et al. Simple all-optical FFT scheme enabling Tbit/s real-time signal processing, Optics Express, April 2010; 18(9) 9324-9340.

11. Wang X.; Ho P., and Wu Y. Robust Channel Estimation and ISI Cancellation for OFDM Systems with Suppressed Features, IEEE Journal

on Selected Areas in Communications 2005; 23(5) 963-972.

12. Shieh W., Bao H., and Tang Y. Coherent optical OFDM: theory and design. Optics Express 2008; 16(2) 841-859.

13. Armstrong J. OFDM for Optical Communications, IEEE Journal of Lightwave Technology, February 2009; 27(3) 189-204.

14. Kikuchi K. Coherent optical communication systems, In: Kaminov, I. P.; Li, T. & Willner, A. E. (Eds.) Optical Fiber Telecommunications VB: Systems and Networks, Academic Press, Amsterdam, London, New York: Academic Press; 2008. p91-129.

15. Da Silva E., Pataca D.M., Ranzini S. M., de Carvalho L.H.H., Juriollo A.A., da Silva M.L., Oliveira J.C.R.F. Transmission of 1.15 Tb/s NGI-CO-OFDM DP-QPSK superchannel over 4520 km of PSCF with EDFA-only amplification. Journal of Microwaves, Optoelectronics and Electromagnetic Applications, 2013; 12(SI-2) 96-103.

16. Cincotti G., Moreolo M.S. and Neri A. Optical Wavelet Signals Processing and Multiplexing, EURASIP Journal on Applied Signal Processing, 2005; 10, 1574-1583.

17. Rao R.M.& Bopardikar A. S. Wavelet Transforms. Introduction to Theory and Applications. Reading, Massachusetts: Addison-Wesley; 1998.

18. Daubechies I. Ten Lectures on Wavelets. Philadelphia, Pennsylvania: Society for Industrial and Applied Mathematics; 2006.

19. Sarkar K.T., Salazar-Palma M., Wicks M.C. Wavelet Applications in Engineering Electromagnetics. Boston: Artech House; 2002.

20. Nerma M. H. M., Kamel N. S., Jeoti V. An OFDM based on dual tree complex wavelet transform (DT-CWT), Signal Processing: An International Journal (SPIJ), 2009; 3 (2) 14-26.

21. Ben-Ezra Y., Brodeski D., Lembrikov B.I. High Spectral Efficiency OFDM Based on Complex Wavelet Packets. In: ICTON 2014: Proceedings of the 16th International Conference on Transparent Optical Networks, 6-10 July 2014, Graz, Austria, We.A1.4 p1-3.

22. Kuschnerov M., Hauske F. N., Piyawanno K., Spinnler, B., Alfiad, M.S. Napoli, A., and Lankl, B. DSP for coherent single-carrier receivers, Journal of Lightwave Technology, 2009; 27(16) 3614-3622.

23. Kingsbury N. Design of Q-shift complex wavelets for image processing using frequency domain energy minimization. In: Image Processing 2003. ICIP 2003. Proceedings of International Conference on Image Processing, 14-18 September 2003, Barcelona, Catalonia, Spain, 1, 1-1013-16.

24. Kingsbury N. A dual-tree complex wavelet transform with improved orthogonality and symmetry properties. In: Image Processing 2000. ICIP 2000. Proceedings of International Conference on Image Processing, 10-13 September 2000, Vancouver, BC, Canada, 375-378.

25. Selesnik I. W. Hilbert transform pairs of wavelet bases, IEEE Signal Processing Letters, 2001; 8(6) 170-173.

26. Ma Y., Yang Q., Tang Y., Chen S., and Shieh W. 1-Tb/s single-channel coherent optical OFDM transmission over 600-km SSMF fiber with subwavelength bandwidth access. Optics Express 2009; 17(11) 9421-9427.

27. Ip, Ezra, Lau, A.P.T., Barros, D.J.F., Kahn, J.M. Coherent detection in optical fiber systems. Optics Express, January 2008; 16(2) 753-791.

CITATION

CHAPTER 1

Majdabadi, A. , Marjani, S. and Sabaghi, M. (2014) Threshold Characteristics Enhancement of a Single Mode 1.55 μm InGaAsP Photonic Crystal VCSEL for Optical Communication Systems. Optics and Photonics Journal, 4, 296-303. doi: 10.4236/opj.2014.410029.

CHAPTER 2

Ben Zid Maha and Raoof Kosai, Multi User MIMO Communication: Basic Aspects, Benefits and Challenges; DOI: 10.5772/57133

CHAPTER 3

Giovanni Geraci and Jinhong Yuan; Physical Layer Security for Multiuser MIMO Communications; DOI: 10.5772/57130

CHAPTER 4

taner cevik and serdar yilmaz; an overview of visible light communication systems; doi : 10.5121/ijcnc.2015.7610

CHAPTER 6

Yan Zhang, Feng-fan Yang, and Weijun Song, "Performance Analysis for Cooperative Communication System with QC-LDPC Codes Constructed with Integer Sequences," Discrete Dynamics in Nature and Society, vol. 2015, Article ID 649814, 7 pages, 2015. doi:10.1155/2015/649814

CHAPTER 7

Rongxing Duan and Jinghui Fan, "Reliability Evaluation of Data Communication System Based on Dynamic Fault Tree under Epistemic Uncertainty," Mathematical Problems in Engineering, vol. 2014, Article ID 674804, 9 pages, 2014. doi:10.1155/2014/674804

CHAPTER 8

Edward G. Kostsov, Sergey V. Piskunov and Mike B. Ostapkevich (2012). 3D ICS with Optical Interconnections, Optical Communication, Dr. Narottam Das (Ed.), ISBN: 978-953-51-0784-2, InTech, DOI: 10.5772/47872

CHAPTER 9

Y. Ben-Ezra and B.I. Lembrikov (2014). Application of Complex Wavelet Packet Transform (CWPT) in Coherent Optical OFDM (CO-OFDM) Communication Systems, Advances in Optical Communication, Dr. Narottam Das (Ed.), ISBN: 978-953-51-1730-8, InTech, DOI: 10.5772/59054.

CHAPTER 10

Y. Ben-Ezra and B.I. Lembrikov (2014). Application of Complex Wavelet Packet Transform (CWPT) in Coherent Optical OFDM (CO-OFDM) Communication Systems, Advances in Optical Communication, Dr. Narottam Das (Ed.), ISBN: 978-953-51-1730-8, InTech, DOI: 10.5772/59054.

INDEX